Design Science and Innovation

Series Editor

Amaresh Chakrabarti, Centre for Product Design and Manufacturing, Indian Institute of Science, Bengaluru, India

The book series is intended to provide a platform for disseminating knowledge in all areas of design science and innovation, and is intended for all stakeholders in design and innovation, e.g. educators, researchers, practitioners, policy makers and students of design and innovation. With leading international experts as members of its editorial board, the series aims to disseminate knowledge that combines academic rigour and practical relevance in this area of crucial importance to the society.

Jon Lang

The Scope of Architectural Theory

Springer

Jon Lang
School of the Built Environment
University of New South Wales (UNSW)
Sydney, NSW, Australia

ISSN 2509-5986 ISSN 2509-5994 (electronic)
Design Science and Innovation
ISBN 978-981-96-3689-1 ISBN 978-981-96-3690-7 (eBook)
https://doi.org/10.1007/978-981-96-3690-7

This Springer imprint is published by the registered company Springer Nature Singapore Pte Ltd.
The registered company address is: 152 Beach Road, #21-01/04 Gateway East, Singapore 189721, Singapore

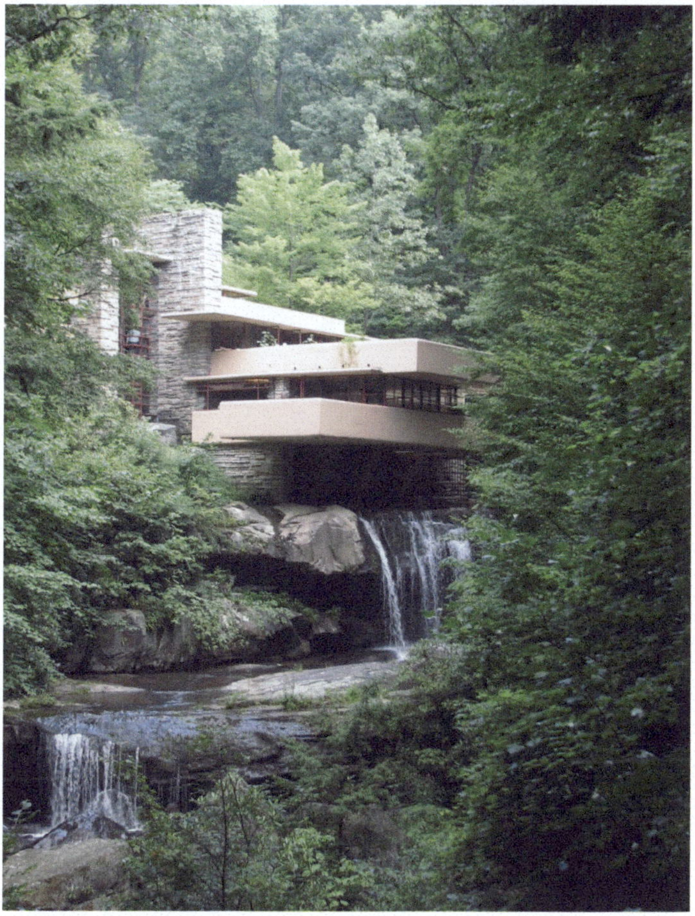

Falling Water, Mill Run, Pennsylvania (1935–1937)
Edgar J. and Liliane S. Kaufman clients; Frank Lloyd Wright architect
Photograph by the author

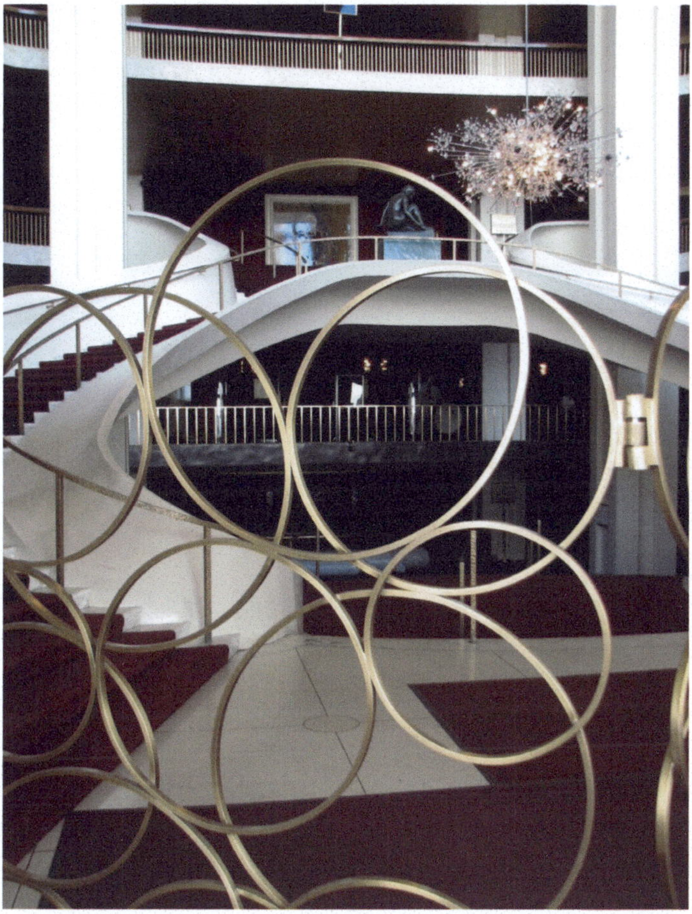

Metropolitan Opera House, Linclon Center for the Performing Arts, New York (1966)
Metropolitan Opera House Association, clients; Wallace K. Harrison of Harrison and Abromowitz, principal architect
Photograph by the author

The contemplation of things as they are, without error or confusion, without substitution or imposition, is in itself, a nobler thing than a whole harvest of inventions.

Francis Bacon in The Collected Works of Francis Bacon edited by Robert Ellis, Douglas Heath, and James Spelling (1996).

Architects are not consciously disdainful of how things work. On the contrary, they have gone to great length to learn what [architectural] saints . . . have said about how places ought to work and what is good for people . . . When contradictory reality intrudes, they shrug it off.

Jane Jacobs in The Death and Life of Great American Cities (1961).

For Susannah Dorothea Lang
An Architect

Preface

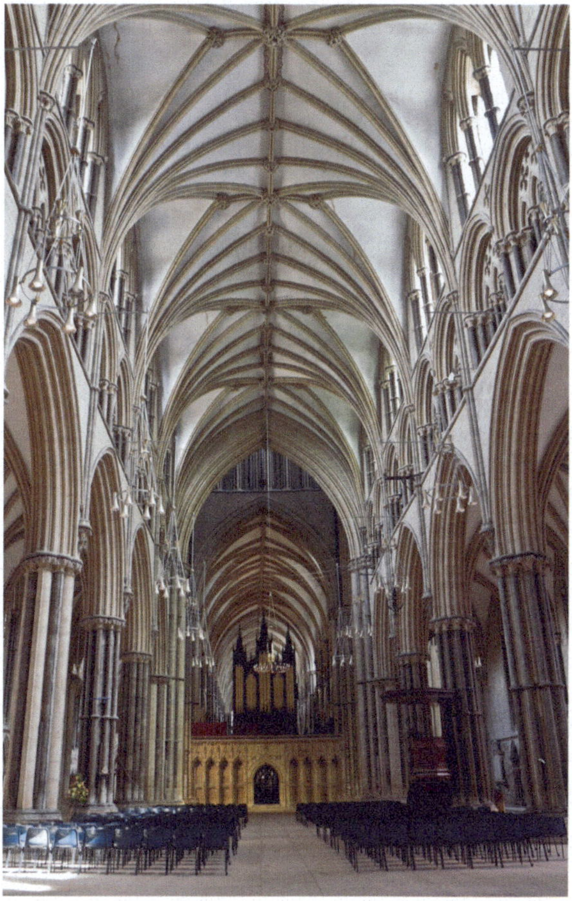

Lincoln Cathedral
(Cathedral Church of the Blessed Virgin Mary of Lincoln)
Originally built in 1072 by Bishop Remigius de Frécamp but rebuilt after an earthquake between 1192 and 1210 and added to between 1307 and 311
Photograph by Michael Beckwith/Wikimedia Commons

Architectural theory needs to inform and be informed by the nature and diversity of architectural practices. If it is, the intellectual gap between academia and the working world will be reduced. The scope and nature of existing approaches to architectural theory needs to be rethought. The book pulls together the thoughts of architects and scholars about how theory and practice can be better integrated.

This book is about architecture. Its assumption is, as Mariana Van Rensselaer put it, that "everything that was ever built is a good work of architecture or a bad one" (Van Rensselaer 1893). The stance thus differs from that taken by Nikolaus Pevsner (1902–1983), the eminent art and architectural historian, who observed:

> A bicycle shed is a building; Lincoln Cathedral is architecture. Nearly everything that encloses space on a scale sufficient for a human being to move in is a building, the term architecture only applies to buildings designed with a view to aesthetic appeal (Pevsner 1948).

Although his position has staunch support (Davies 1994, Scruton 2013), every building has an aesthetic quality even if it was designed without an explicit intellectual idea guiding it.

All buildings are based on their creators' assumptions about how the world works and some view of the benefits their works will create. In other words, architects base their designs on some theory of how buildings and/or urban designs function and some theory of what makes a superior design. In creating a design, architects follow a series of steps. Whether they go through these steps unselfconsciously by habit or self-consciously, it is based on some theory of the nature of the design process and some view of how it should be conducted.

All schools of architecture have theory of architecture courses introducing prospective architects to the nature of design products. These courses inform students about the history of the development of architectural works and/or what individual architectural luminaries consider to be good designs. Very few architectural programs offer courses on design methodology—the study of the processes of analysis, creation, and evaluation involved in the creation of designed products. Students internalize a model of the design process through their experiences in studio courses. It is a big shock to graduates when they enter practice to find out it is not what they expected from their architectural education. That education lacked a theoretical framework describing and explaining the range of architectural practices.

In my teaching I have attempted to rectify this situation by leading courses on design methodology and on functionalism. I have offered studios that get students to explore the range of design paradigms that have their proponents. Much that I have presented has been theoretically naïve. Continuing research has led to greater understandings that have informed my present efforts to encapsulate the theoretical background that can help architects and architectural students understand the work they see around them as well as their own endeavours. This book presents my understanding of the scope of architectural theory that will strengthen the intellectual underpinnings of architects' efforts.

Along the way I authored a book, *Creating Architectural Theory* (1987), that aroused some interest. In it I described some of the efforts of scholars and architects to be explicit and precise about the functions that architecture can serve. It drew on diligent research about what different patterns of built form afford different people in terms of their activities and aesthetic experiences. In the book I also presented the contemporary understandings of the design process—design methodology.

In focussing on the development of functional theory and program-based design processes I failed to fully acknowledge that most architects, in formulating designs, rely on adapting existing building types to the situation at hand. They imitate precedents and generic solutions. Sometimes they simply copy them. This book includes a chapter of typologies as part of the theoretical knowledge base of architects. The need now is to better understand what existing types might afford across the spectrum of the potential functions of buildings and urban designs. It also more clearly distinguishes among different approaches to conducting the design process. Doing so is particularly important as Artificial Intelligence (AI) algorithms impact professional practice.

Architectural schools have largely brushed aside the systematic study of these two aspects—the understanding of the functions of the built world and design methodology—of designers' knowledge base. As Jane Jacobs (1916–2006) an American journalist and architectural critic, recognized years ago, schools have focussed instead on developing students' knowledge about what, as she put it, "architectural saints" consider to be good architecture (Jacobs 1961). Implicitly, architects have accepted that a deep understanding of what has been shown to work or not work hampers their search for creative designs. It is, after all, difficult to produce designs of novel geometries if you know that they will not function well on many dimensions of analysis. Luckily, people are adaptable and accommodate their lives to the affordances of what is on offer.

Architects talk little about how they produce their ideas. Yet, over the past seven decades much thought and some diligent research has increased our knowledge about the design process. Many architects fear that being self-conscious about the design process will inhibit their thinking. There is no evidence to support this belief. Architects need to know how to go about designing if they are to retain or develop a useful role in a complex, multi-cultural world. Otherwise, the profession may end up being, as the late Zaha Hadid (1950–2016), the 2004 Pritzker Prize laureate put it, one of "cake decorators" (Hadid 1982). Alternatively, AI may replace architects' creative design role.

Some architects are highly articulate about their views on what constitutes good architecture, others less so. One must infer from their works what their beliefs are. Some architects practice what they profess. For others, a gap exists between what they say and what they do. What they say they believe and what they actually believe may differ. Alternatively, they may not have been able to put their views into practice because of financial, legal, or social constraints. In *Creating Architectural Theory*, I called these beliefs *normative theories* of architecture. On reflection, now over thirty years later, it is better to regard them simply for what they are, architectural ideologies—*prescriptive theories*. These theories are fundamental to understanding the characteristics of the self-consciously designed built environment. Understanding the ways of thinking and the attitudes they represent is equally important.

Ideological positions on how the design process should be conducted also exist. There are fewer such advocacies than advocacies for specific building configurations and aesthetic qualities. There is, however, a body of teachable knowledge about the nature of designed products and the processes for making them. This knowledge base is vitally important in understanding the opportunities afforded by developments in AI.

The Structure of the Argument

This book consists of a Prologue, three parts and an Epilogue. The purpose of the *Prologue* is to present a model of the scope of architectural theory, and the argument for why a solid body of theory is important if the discipline and profession of architecture is to make a significant contribution to life on the planet Earth. *Part I* covers the nature of declarative, or descriptive, architectural theory. *Part II* reviews the prescriptions, the ideological positions of architects, and *Part III* their work. My goal is to provide an intellectual structure that enables us to organize our knowledge. It is also to provide thoughts on the strengths and shortcomings of our present understandings. Elaborations and details of the argument I present can be found in the writings identified as the sources of information. In a way this book is a guide to the architectural theory literature. The *Epilogue* raises questions about the self-images that architects present to the world and their attitudes to the concerns facing them in their work. Resolving unresolved issues facing the discipline will shape the nature of architecture and thus the scope of architectural theory in the future.

The content of architectural theory in the future will depend on the nature of architectural practice in a changing technological world as well as the future socio-political context in which architects work. The role that AI will play will determine much, but the essence of architectural theory will remain unchanged.

Wollstonecraft, Sydney, Cammeraygal Country, Australia/Terrigal, Darkinjung Country, NSW, Australia
November 2024

Jon Lang

Prologue: The Domain of Architecture and the Scope of Architectural Theory

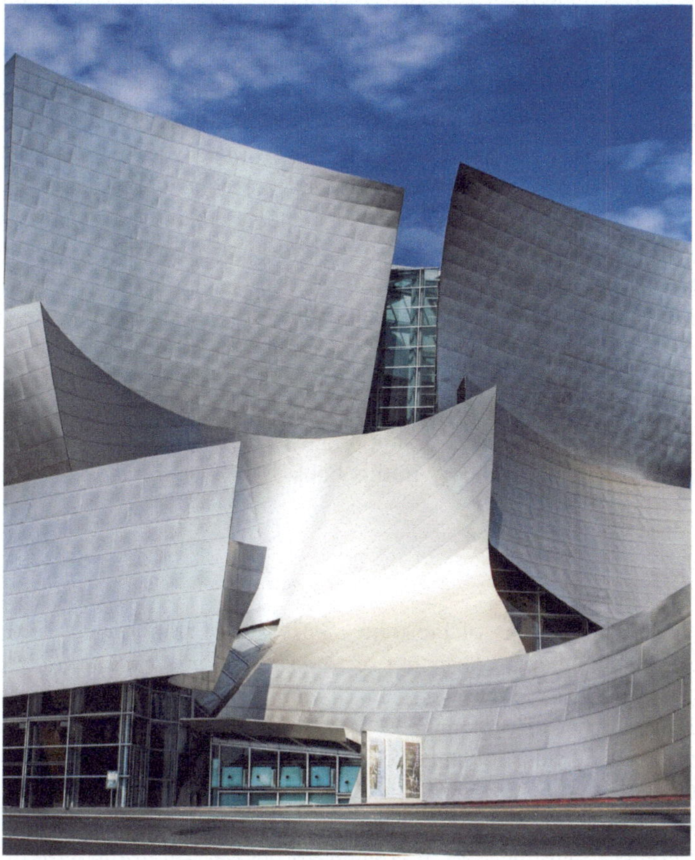

Walt Disney Concert Hall, Los Angeles (1999–2003)
Los Angeles Music Center, client, Frank Gehry, architect and Cosentini Associates, structural engineers
Photograph by Carol Highsmith/Wikimedia Commons

Architectural theory has focused on the images of ideal worlds and the design forms and patterns advocated by leaders in the field. These theories are based on (1) some image of how the world works, its problems, and (2) images of good enough, if not ideal, futures. The former consists of a model of how things work; the latter consist of prescriptions of how it should work. The former is the domain of *declarative theory*, the latter of *prescriptive theories*. They both have generally dealt with *substantive* concerns—the nature of buildings as habitations and displays.

It is equally important for architectural students and architects to understand different ways of thinking about the design process. The concern is with *procedural theory*—the domain of concern of design methodologists. A distinction can also be made between the way we work—*declarative procedural theory* and the way different advocates say we should be working—*prescriptive procedural theory*. This book is a thesis on the scope of architectural theory, both substantive and procedural, and both declarative and prescriptive.

In practical fields such as architecture a gap often exists between theory and practice. Theory is regarded as one thing and the nature of practice as another. Theory, to be of use in understanding the nature of architecture, must, however, inform practice and, in return, it must be informed by practice (Hutchings and Jarvis 2012). What, however, do we mean by theory?

If George Orwell (1903–1950), an English novelist and essayist, was authoring his essay *Politics and the English Language* today, he might well have included *theory* as one of the words that contemporary usage has rendered meaningless (Orwell 1954). The term is used in many ways. As this Prologue outlines the domain of architectural theory it is important to be clear about what is meant by theory and by architectural theory in this book.

Theory

Two widely used meanings of the term *theory* are of importance. In the first, a theory is a mental schema consisting of *descriptions* and *explanations* of a phenomenon or a process (Sklar 2000). In *Creating Architectural Theory*, I referred to such descriptions and explanations as *positive theory*. Critics are uneasy with that nomenclature because it can be seen to assume a positivist philosophical position. That view assumes that no truth exits beyond that which can be verified or falsified as espoused by the Scottish philosopher David Hume (1711–1776) (see Ricœur 1977). No such meaning was intended but the use of the term *declarative theory* avoids any such misunderstanding.

Declarative theories of *phenomena* or *processes* are ideally based on systematic studies following the rules that govern scientific research particularly those of operationally defining terms and repeated and controlled observations (Schuyt 2019, Brewer 2020). Much in architecture is, however, based on the casual observations that we make of the world as part of our everyday lives. We tend to generalize from single observations to a whole class of phenomena. Less honestly, theories may simply be fabrications based on the values of a person. Such views are used to legitimize an ideological position for the actions one takes.

Much about our knowledge remains at a hypothetical level. A *hypothesis* is an untested assertion about a phenomenon or process being the way it is. As such, the term is often used as a synonym of theory but a hypothesis is really an assertion awaiting testing. In a sense, everything that an architect designs is a hypothesis about what works that awaits being assessed in operation when built. Much architectural theory certainly consists of untested hypotheses. Many buildings and urban designs based on it have, fortuitously, turned out well in whatever way that outcome is measured. Other schemes have not achieved their stated objectives. We need to learn from both successes and failures.

The veracity of the description and explanation of a phenomenon or process can be disproven but cannot be proven. A theory stands until it is refuted or one that describes and explains more is devised. The approach to describing and explaining the domain of architectural theory presented here can stand until a better way is found.

In the second use of the term, a theory is comprised of the principles that *prescribe* the views and the actions that are regarded as good practice in a field. In architecture, they are value-laden *paradigms*, of what constitutes a good world or a good object within it or how the design process should be conducted. As such, a theory is an *ideological* position—a rhetorical view stripped to its essential—that shapes intentional acts such as designing a building. In the past I referred to such positions as *normative theories* but the ambiguity of the word 'normative' suggests that referring to them simply as *prescriptions* seems wiser.

Scholars use abstract *models* to display the structure of theories and processes. Models can be in mathematical, graphic, and/or verbal forms. Many aspects of theories of structures in architecture are represented in mathematical formulae. Flow diagrams are used to present the framework of procedural algorithms, and many ideas are presented in verbal form. Drawings, static or dynamic, and physical models at a much-reduced scale are used to examine and illustrate ideas for buildings and urban designs. Good models describe a lot as simply as possible and enable an individual to make accurate predictions of how a product will function when implemented. As our knowledge is advanced, we can improve on the models that we use including the model of theory presented here.

Declarative Theory

The declarative theory of any practical field provides the knowledge base for the actions that take place within it. Each one of us possesses such a body of knowledge about how the world works in our heads. It is based on our daily experiences and through our formal educations. We learn what is acceptable and what is not within the cultural norms that establish the moral codes that shape our behaviour. Such personal theories about the world are biased by our personal lives and may serve us well as a basis for making decisions for ourselves. The questions are: How well do they serve in making decisions for others? How well can we generalize from our own experiences to the world of others? They are after all limited by the lives we lead. The aim of scientific, quasi-scientific, and phenomenological research is to generate widely accepted truths that hold until further research suggests a more accurate way of describing the world (Kuhn 1962, Sklar 2000).

Theories not only consist of descriptions of phenomena but of *explanations* of why those phenomena are the way they are. Explanations are important. Without them the understanding of phenomena is limited. In problem-solving fields, such as architecture, there is a danger of basing decisions on descriptions of situations—on the symptoms of a problem—and not its causes. This situation prevailed in architects' prescriptions for slums. Tackling the underlying causes was neglected. Architects jumped from descriptions to prescriptions.

Declarative theory, at its best, accurately describes and explains the present state of circumstances, events, and processes of change. It provides the basis for predicting the future although predictions must be made cautiously because unforeseen intervening variables can and do alter current states and trends. Our declarative theories are inevitably incomplete and evolve as more research is completed and from feedback received from the outcomes of actions in practice.

Constructing theories, like designing anything, requires creative thought. It involves the conjecturing and testing of suppositions about what is worthwhile and what is not. We often have competing explanations for why commonly accepted descriptive states exist. One of the goals of ongoing research is to test alternative explanations and resolve the conflicts among them.

Prescriptive Theories

Prescriptive theories are created deliberately. They are shaped to maintain the status quo or to enhance or counter trends. Sound declarative theory increases our ability to make wise, if uncertain, prescriptions for the future in circumstances about which we are unsure. Every practical field needs such an intellectual foundation.

Views on what constitutes a good world and good ways to decide how to shape it are value laden. Although a discipline may seek to have a unified body of declarative theory as its intellectual base, it is almost certain that there will be different, possibly radically different, views on what its implications are for making decisions. In many fields there are thus competing ideologies and competing prescriptive theories advancing courses of action. The world of architecture is full of overlapping and opposing prescriptions (see Conrads 1970, Hays 1998, and Taschen 2022 for samples).

Prescriptive theories consist of statements shaped to persuade people to support a specific position that benefits a person or group of people. They are often presented in manifestoes that attempt to persuade people to act in particular ways. Most buildings are based on implicit rather than explicit prescriptive statements.

They are designed *pragmatically* drawing on some declarative theoretical base and implicitly on a prescriptive theory specifying ends and means.

Some prescriptive design theories, such as those based on the *Shastras* in India (Chakrabarti 1998, Bubbar 2005, Sachdev 2011) or *Feng Shui* in China (Anderson 2014) can be highly dictatorial; others are less so. Thoughtful ideological theories not founded on religious precepts are based either on *empirically* observed phenomena and processes that work or on *rationally* derived beliefs and hypotheses of how phenomena or processes should work. Pragmatist, Rationalist, and Empiricist views on good approaches to dealing with the world differ.

In architecture there have been and are competing paradigms although one may hold sway at a time. The paradigm holds until an alternative that is regarded as better is formulated. A paradigm shift then occurs. It may result from incremental changes but is just as likely to be dramatic. The emergence of Modernist architecture was dramatic. Its proponents were strong advocates for its prescriptions to be accepted.

The Scope of Architectural Theory

Some architects consider themselves to be experts in designing anything from a spoon to a city. They have indeed produced thoughtful spoons that one can use to eat with but others are simply objects to look at and admire for their novelty. The design of products such as teaspoons falls outside the domain of concern here although much that is discussed is pertinent to the design of any product. The concern here is with the built environment of human life, but only in passing, with the world of other species, animate or inanimate. The domain of direct concern is narrowly focussed on architects' roles in designing the human habitat.

Declarative Theory: Procedural and Substantive

An understanding of what constitutes architectural theory is moulded by what one accepts as the domain of architecture as an intellectual discipline and a field of practice (Johnson 1994). Most commentaries on architecture are about the qualities of individual buildings and how they represent their author's oeuvre and/or fit current paradigms of what constitutes high quality. These commentaries are a necessary but an insufficient component of architectural theory. To enhance professional work and architectural education the need is to understand the nature of the whole design process and the nature of buildings and urban designs.

Procedural Theory: Understanding the Designing Process

The study of how designs are created is known as *design methodology*. The discussion of methods of analysis, synthesis, and evaluation has been led by scholars seeking both a *declarative theory* of designing and a *prescriptive* one—knowledge of how we *do* go about designing and the different positions on how we *should* go about it. They sought a science of the design process as promulgated in *The Sciences of the Artificial* (Simon 1969) by Herbert Simon (1916–2001), an American political scientist and Nobel laureate.

Architectural works can be configured *unselfconsciously* by habit or *self-consciously* to meet specific goals. A variety of animals, birds, reptiles, and insects make habitats for themselves. The environments they create result from instinctive behaviour although the young of some species clearly learn from the example set by their parents (von Frisch 1974, Hansell 2004, Arndt and Tautz 2014). Architectural design is a self-conscious, intentional activity although much is carried out unselfconsciously (Burnette 2009). Research has enhanced our understanding of the self-conscious design process.

Findings and hypotheses about how architects design are widely scattered and little considered in architectural schools. Students, instead, are socialized into accepted processes of analysis, synthesis, and evaluation by doing them. They internalize a model of how designs are generated. The model is not articulated so is not open to discussion. Students learn to design following a procedural paradigm without fully grasping its structure or being aware of alternative ways of organizing their efforts. The observation is that architects do well enough without an explicit declarative procedural base for the work so why worry about it. Not everyone agrees.

Design methodologists have been seeking an understanding of how design programs, or briefs, are created, design possibilities generated, and then evaluated and implemented (Jones 1992, Vilchis Esquivel 2014). They seek to understand how designers think (Rowe 1987, Lawson 2005) and how they reflect, or not, on their work (Schön 1983). It is generally recognized that each step in designing a building is one of repeated analysis, synthesis, evaluation, and the choosing among possible courses of action. The question is: What is the nature and sequence of these processes?

Our knowledge of declarative procedural theory has advanced considerably since the 1970s. Several scholars have studied the approaches to designing that different architects follow in practice (Eastman 1970, Schön 1983, Gutman 1988, Lawson 2005, Yaneva 2009). Accompanying the development of procedural theory about how to go about designing is a better understanding of the knowledge base on which architects can draw. First, there is a development in our understanding of the *functions* that different patterns of the built environment can fulfil for humans and other species and, secondly, an enhanced understanding of the building *types* and *typologies*.

Functional Theory: Understanding the Purposes of Architecture

Writings on the concerns of architecture go back, at least, to *De architectura: libri decem* written by Marcus Vitruvius Pollio (*ca* 75–15 BCE), a Roman military engineer. His writings have been translated into English as *The Ten Books on Architecture.* Vitruvius suggested that buildings must serve three basic purposes: *utilitatis*, *venustatis* and firmitatis—utility, beauty, and strength. Sir Henry Wotton (1568–1625), an English philosopher and politician, took these ends and, with some hesitation, wrote:

> In *Architecture* as in all *Operative* Arts,
> The *end* must direct the *Operation.*
> The *end* is to build well.
> Well building has three Conditions:
> *Commoditie, Firmenes*, and *Delight* (Wotton 1624).

All later specification of the concerns of architecture are adaptations of Vitruvius' observations.

Traditionally, architects have considered the instrumental or task function (*commoditie*) of buildings as something separate from their aesthetic function (*delight*) but it will be argued here that both constitute part of a hierarchy of functions that buildings serve. *Firmenes* deals with structural and constructional soundness and thus with the technical side of architecture. It has a well-established declarative theoretical basis. It will only be considered here to the extent that it impinges on the way buildings function as objects, as inhabited environments, and as environment-making objects.

Understanding "*Commoditie*" and "*Delight*"

A model of functions based on the theory of human motivations developed by Abraham Maslow (1908–1970), a humanist psychologist, offers the most in structuring our understanding of the potential utility of patterns of built form. Although open to criticism because it lacks scientific rigour, Maslow's model of human motivations, originally devised in the 1940s, has stood up to scrutiny over the years (Cherry 2022). It can serve as an organizing model for presenting the functions served by patterns of built form until a better one is developed.

Maslow hypothesized that people have a hierarchy of needs that they are motivated to fulfil. The most basic is to survive. At the other end of the hierarchy, the most advanced, is for a person to be self-actualizing. The set of needs motivates people to take actions on many dimensions of their lives. One dimension is to devise and inhabit physical settings with specific characteristics and to possess objects that enhance their ways of life and/or simply display their tastes and social status. How these settings and objects are created must be seen within a cultural context and moral order. Functional theory in architecture is thus concerned with describing and explaining what different patterns of objects and environments can afford people.

The goal in the ongoing development of functional theory is to provide architects with an intellectual framework for understanding their own work and what it accomplishes. It is also to serve as the basis for asking questions about what *should be done* to meet the goals of a range of clients, sponsors, and users who have different interests and motivations to achieve the ends they seek. Functional theory further informs architects of the extent and the limitations of the social impact of their work. Implicitly, it warns us to neither claim too much for it nor to belittle the social consequences of designs. This imperative is difficult to heed because architects are businesspeople. Boosting the importance of their work and emphasizing its good qualities is often necessary to clinch a deal.

Typology

Typology is the study and categorization of phenomena according to their similarities on one dimension or another (Croft 2002). For architectural theory the concern is with buildings and urban designs and the instrumental functions they serve and/or their functions as aesthetic objects to be examined and admired (or not). Buildings and urban designs are categorized into types by their geometric shapes or organizational logic and/or such distinctions as public and private spaces (A. Colquhoun 1967, Moneo 1978, Franck and Schneekloth 1994, Madanipour 2003). Most frequently they are categorized by the instrumental purposes they serve. The architectural literature is full of buildings type studies: hospitals, schools, museums, theatres, retail shops, etc. (see, for instance, Sherwood 1978 on modern housing types).

Architects rely heavily on an understanding of existing building types and *generic designs*—conceptual solutions to classes of standard problems—when creating designs for familiar instrumental building types (Argan 1963, Colquhoun 1967, Güney 2007, Heyda 2019, Graves 2020). In doing so they work within one or other contemporary paradigm. Every now and then new generic designs are introduced on to the scene by innovative architects who manage to convince others that their new paradigms are superior to existing ones. The success of the Rationalists among the Modernists in convincing other practitioners and academics of the rectitude of their design paradigm is a prime case in the recent history of architecture. The Neo-Traditionalists are having some similar success in the early twenty-first century. The Postmodernists convinced fewer architects of the virtue of their ideas.

Prescriptive Theories: Substantive and Procedural

Architects, explicitly or implicitly, take stances on what constitutes good architecture and what are good design processes. In education and in practice the focus has been on the former. The latter and its impact on the former are seldom rigorously discussed. We have the knowledge to do better.

Substantive Theories: Good Worlds and Good Architecture

The beliefs of individual architects about what constitutes good architecture is a fundamental part of their knowledge base. Students learn about the paradigm that a specific architect follows and what qualities establish that architect's brand of work. They learn less about the characteristics of the people who commission architectural works. The focus is on the architect's ideas and how they shape their designs. Organizing those thoughts and designs so that they can be understood by others has been done in many ways. The one that explains the most is by organizing them first into the competing philosophical traditions into which they fall.

Distinguishing among architects who are *Pragmatists*, *Rationalists*, and *Empiricists* and how each line of thought acts in parallel and/or in competition with the others is a fruitful way of thinking about architectural works and the ideologies on which they are based (van Schaik 1985). All architects fall primarily into one or other category, some more clearly than others. Each's prescriptions and work, however, show characteristics of all three philosophical traditions. Some pragmatic thought is certainty required to get a project implemented. In addition, an architect's approach may evolve from thinking in one way to another. This last observation seems to be particularly true of those architects and critics who were once staunch Rationalists but became doubters. The boldness of Rationalists' ideas and works and their departure from the norms of practice is, nevertheless, much admired by critics and in architectural academia. Recently, the creative efforts of Empiricists, who base their designs on precedents, have been more closely scrutinized. The work of Pragmatists is largely neglected even though many, if not most, architects work for large commercial firms producing highly pragmatic work.

The Rationalists make assumptions about what are good worlds and what ought to be based on their intuitions (or insights), logical thinking, and deductive reasoning (Sharp 1978, Peckham and Schmiedeknecht 2013). Their ideas are based on what to them are ideals. They set precedents for pragmatic architects to imitate. The Empiricists, among architects, base their thinking on experiences with what works in the everyday world and inductive thinking and not so much on introspections. Their work, nevertheless, like that of the Rationalists, is backed by a sense of idealism, sometimes messianic idealism.

Procedural Ideologies

Declarative procedural theory yields a general understanding of the structure of the design process but that structure can be operated in different ways. How should we go about designing? What do different architects advocate? Two schools compete for attention. Their positions were clearly articulated by Colin Rowe (1910–1999), an architectural historian and critic (Rowe 1983). One is that architects should base their designs on the current paradigms developed by leaders in the field. It is a form-led approach to design. The second position is that architects should base their work on carefully crafted, uniquely designed, programs/briefs that specify the instrumental requirements to be met and the aesthetic ends that should be sought when designing a building. Following this approach, the patterns required to solve each aspect of the program are worked out and adjusted to achieve an integrated solution. The design of the program is regarded as the truly creative act.

Implicit in these observations is that every design is based on an advocacy for the interests of one group of people rather than another. Almost all building designs deal with a multiplicity of stakeholders—owners, financial sponsors, the cognoscenti, future users, plus the public interest. In practice, the aesthetic qualities of the designs that are admired by the profession and that feature in architectural journals generally cater to the cognoscenti's tastes. This line of thought leads to architecture being something for architects to admire. If others do not understand it, it does not matter.

Although all designs are advocacies for one set of ends or another, *Advocacy Designers* are those who argue that a broader group of people, particularly the less privileged, and their ends should be the focus of attention when designing a new building and certainly when creating urban designs (Appleyard 1982, Hatch 1984, Ornstein 2022). The concern for designing with users in the forefront of an architect's mind is now often dismissed as a 1980's concern. The issue of for whom architects are designing is, nevertheless, still with us (Filippe and Hooper 2019, Pelletier 2019, Troy 2021). Procedures for designing with a multiplicity of stakeholders are constantly being promoted.

Whatever approach is advocated, questions arise about how best to go through the phases of the design process. Should architects go through them with a speedy depth-first approach to arrive at a design and then incrementally improve it? Should they then try to develop other possibilities in the same manner? Alternatively, should architects strive to carry parallel lines of thought through to achieving several designs to evaluate? Prescriptive theory deals with the positions taken on such concerns.

Commentary

Scholars have given considerable thought to the nature of functional theory, typology, and design methodology over the past seventy years. Some architects have learnt from the results but others treat them with diffidence. This situation has resulted from the failure of researchers to communicate their findings in a manner that interests architects and/or researchers' failures to address issues of interest to practitioners.

Three widely held beliefs about declarative theory continue to shape architectural practice and education. The first is that a person's experiences and personal beliefs combined with common sense hold the profession in fine stead. They suffice. They are good enough. Common sense and intuitive thinking are certainly necessary in obtaining good designs but they are insufficient. Our common sense has let us down too frequently and many phenomena work counter intuitively. We have learnt those lessons the hard way. In addition, buildings when completed are seldom systematically examined for what they offer in use once the honeymoon period of their lives has ended. Photographs of buildings uninhabited do not teach us much about how well they function except as objects to contemplate.

The second belief is related to the first. It is that a having a strong, well-researched body of declarative theory (other than a sound declarative theory of structures) inhibits creative thinking and the production of innovative geometric forms. It is, indeed, difficult for an architect to create a geometrically innovative design and to defend it knowing that it will not function well in what it affords for the activities and symbolic interpretations of those who commission it or those who are exposed to it as part of their everyday lives.

A third reaction is based on an accurate, but unhelpful, observation. Architects have created fine buildings and urban complexes without having a solid body of explicit declarative theory. This situation means that matters can be discussed without a full understanding of the issues of concern. As the demand for better public accountability grows, so a strong declarative base for making, explaining, and defending design decisions becomes imperative.

Understanding procedural theory assists architects in working with clients. It enables them to explain how they go about creating building programs, designs, and how they evaluate them. In doing so architectural theory needs to learn from practice and help in informing practice. To be able to fully understand what a building or urban design offers and might offer requires a strong declarative theoretical base. Such a base will never be complete, but we understand much more than we acknowledge. Understanding our own processes of designing and what procedural options are open to us enables us to self-consciously decide how best to work in different circumstances. Using such bodies of knowledge enriches both practice and makes architectural students more aware of the full range of potential future professional roles available to them.

References

Hutchings M, Jarvis P (2012) The relationship between practice, theory and research. In: Practice-Based Education: Practice, Education, Work, Society 6. Rotterdam: Sense Publishers, 175–186. https://doi.org/10.1007/978-94-6209-128-3_13. Accessed 3 Sept 2022

Kuhn TS (1962) The structure of scientific revolutions. Chicago, IL: University of Chicago Press

Lawson B (2005) How designers think: the design process demystified. Revised edition. Oxford: Architectural Press
Schön DA (1983) The reflective practitioner: how professionals think in action. New York, NY: Basic Books
Simon HA (1969) The Sciences of the Artificial. Cambridge, MA: MIT Press
Sklar L (2000) The nature of scientific theory. Philosophy of Science: Book 2. London and New York: Routledge
van Schaik L (1985) Rationalism and contextualism. Architecture SA (May–June), pp 52–56

Acknowledgements Over the years I have learnt much from architectural students and practicing architects. The story presented in this book relies heavily on their ideas and the research of others. I have assembled them to flesh out the story I offer here. Caroline Nute provided invaluable editorial assistance in getting that story straight. The responsibility for its limitations is mine alone.

The names of the people who inspired and informed this book appear throughout it and in the References and Bibliography at its end. Many largely forgotten books and articles published during the twentieth century have been recently republished. Once, brushed aside, their message remains valid. In referencing them I have included the first date of their publication to indicate the era in which they first appeared in print.

I have acknowledged the sources of all information to the best of my ability. I apologize to those whose work I may have failed to acknowledge or have acknowledged incorrectly. I hope no apologies are needed. I, unapologetically, have plundered my own writings particularly *Creating Architectural Theory* (1987), *The Routledge Companion to Twenty and Early Twenty First Century Urban Design* (2021), and *Functionalism Revisited* (2010) that Walter Moleski (1936–2019) and I wrote together. I have used material from those works and other of my writings to structure our contemporary thinking about the scope and nature of architectural theory. I thank Taylor and Francis for allowing me to do so.

The illustrations come from many sources. Unless otherwise acknowledged in Credits at the end of this book, the photographs, diagrams, and tables are my own work or I hold their copyright or the copyright has expired. Some have been so widely published it has been impossible to ascertain who holds their copyright. In addition, colleagues and students have given me photographs over the years. Where I have failed to record their sources, I have simply stated that they are part of my collection. I am extremely grateful for all these gifts and regret my failure to keep track of who provided me with them. Please contact me at jonl@unsw.edu.au about any missing or erroneous attributions of copyrights. I am anxious to rectify them.

Competing Interests The author has no competing interests to declare that are relevant to the content of this manuscript.

Contents

Part I Procedural and Substantive Declarative Theory: The Way Matters Are

1 The Built Environment 5
1.1 The Cultural Context 5
1.2 The Nature of Environments and the Nature of Objects 6
1.2.1 The Concept of *Affordance* 7
1.2.2 The Concept of Competence and the Perceiving of Affordances 9
1.2.3 The Concept of Costs and Rewards in Evaluating Environments 10
1.3 Commentary 10
References 12

2 Procedural Theory: The Nature of the Design Process 13
2.1 Unselfconscious and Self-conscious Design 14
2.2 The Nature of Architectural Problems 14
2.3 Generic and Specific Models of the Design Process 16
2.4 The Phases of the Design Process 19
2.4.1 The Intelligence Phase: Perceiving Problems/Opportunities 19
2.4.2 The Intelligence Phase: Designing the Program 20
2.4.3 The Design Phase and Nature of Designing 23
2.4.4 The Choice Phase: Evaluating Potential Designs and Selecting One 25
2.4.5 The Implementation Phase 27
2.4.6 Evaluating Completed Designs 27
2.5 Three Approaches to Going Through the Phases 28
2.5.1 The Design-Led Approach 29
2.5.2 The Type-Led Approach 29
2.5.3 The Problem/Program-Led Approach 31
2.6 Commentary 31
References 32

3 Functional Theory 33
3.1 Functionalism 34
3.2 The Basic Functions 34
3.2.1 The Accommodation of Activities 35
3.2.2 Salubrious Environments 40
3.2.3 Safety and Security Functions 41
3.2.4 Community and Social Status 44
3.2.5 Delight and Experiential Aesthetics 45
3.2.6 Taste Cultures 50
3.3 The Advanced Functions 52
3.3.1 The Cognitive Function of Environments and Objects 52
3.3.2 Intellectual Aesthetics and the Cognoscenti 52
3.4 The Catalytic Function of Buildings: Multiplier and Side Effects 53
3.5 The Concept of Fit 55
3.5.1 Fixed-Feature Spaces 56
3.6 The Functions of Buildings for Architects 56
3.6.1 Central Ideas and Brands 58
3.7 Commentary 58
References 59

4 Types and Typologies 61
4.1 Architectural Typologies 61
4.1.1 Building Typologies and the Potential Functions of Architecture 65
4.2 Typology and Types Today 65
4.2.1 Urban Form Types 66
4.2.2 Elements of Urban Form 66
4.2.3 Changing Technologies and the Emergence of New Types 69
4.3 Urban Design Types 72
4.4 Building Types 72
4.4.1 Form Types 73
4.4.2 Structural Types and Construction Types 73
4.4.3 Instrumental Functional Types 75
4.4.4 Building Component Types 77
4.4.5 Lighting Types 77
4.5 Invented Ideal Types and Generic Solutions 81
4.6 Commentary 83
References 86

Part II Prescriptive Theories: Views of the Way Matters Ought to Be

5 Ends and Means: A Prolegomenon to the Study of Prescriptive Theories of Architecture ... 91
5.1 Clients and Their Ends ... 92
5.1.1 Buildings and Urban Designs as Public Art—Displays—And as Habitable Environments ... 95
5.2 Pragmatists, Empiricists, and Rationalists ... 97
5.2.1 The Pragmatists ... 97
5.2.2 The Empiricists ... 99
5.2.3 The Rationalists ... 101
5.3 Rationalism and Empiricism and Postmodern Architecture ... 102
5.4 Pragmatism, Rationalism, and Empiricism in Architecture Today ... 103
5.5 Advocacies for Specific Ends ... 104
5.5.1 Climatically and Ecologically Sound Environments ... 104
5.5.2 Culturally Appropriate Environments ... 106
5.5.3 Barrier-Free Environments ... 107
5.5.4 Educative Environments ... 107
5.6 Colour Advocacies ... 108
5.7 Designing for the Public Interest: Design Controls and Architecture ... 109
5.7.1 The Interests of Other Species ... 110
5.8 Means ... 111
5.8.1 Designing for *Firmness* ... 111
5.8.2 The Architecture of Structural Dexterity ... 113
5.9 Commentary ... 114
References ... 114

6 Architectural Prescriptions: A History of the Development of Ends and Means ... 115
6.1 Unselfconscious Design: Vernacular Architectures ... 115
6.1.1 Clients and Designers ... 116
6.1.2 The Work: Types and Functions ... 118
6.2 Self-conscious Design ... 119
6.2.1 Classical Architecture ... 119
6.2.2 Medieval Architecture ... 122
6.2.3 The Renaissance ... 128
6.2.4 Revivalism: Neoclassic, Neogothic, and Beaux Arts Architecture ... 135
6.3 Self-conscious Design: Twentieth-Century Modernism ... 140
6.3.1 The Pragmatists ... 141
6.3.2 The Empiricists ... 143
6.3.3 The Rationalists ... 150

6.4 Self-conscious Design: Post-modernism 157
6.4.1 The Architects 160
6.4.2 The Clients 160
6.4.3 The Work 161
6.5 A Postscript: Self-conscious Design in the Early Twenty-First Century 163
6.6 Commentary 165
References 167

7 Designing Ends and Means: Prescriptions for the Design Process 169
7.1 Operating the Design Process: Design-First and Program-Oriented Approaches Revisited 170
7.1.1 The Design-First Approach 170
7.1.2 Program/Problem-Oriented Design 171
7.2 Advocacies for Getting All the Stakeholders Involved in Developing Designs 174
7.2.1 A Note on Advocacy Design 175
7.3 Advocacies for Systematizing the Designing Process 177
7.3.1 Computer-Based Design-Generating Algorithms 178
7.4 Commentary 181
References 181

Part III Architects, Their Prescriptive Theories, and Their Designs

8 Understanding Architects' Prescriptions and Their Designs 187
8.1 Architectural Firms and Individual Architects 188
8.1.1 Professed Positions and Practised Work 189
8.2 Three Pragmatists 190
8.2.1 Howard Van Doren Shaw 191
8.2.2 R. Seifert and Partners and Richard Seifert 195
8.2.3 Alejandro Aravena Mori and Elemental Consultores S.A 199
8.3 Three Rationalists 205
8.3.1 The Le Corbusier and Pierre Jeanneret Partnership and Le Corbusier 205
8.3.2 Bernard Tschumi Architects and Bernard Tschumi 211
8.3.3 Zaha Hadid Architects (ZHA) and Zaha Hadid 217
8.4 Three Empiricists 222
8.4.1 The Angles Architects and Darshan Kumar Bubbar 222
8.4.2 Abdel-Wahed El-Wakil, Architect 227
8.4.3 DPZ CoDesign and Andres Duany and Elizabeth Plater-Zyberk 232
8.5 Commentary 237
References 238

Epilogue 241

Credits 255

References and Bibliography 261

Bibliography 263

Index 289

About the Author

Jon Lang is an international consultant on architecture and urban design and is an emeritus professor of architecture at the University of New South Wales. Jon was educated in India, England, South Africa, and the USA. He was formerly a director of ERG, Inc., in Philadelphia, Pennsylvania, and headed the joint M.Arch./MCP program at the University of Pennsylvania. He has served as a juror, including as chief juror, on international architectural competitions. He has authored books on architectural and urban design theory and practice and modern architecture in India. In 2010, he received the Reed and Malik Medal from the Institution of Civil Engineers in London.

Books by the Author

Designing for Human Behaviour: Architecture and the Behavioural Sciences edited with Charles Burnette, Walter Moleski, and David Vachon. Stroudsburg, PA: Dowden Hutchinson and Ross, 1974.

Creating Architectural Theory: The Role of the Behavioural Sciences in Environmental Design. New York: Van Nostrand Reinhold, 1987. This current endeavour draws heavily on the line of thinking presented in that book.

Urban Design: The American Experience. New York: Van Nostrand Reinhold, 1994.

The Architecture of Independence: The Search for Identity—India 1880–1980 authored with Madhavi Desai and Miki Desai. New Delhi: Oxford University Press, 1997. Second edition published by CEPT University Press in Ahmedabad, 2022.

Urban Design: A Typology of Processes and Products Illustrated with over 50 Case Studies. Oxford: Architectural Press, 2005. The second edition was published by Routledge, London and New York, in 2017.

A Concise History of Modern Architecture in India. New Delhi: Permanent Black, 2010.

Functionalism Revisited: Architectural Theory and Practice and the Behavioural Sciences authored with Walter Moleski. Farnham: Ashgate, 2010. The intellectual framework of *The Scope of Architectural Theory* draws heavily on that presented in the book which, in turn drew on the building programming work of the Environmental Research Group, Inc. in Philadelphia.

The Bungalow in Twentieth Century India: The Cultural Expression of the Bungalow in the Domestic Architecture of Colonial and Postcolonial Society authored with Madhavi Desai and Miki Desai. Farnham: Ashgate, 2012.

Urban Squares as Places, Links and Displays: Successes and Failures authored with Nancy Marshall. New York: Routledge, 2016.

The Great Houses of Calcutta: Antecedents, Precedents and Portents authored with Joanne Taylor. Delhi: Nyogi, 2016.

The Routledge Companion to Twentieth and Early Twenty-First Century Urban Design. A History of Shifting Manifestoes, Paradigms, Programs, Generic and Specific Designs. London and New York: Routledge, 2021. The book presents an exploration of architects' prescriptions for design within the intellectual framework provided by pragmatic, empirical, and rationalist philosophies.

Part I

Procedural and Substantive Declarative Theory: The Way Matters Are

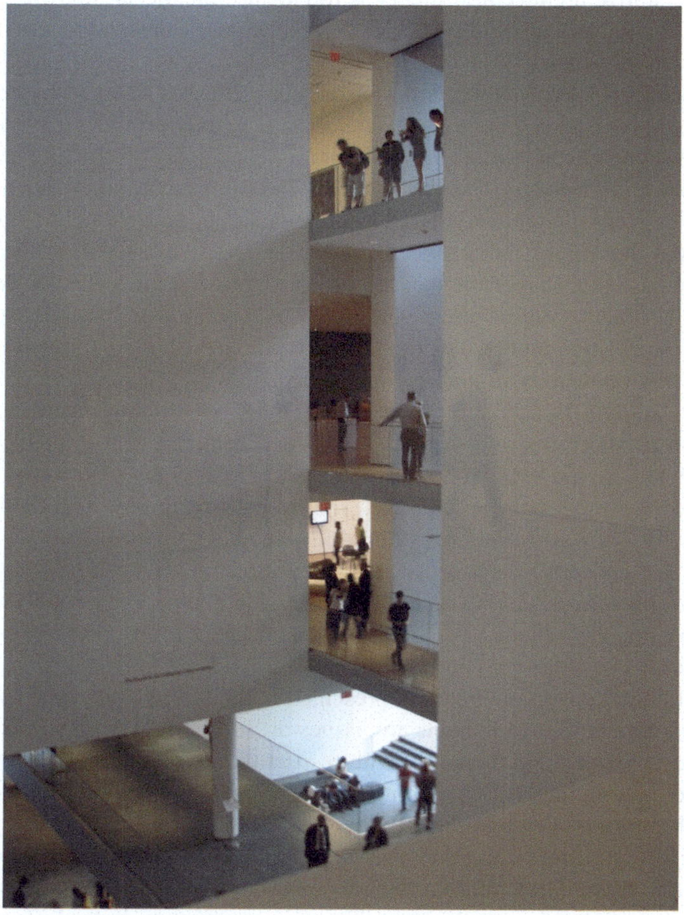

Museum of Modern Art, New York
MOMA Trustees client, Philip Gordon and Edward Durrell Stone, architects
Photograph by the author

Part I begins with descriptions of the nature of the built environment and the nature of the designing process. Most architects follow a typological approach to design.

The need is to understand what functions patterns of the built environment can serve and the functions implicitly served by specific building types and generic solutions.

The world is complex. The tendency is for people to be selective about the concerns they address. It is easier for us to deal with the ones that we feel comfortable in addressing. Consciously or subconsciously, architects are encouraged by leaders in the field to challenge building programs and to redefine problems in their own terms to accommodate their own brands of design. The danger is that in doing so, important issues are neglected. Drawing on abstract theoretical knowledge enables architects to deal with a higher level of complexity than they are wont to do. The role of declarative architectural theory is to enhance architects' knowledge of (1) the nature of the artificial built world and (2) the ways the design process has been, is, and can be conducted.

To structure our knowledge, Part I of the book consists of four chapters. Each chapter could have been a book so the coverage here only provides a synopsis of a topic—the scope of concern. The goal of all four chapters is to structure and access our cumulative knowledge. The first of them, The Built Environment, identifies the domain of concern of architects. The second reviews our understanding of the ways we go about designing. Based on this understanding, the discussion turns to the knowledge base that shapes the making of decisions. The third chapter deals with the potential functions of different patterns of built form. The fourth discusses the functional knowledge embedded in building typologies and generic solutions that architects rely on heavily in making design decisions.

Theory building is an ongoing creative activity. Continuing research and the recognition of successes and failures in practice continue to inform theory. The four chapters here are thus works in progress and reflect the reality that the content of theory always will be tentative. The ever evolving physical, cultural, and social state of the world challenges our ability to stay abreast of change. Many of the matters of concern to architects are difficult to study rigorously. Variables are bundled together so designing research studies that control for possibly intervening variables pose problems. We can assume and, indeed have done so too easily, that correlated variables are causally linked.

Several limitations to the material presented in the four chapters need to be recognized. Important though budgets are, issues of 'form following finance' are only considered in passing and then only with respect to the limitations they impose on clients and their architects implementing their ideas. It must, nevertheless, be recognized that no building is erected based on an infinitely elastic money supply even though the Sydney Opera House with its 1400% cost overrun, may appear to have been so supplied (Willis 1995, Wainwright 2019). All buildings need to meet Wotton's *firmnesse* concern too. Theories of structure both declarative and prescriptive are only addressed to the extent that they impinge on what architects dream about and can accomplish.

Despite these misgivings, there is much that we do understand and can apply to creating a sound declarative theoretical base for architecture and architectural education. The objective here is, in sum, to provide a structured foundation for making our cumulative knowledge retrievable and so useable. The overall goal is to enhance the utility, liveability, and enjoyment of buildings and, more generally, the built environment of human settlements.

1 The Built Environment

Human beings (along with other species) have created an artificial world of settlements, buildings, streets, squares, and parks on the planet Earth. Agricultural and gardening practices have rearranged the planet's vegetation. This artificial world consists of all the changes that have made to the *terrestrial*, or geographic, environment of a locale in response to the evolution of species' capabilities, activities, and, for humans at least, aesthetic values. This world is not static but is constantly being altered, for better or worse, by people's constructive and destructive actions, and by natural chemical processes abetted by wind and rain.

Many changes in the world result from the *unselfconscious* piecemeal and habitual actions of people conducting their daily lives. Others are the product of *self-consciously* planned changes that are based on explicit, predetermined courses of action. Interior designers, architects, landscape architects, and urban designers, and a host of engineers have taken on specific roles in helping societies, and groups and individuals within them, to determine the best way to make the changes that they seek. Inevitably, the changes that are made, whether unselfconsciously or self-consciously, benefit some people more than others. Often, they have had a detrimental effect on the lives of other species, animate and inanimate, that inhabit the planet that we share and the functioning of the planet itself. The activities and values of people and the environments and objects they have created and the way they have done so are part of the diverse cultures that exist on our planet.

1.1 The Cultural Context

People differ. Physiological differences exist among them because over the centuries they have adapted to habitats that have unique qualities. Social networks have evolved in these different settings. People have been and are socialized into

J. Lang, *The Scope of Architectural Theory*, Design Science and Innovation,
https://doi.org/10.1007/978-981-96-3690-7_1

different cultural groups by their formal educations and informally as part of their everyday interactions with other people, other species, and different places. The built environment forms part of the culture of a society within which sub-societies exist each with its own cultural qualities, some unique but others universal.

The culture of a society consists of a *tangible*, or *material*, component consisting of artefacts created within it, an *intangible* set of languages and rituals, and an *abstract* component of behavioural norms and values. The three are closely intertwined (Suswanto 2000). Moral codes, ways of life, individual roles, social networks and aesthetic values are part of that overall cultural environment. The professions that have established themselves in the world, including that of architecture, have their own cultures that exist within those of the broader society of which they are a part. Their members are socialized into cultural norms by formal educations or through apprenticeships.

Changes take place in cultures by the implementation of public policies that are designed to alter behaviour patterns and attitudes, technological developments, and by individual actions and advocacies. Today, some observers worry that the globalization of the world's capital investment mechanisms and the ease of communication across the planet will result in a reduction in the diversity of cultures. They decry the development of globally similar patterns of behaviour and values; others embrace such changes.

Architectural artefacts are shaped by the creative energy of their makers but also by the prevailing values of the social, political, and economic worlds in which they are created. The design paradigms—the acceptable designs and way of designing of the Modernists among architects—were formulated within an atmosphere of strong socialist tendencies. Today they are being informed by neoliberal economic attitudes. Design paradigms are also shaped by the changing values and fashions acceptable to the profession itself, the demands of its clients, and the desire of individual architects to express themselves as artists.

The desire to ally architecture with the fine arts goes back to the profession striving to attain its own unique identity. Architects wished to distance themselves from engineers. They saw themselves, and their clients, as people with the "right stuff" and "good taste" (Wolfe 1981). A theme running throughout this book is how different architectural movements and individual architects have dealt with and are dealing with the twin tugs pulling on their hearts and minds when creating buildings and urban settings as inhabited sculptures—between designing environments for everyday life and creating objects to contemplate.

1.2 The Nature of Environments and the Nature of Objects

Environments surround. They are places composed of the arrangement of surfaces, natural or artificial, that are made of a variety of materials possessing different textures and pigmentations. The surfaces have different degrees of rigidity, strength, and transparency or translucency. They are organized to make spaces that afford a diversity of human activities. Architects think of buildings as a set of *spaces*. This view was promoted by Sigfried Giedion (1888–1968), a Modernist architectural historian and critic in his book *Space, Time and Architecture* (1963). In daily life we, consciously or subconsciously, consider them to be a set of *behaviour settings* (Barker 1968).

The world and the locales within it consist of a nested set of evolving behaviour settings enclosed to greater or lesser degrees by surfaces of different qualities. They are furnished with fixed, semi-fixed and movable objects that support people's desired activities and comfort levels within the resources available to them. The arrangement of surfaces is constantly being changed by people to better accommodate their lives. The changes are also made for people to admire for their geometric and sculptural qualities. Architects are much involved in making many of these alterations although most such changes are made by a variety of individual and groups seeking to better their circumstances.

If environments surround to create places, *objects* are seen from the outside as elements in our surroundings—as features that potentially have some instrumental utility and/or as displays. They are seen as figures against a background. They may be part of the natural world or be shaped by human hands. Objects vary in size from something like a cut and polished diamond, to a sculpture, to a building, to a landscape, to a whole city and, if one is in outer space, the whole planet. It does require a special mode of attention to consider environments as objects but sometimes we do stop and contemplate them as such. We certainly photograph them as objects.

The organization and structure of the natural and artificial environments creates the settings—the milieus—that afford the interactions between people and the terrestrial world and among people. The existing patterns of built form afford or do not afford the functions that we want them to serve. These functions vary from the most basic one of providing of shelter to more complex ones of identity and, even more advanced, the self-conscious analysis of aesthetic concerns whose logic is meaningful only to those in the know, the cognoscenti. We change the existing patterns to better afford our purposes (Fig. 1.1).

1.2.1 The Concept of *Affordance*

The word *affordance* was coined by James Gibson (1904–1979), an American perception psychologist. He defined the word as follows:

> The affordances of the environment are what it offers the animal, what it provides or furnishes, either for good or ill. The verb to afford is found in the dictionary; the noun affordance is not. I have made it up. I mean by it something that refers to both the environment and the animal in a way that no existing term does. (Gibson 1979, 127)

The concept of affordance is now commonly used by architects. The term clarifies the relationship between people and the built environment. Architects have

Photograph by the author

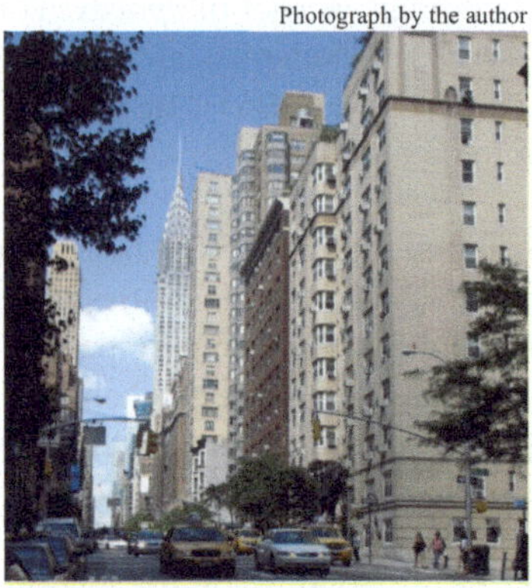

Photograph by Gren/Wikimedia Commons

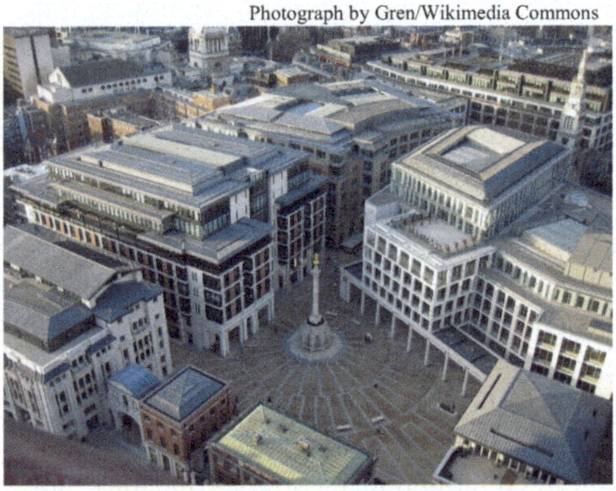

Buildings creating environments – as place makers, 4th Avenue, New York (left) and Paternoster Square, London (right)

Photograph by the author

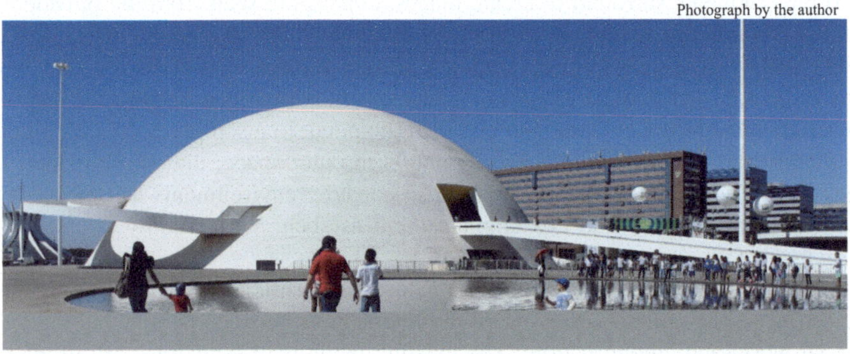

Buildings seen as objects – as displays in space: The National Art Gallery and government buildings, Brasília

Photograph by the author

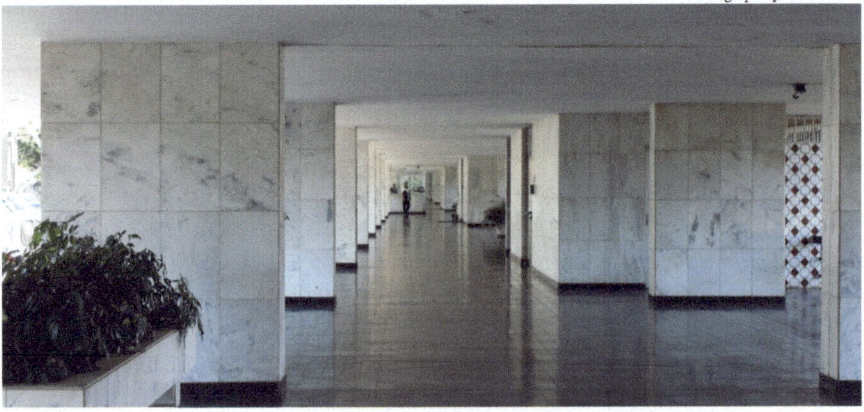

A built space as an object or as an environment depending on the viewer's focus of attention

Fig. 1.1 Built forms as environments and as objects

traditionally used other terms such as 'capability', 'environmental availability', 'environmental opportunity' and 'invitational quality' to convey a similar idea.

Different patterns of the environment and their various physical qualities afford humans ambient conditions that provide comfort levels, behavioural opportunities, interpretations of meaning, and aesthetic appreciations. Whether we recognize and pay attention to the affordances depends on our knowledge, competencies, physical and mental, and our motivations. As Salvador Dali (1904–1989), the Spanish Surrealist artist, observed "A chair may serve many purposes though not necessarily that of sitting down on" (cited in Westerman 2019). It can be a stool, a weapon to wield, or simply be an object—an aesthetic display—for people to contemplate.

The built environment and the objects it contains afford a variety of human experiences. Perception is multimodal. When designing architects tend to pay more attention to the affordances of what people see than what they hear, feel, or smell because design proposals are modelled in drawings. Our experience of the behaviour settings we inhabit depends, however, on the range of perceptual information that we obtain from our surroundings, and the people and the activities taking place within them (Rasmussen 1959; Gibson 1979).

1.2.2 The Concept of Competence and the Perceiving of Affordances

Recognizing the affordances of the different patterns of the built environment depends on the motivations and competencies of a species and/or one of its individual members. Our actions, ways of thinking and, more generally, ways of life are constrained by our physical and intellectual abilities. Our competences develop as we grow up and experience the world and receive formal instructions. Many competencies decline with ageing. In illnesses competence levels sink. The frailer people are, the more the nature of the built environment affects the range of their activities. People recognizing the affordances of an environmental pattern for others and not being able to avail themselves of those affordances may feel frustrated and envious, even angry.

Our competencies in dealing with our surroundings are limited by how well our perceptual systems function. The blind pay attention to and recognize the affordances provided by other than their visual systems. For them, the sonic and tactile information drawn from the environment is full of meaning while the fully sighted hardly pay attention to it (Southworth 1969; Tovar 2023b). The sounds, textures, and smells of the world, nevertheless, always impinge on and shape our experiences.

As shown in Fig. 1.3 (left) the lower our competencies the greater the difficulty we have in negotiating our way around the everyday world (Lawton 1989). The greater the physical competence of people the more the most challenging of environments, such as climbing sheer rock faces can be tackled. There is a degree of environmental press—the level of challenge presented by the built world—at which an individual feels comfortable. What the diagram also shows is that the less demanding our social and physical environments are for us, the more our competencies are likely to decline to the level demanded of them. This reality has long been a significant issue of contention in the design of places such as housing for the elderly and mental health institutions (see, for instance, Goffman 1961) (Fig. 1.2).

Defining the appropriate level of environmental press to design for in the public environment is a subject of debate. Creating barrier free environments for those in wheelchairs is now expected of architects in many countries. Designing for the blind, in contrast, is poorly considered (Nicolau et al. 2009). Designing for people with dementia is seldom a concern in places other than institutions housing the elderly but the knowledge of how to design dementia-friendly neighbourhoods exists (Passini et al. 1998; Mitchell et al. 2004; Weaver 2023). Individuals often withdraw from participating in life because the environmental press on them is too severe. They lead lonely lives.

The ability to function well in the built environment contributes to a person's sense of wellbeing and self-esteem. While individuals may raise their sense of self-esteem in overcoming obstacles, if other people impose those obstacles the implicit message is that those others do not care.

1.2.3 The Concept of Costs and Rewards in Evaluating Environments

In the everyday world our behaviour is guided by our motivations and our perceptions of the consequences, positive and negative, of engaging in specific behaviours. We thus evaluate the milieu (and, more likely, the whole behaviour setting in which we are engaged) in terms of the costs and rewards of being in a place. We have much higher expectations of those places that we inhabit or visit that are expensive, either in monetary or psychological terms, than of those that are cheaper (see Fig. 1.3 (right)). An ideal world would be one in which the costs are zero and the rewards infinite. Striving for such a world is a worthwhile goal.

Our expectations of places are also shaped by what we are used to seeing and using—our habituation levels (Cherry 2023b). The totally new can be a shock (Hughes 1980). Relatively small changes from the norm can be pleasing but major changes take some getting used to. Structures that we once deplored become part of our daily worlds. The Eiffel Tower in Paris is an example. Parisian hated it when it was being constructed but it is now a much-loved landmark in the city.

1.3 Commentary

Architects have taken on the role of designing changes to aspects of the terrestrial environment in response to the changing needs and demands of societies and individuals—clients for services—within them. They design environments and objects for a subset of human populations to use and enjoy. The construction industry

Photograph by the author

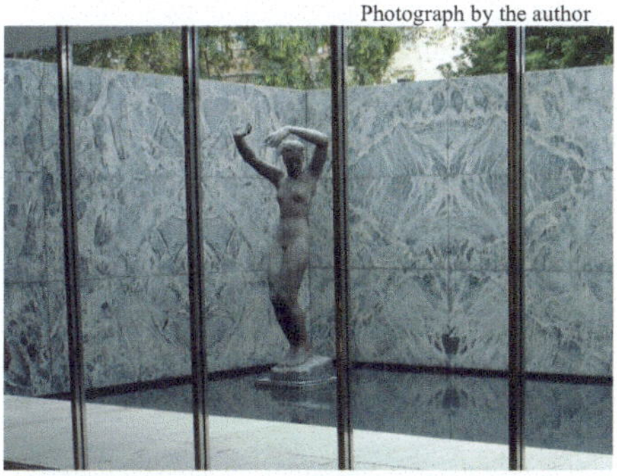

The Environment as an object to contemplate

Drawing by Walter Moleski: Dept of the Army (1976, 3-9)

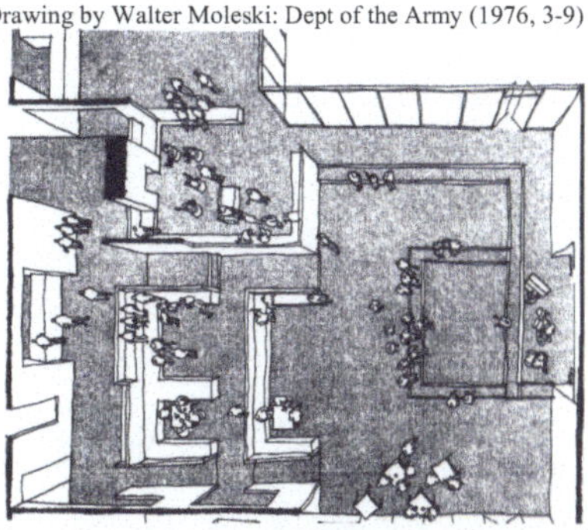

The environment as a nested set of behaviour settings

Photograph by the author

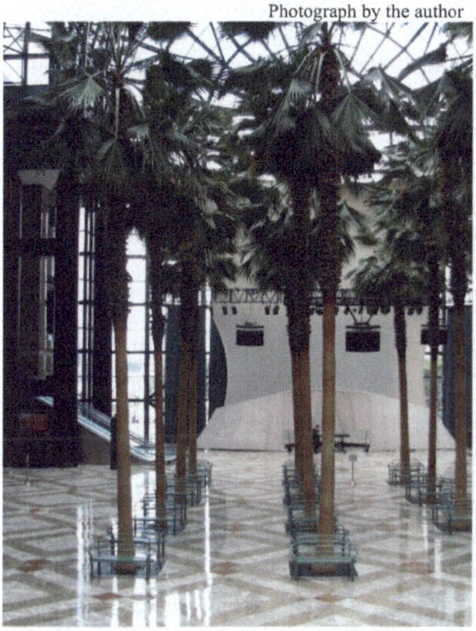

Photograph by the author

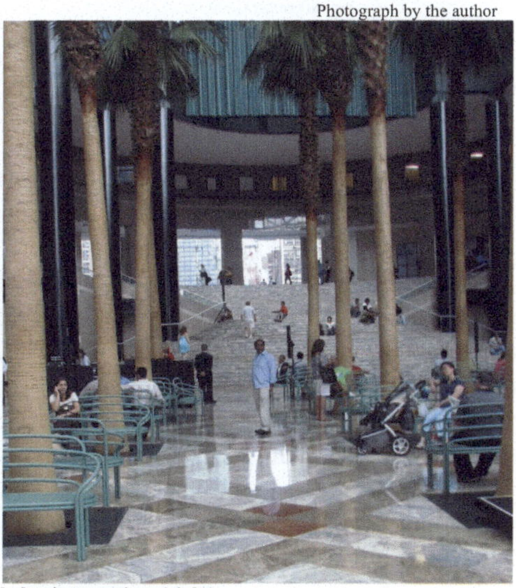

A potential behaviour setting (left) and an effective behaviour setting (right)

Photograph by the author

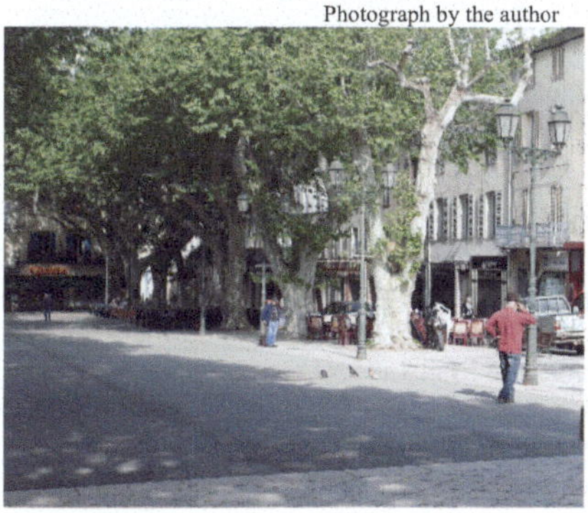

Photograph by the author

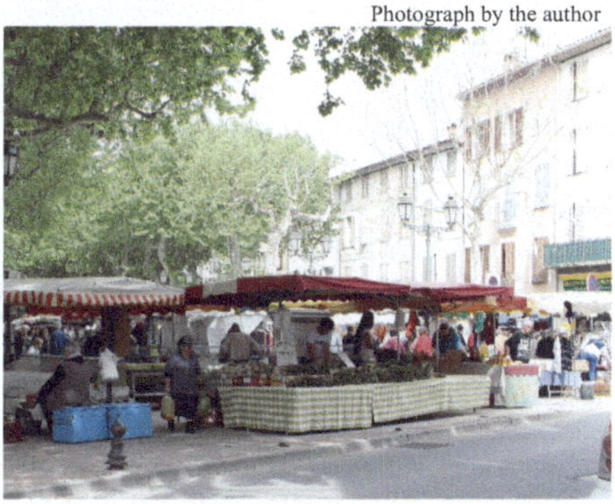

One effective behaviour setting (left) and an another within the same milieu (right)

Fig. 1.2 Environments and behaviour settings

Diagrams adapted from Lang and Moleski (2010, 42 and 58) by the author: Courtesy of Taylor and Francis

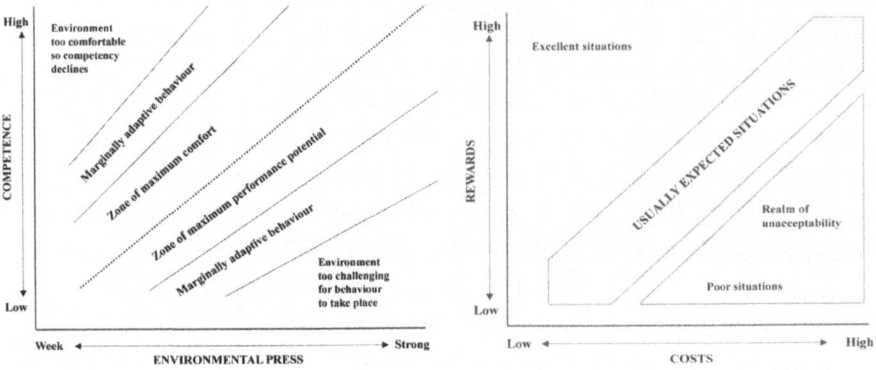

Fig. 1.3 Competence and environmental press (left) and perceptions of environmental quality (right)

and the architectural profession are thus part of a network of businesses that provide services to clients. To survive, the businesses must have a steady flow of commissions.

George Bernard Shaw (1856–1950), an Irish playwright, may have dismissed professions as "a conspiracy against the laity" (in *The Doctor's Dilemma,* I) but, in combination, the design professions have had great success in creating well-loved places, the links between them, and the objects they contain. They have also created places and objects that disappoint. Some have failed to meet the goals that were supposed to achieve. Sometimes the goals have simply been misguided.

What has been accomplished has depended on the way designers and their clients look at and examine the world, the problems they address, the opportunities they observe, and the processes of analysis, synthesis, and evaluation—the design methods—that they employ. The next three chapters examine these concerns.

References

Barker RG (1968) Ecological psychology: concepts and methods for studying human behavior. Stanford University Press, Stanford, CA

Gibson JJ (1979) The ecological approach to visual perception. Houghton Mifflin, Boston, MA

Heft H (2001) Ecological psychology in context: James Gibson, Roger Barker, and the legacy of William James's radical empiricism. Lawrence Erlbaum, Mahwah, NJ

Lang J, Moleski W (2010) Functionalism revisited: architectural theory and practice and the behavioral sciences. Ashgate, Farnham. Republished by Routledge in 2016

McClure WR, Bartuska TJ (eds) (2007) The built environment: a collaborative inquiry into design and planning, 2nd edn. Wiley, Hoboken, NJ

Procedural Theory: The Nature of the Design Process

2

Architecture is one of several fields where individuals and teams design environments and objects that are going to function in the future. They do so on behalf of a client. A knowledge of procedural theory enables decision makers to understand and be articulate about what they are doing. It also provides the intellectual base for them to configure the process of design suitable for the specific tasks that face them rather than simply relying on habitual processes. As Artificial Intelligence (AI) developments increasingly shape design practices, an understanding of how architects think will help them shape their future creative roles.

Although thinking about the nature of the design process has a history going back to the writings of the French philosopher and scientist René Descartes (1596–1650), *design methodology* only became established as a recognized field of study in the 1970s. It had been given a boost by Herbert Simon's *The Sciences of the Artificial* (1969). While recognizing the importance of intuitive and reflective thinking, Simon sought to understand the decision-making process through intellectually rigorous research (Huppatz 2015; Schaathun 2022). His goal was to develop a body of knowledge that is teachable and that can help both professionals and students understand the mental activities involved in producing products. In striving for that goal, Simon recognized the limits of human rationality.

All architects follow some procedural model of design in their work. Many of them prefer to see designing as a purely intuitive process shrouded in mystery. We understand much about the strengths of working intuitively (Eastman 1970; Lawson 2005; Rowe 1987; Lockard 2017). Dealing with complex, multidimensional problems does, however, require more systematic ways of working.

Much of our current understanding of the design process comes from architects reflecting on their own working procedures (Archer 1970; Bahrami and Dagli 1993; Roozenburg and Dorst 1998; Çalişkan 2012). Studies on the nature of practice (Schön 1983; Gutman 1988; Cuff 1992), observations of architects at work (for

J. Lang, *The Scope of Architectural Theory*, Design Science and Innovation,
https://doi.org/10.1007/978-981-96-3690-7_2

example, Yaneva 2009 on OMA), and some experimental work have advanced the quality of procedural theory. It has also been enhanced by the research conducted in allied fields such as operations research. It was a response to the insights that contemporary systems approaches afforded scientists and engineers in dealing with unique, complex problems, such as those in NASA's space program.

2.1 Unselfconscious and Self-conscious Design

In his book *Notes on the Synthesis of Form*, Christopher Alexander distinguished between *unselfconscious* and *self-conscious* design processes (Alexander 1964). Unselfconscious processes are a characteristic of societies where indigenous traditions dictate building forms. Those forms have evolved from incremental adjustments made over a long period of time. Vernacular architectures respond insightfully to the climatic demands and cultural traditions of a location using the, usually limited, resources available (Rapoport 1969; Oliver 2003). Buildings are designed by individuals or groups of people following procedures honed into habits. Learning to build takes place by copying, not imitating, the work of people with expertise. Changes occur pragmatically by recognizing what works and what does not.

Self-conscious design is different. It involves the creation and evaluation of possible courses of action prior to implementing them. It is a process of "simulating what we want to make before we make it as many times as may be necessary to feel confident in the result" when implemented (Booker 1966 cited in Salama 1995). It is regarded as a creative process that is learnt through formal institutionalized educational procedures as much as by practice. Gerhard Fischer (b. 1945), a University of Colorado professor, summarized the difference between unselfconscious and self-conscious computer software design processes in Table 2.1 (Fischer 2004). It applies just as much to architecture and to the design process.

Self-conscious architectural design has a long history. The settlements and buildings of the Indus Valley, those of the Incas and the Aztecs in South and Central America, and the buildings of South and East Asia based on cosmological considerations were all purposeful, pre-determined designs. Throughout history architects have considered the process that they go through in creating designs with varying degrees of diligence. In the western world, it was René Descartes who first systematically analysed the processes of reasoning. Over time it became clear that an understanding of the nature of problems that architects face is fundamental to understanding the design processes that architects follow.

2.2 The Nature of Architectural Problems

A problem can be said to exist when, in somebody's mind, there is a discrepancy between the present situation and a desired one, real or imaginary. Architects often face puzzling, troubling, and uncertain assignments in resolving problems and/

Table 2.1 The distinction between unselfconscious and self-conscious design

	Unselfconscious design	Self-conscious design
Definition	Process of slow adaption and error reduction (situational knowledge)	An explicit, externalized description of a design exists (theoretical knowledge)
Original association	Primitive societies. Handmade things Design for self	Professionally dominated design Design for others
Primary goal	To solve own problems	To solve the problems of others
Examples	Organically grown cities: London, Paris Linus	Designed cities: Brasília, Canberra, Islamabad Microsoft Windows
Strengths	Many small improvements. Artefacts well-suited to their functions Copes with ill-defined problems	Activities can be delegated Division of labour becomes possible
Weaknesses	No general theories exist or can be studied because designing is internalized	Many artefacts are ill-suited for the functions expected of them
Requirements	Owners of problems must be involved because they have relevant unarticulated knowledge	Externalized descriptions must exist
Evaluation criteria	Personally meaningful. Pleasant and engaging experience, self-expression	High production value: efficient process, robust, reliable
Relation with context	Both problem framing and solving take place within the bigger context	Context required for the framing of the problem

Adapted from Fischer (2004) by the author: Courtesy of Gerhard Fischer

or exploiting opportunities. They address *wicked*, or *meta*-problems. Horst Rittel (1930–1990), a design theorist and university professor, was the first to describe architectural problems as such. West Churchman (1913–2004), a systems scientist, gave the characterization academic weight in his writings (Churchman 1967; Coyne 2005).

Wicked problems are ill-structured, complex, and have no ascertainable definitive formulation. Many architectural design assignments become particularly wicked when many stakeholders needs and images of good buildings vie for attention. *Tame* problems are those that can be addressed with some probability that they are clearly defined. *Simple* ones have a definitive solution and rules for defining whether they are correct or not. Simple does not mean easy. Some simple problems are very difficult to resolve but a solution is recognizably correct when it is found. There is no way of being certain that a proposed solution to a wicked problem will work well prior to its implemented nor a way for designers to recognize an optimal solution if they have somehow achieved one (Rittel 1971; Skaburskis 2008).

Architectural problems are compound ones consisting of wicked, tame, and simple components. In practice, the way we deal with them is to focus on what is understandable and to assume that the other components are simple problems, whether they are or not. The easiest way to deal with complexity is to simplify the situation or, as one of America's greatest architects, Louis I. Khan (1901–1974; né Itze-Leib Schmuilowsky in Estonia), suggested, to redefine the problem in one's own terms (Dagit 2017). More frequently, we imitate precedents, generic solutions, and building types that appear to work well enough to meet the requirements of the situation being addressed. In doing so we deny the wickedness of the problem. We treat them as simple. Doing so suffices.

A major characteristic of wicked problems is that there is neither a stopping rule for terminating the seeking of a better definition of a problem nor finding a better solution to it. We do not indefinitely strive to get *the* perfect design. We stop when we have devised a *satisficing* one—one that is good enough (Simon 1957). Designers keep on working until no recognizable improvement is made in devising a better solution and/or when doing so becomes too costly and/or the emotional resources for doing so are depleted (Churchman 1967). Project budgets restrict the performance of designers (Asimov 1962). These observations hold for every aspect of the architectural design process from its inception to the evaluation of the implemented design in operation.

2.3 Generic and Specific Models of the Design Process

The rational model of the design process, while inaccurate as a description of how designs are really generated, has great utility as an 'ideal type' (Bazjanic 1974). It becomes the basis for asking questions about what we really do. The model assumes that we have perfect knowledge about the present and can predict accurately how our designs will function in the future. The model thus assumes that the design process can be carried out in a step-by-step linear progression of intellectual activities. The process begins with the preliminary definition of the problem being addressed and the constraints under which the solution must operate. The next is step is one in which designers brainstorm, test and evaluate possible solutions before selecting one to implement and then implementing it (Fig. 2.1).

Morris Asimov (1906–1982), a professor of engineering, in *Introduction to Design*, considered the designing process to consist of five major intellectual steps: *Intelligence, design, choice, implementation*, and *operation* (Asimov 1962). Herbert Swinburne (1912–2001), a much-awarded American architect, provided a more descriptive model of the architectural design process. He divided it into seven steps: *Definition, analysis, synthesis, development, implementation, operation*, and *evaluation.* He identified the people involved at each step. In doing so he recognized that getting a building designed and built is a cooperative task

Adapted from various sources by the author

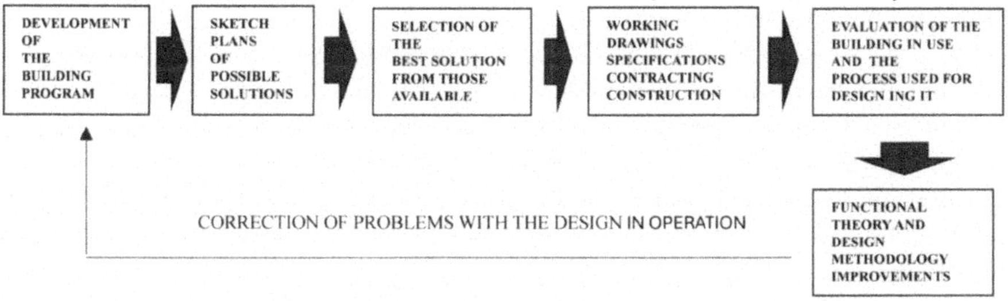

Fig. 2.1 Architectural design as a rational process

(Swinburne 1967; see also Lang 1987, 38–39). Building on Asimov and Swinburne's observations it is possible to devise a generic model of the design process (Fig. 2.2).

The design process begins with the perception of an existing problem or opportunity. The problem being addressed is operationally defined and a program to address it developed. This step is followed by a design phase during which potential solutions are created. Each potential design is evaluated based on a set of criteria. A decision is made as to whether any of them is good enough until a single design to use as the basis for proceeding is chosen from the possibilities. Once that decision has been made the process of implementing the design begins. Doing so involves moving from the symbolic world on the computer screen to the 'real' world of daily life. The implementation phase involves a series of steps that include the production of working documents, the bidding among construction companies for the job and its construction. Once built, the implemented building,

Adapted from Lang (1987) by the author

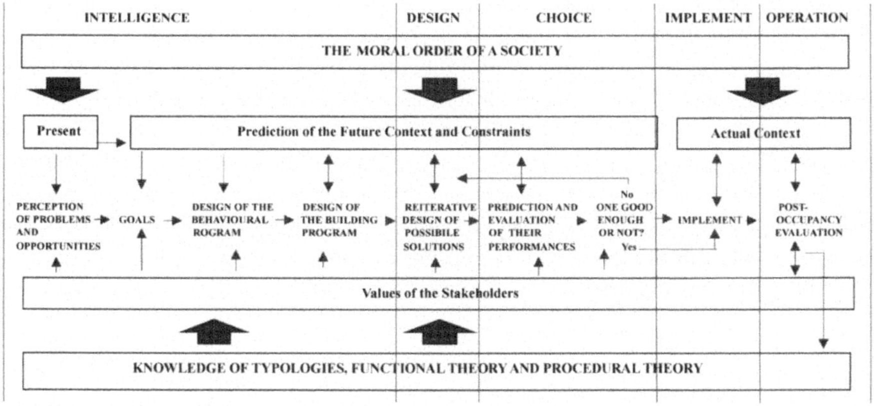

Fig. 2.2 A generic model of the architectural design process

as an environment and as an object, is inevitably evaluated in operation by those who own, inhabit, and/or observe it.

The process is seen as a general linear progression with much questioning of ends and means resulting in feedback loops. All the thoughts and speculations do flow in a sequence but not in a neat linear fashion. It is, more accurately, a spiralling process in which architects backtrack, repeat activities, and constantly test ideas until they create a design that they consider satisfactory, if not the best, in the circumstances. Along the way designers may move away from, rather than towards the goal of problem resolution. They repeat a series of activities again and again, resolving new problems with each repetition. These apparently multi-directional movements together result in one general movement towards the goal of getting a building built (Zeisel 2006, originally 1981). The process is one of simultaneously improving the quality of the problem's definition and the solution to it. Indeed, the program implicit in the design implemented may end up differing considerably from that established at the outset. The problems addressed keep on getting redefined as potential designs are examined.

The thoughtfulness given to each phase of the process varies. Sometimes the whole process is fast-tracked. Figure 2.2 shows that designers are constantly drawing on their knowledge of functional theory, design types and precedents, and their knowledge of how the design process can be conducted. It also shows that design is based, consciously or not, on a moral philosophy of what is right, what is acceptable and what is wrong (Watkin 1977; Verbeek 2008).

Making decisions involves reasoning and arguing and repeatedly developing and testing hypotheses, or conjectures, about what is to be done (Cremer-Petersen and Ahmed-Kristensen 2016). Architects argue with themselves and with their clients about ends and means, about what the program should be, and about the design patterns to be employed. Almost always budgetary constraints form part of the discussion. Any rationality to the process depends on the quality of the functional theory employed and architect's knowledge of the approaches to design available to them. With experience the design process becomes driven by the habitual ways of thinking that shape an architect's brand of work.

The actual sequence will vary from project to project. Figure 2.3 shows the steps involved in an all-of-a-piece urban design as carried out by the architectural firm of Alex Cooper and Stanton Eckstut in the design of Battery Park City, New York (Gordon 1997; Lang 2017, 2024). The first product of the process is a master plan. Guidelines are then written by its designers to attain specific characteristics in the components of the design that are to be built by individual property developers and designed by their architects. From beginning to completion, it is a highly complex and politically charged process.

Adapted from Lang (2017)by the author

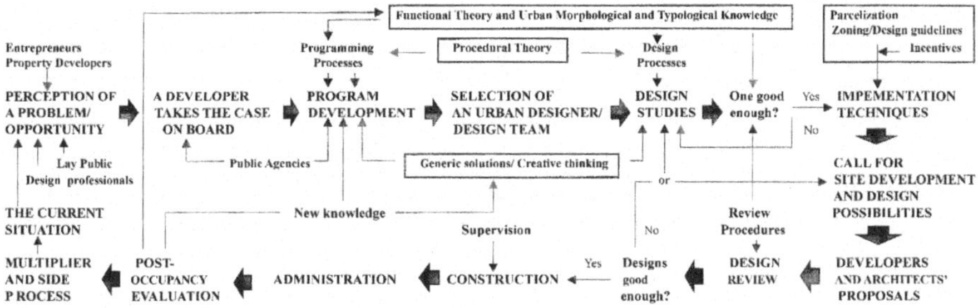

Fig. 2.3 A model of the steps of the all-of-a-piece urban design process

2.4 The Phases of the Design Process

The design process can be further understood by considering the nature of its constituent phases in greater depth. Each phase involves repeated analytical reasoning and the repeated divergent and convergent production of ideas. Each phase requires the making of evaluations and the drawing of conclusions. The phases may well blur together especially when a project is fast-tracked.

2.4.1 The Intelligence Phase: Perceiving Problems/ Opportunities

All designs begin with people who perceive problems that they wish to see addressed. They examine the world in terms of their own interests and they look for opportunities to fulfil their needs and/or desires. Property developers, whether private or public, international conglomerates, families, or individuals, seek opportunities to implement the type of project that will benefit them. In contrast, some problems are identified by individuals on behalf of what they perceive to be in the public interest. Resolving them, they argue, will bring benefits to society. Public interest is, however, notoriously difficult to define.

> The public interest may be presumed to be what [people] would choose if they saw clearly, thought rationally, acted disinterestedly and benevolently. (Walter Lippmann (1889–1974), a political commentator, cited in Wheeler 2006, 25)

Operationally, the public interest means that somebody perceives that a problem exists for some population, human or other species, and lobbies for it to be addressed. Frequently people arguing for something on behalf of the public interest are arguing for what are really their own interests. In architecture it is usually assumed that the public interest is represented by zoning regulations, building by-laws and codes.

Some architects, like science fiction authors, identify hypothetical problems that enable them to offer designs that satisfy their own curiosities. This observation holds particularly for architects exploring the utility of existing, emerging, or imagined structural technologies (Dahinden 1972). Groups such as those working with Buckminster Fuller (Baldwin 1996), Archigram (Sadler 2005), the Metabolists (Pernice 2022), and those advocating Megastructures (Banham 1976) have attracted much interest for the novel boldness of their ideas. Little, if anything, of their proposals has been implemented. The designs, while often dealing with world-wide concerns, are narrowly functional and largely unrelated to how people live or want to live. They deal with hypothetical, imaginary worlds. They, nevertheless, make us think about possible futures.

In dealing with everyday design problems the developer becomes a client seeking architectural services. Clients play an important part in the debates over ends and means during the whole design process (Jenkins 1961; Eaton 1969; Soth 1988). They vary in character and what they demand of the architect. As noted, the client may consist of many stakeholders promoting their own interests. Some wait to see what their architects produce before getting deeply involved; others play a creative role in developing the building program and the design.

2.4.1.1 Participatory or Co-design

All designs are the result of some level of cooperation between clients and architects. Participatory design, also known as co-design, is, however, an approach to cooperatively defining design problems and generating design ideas. Studies in several fields suggest that decision makers generate more ideas and more innovative ones when working closely with others (Mitchell et al. 2015). Property developers, architects, and potential end users are actively involved in making the decisions that affect them (Hensley and Kitch 2023). The use of participatory design methods has proven to be particularly important when architects are designing for people of cultural backgrounds that differ from their own. Participatory design often takes place in the form of a charrette, an intense period of stakeholders working together (Batchelor and Lewis 1984; Pernice 2013; see Chap. 7).

2.4.2 The Intelligence Phase: Designing the Program

Identifying the problems that need to be addressed with clarity is the most important step in designing a room, a building, or a complex of buildings. Identifying precisely what needs to be done is a discursive process with the discussion of ends and means going back and forth among the stakeholders. It involves arguing and reasoning (Hansen et al. 2009; Cremer-Petersen and Ahmed-Kristensen 2016). Increasingly, facility programmers who understand how to ascertain the nature of an organization's desired activity systems and their spatial implications are employed to create the program. In many situations, the potential range of possible activities to be housed and the aesthetic values to be met are so numerous that only the most important are considered. As the problems are inherently *wicked*, the requirements are simplified and operationally defined in a program.

The design of a building program is always conducted under uncertainty about the range of concerns that should be addressed and about the future social and economic environment in which the building will operate. Most programs, implicitly or explicitly, are based on the prediction that tomorrow will be the same as today or that current trends, linear or cyclical, will continue. Few programmers look ahead to deal with the likely catalytic impact of a proposed building on its surroundings unless the neighbours raise concerns. In addition, as the instrumental use of a building is likely to change over time, how much and how to design for adaptability becomes an issue open for discussion (Andrade and Bragança 2019). Conclusions are drawn after much conjecturing, arguing, and evaluating of possibilities by the stakeholders.

2.4.2.1 Types of Programs

Two types of building program are used to state what the nature of a building should be before it is designed. One—*prescriptive*—specifies what spaces and their sizes a building or urban design should contain; the second—*performance* programs—specify how a building and its components should perform. Performance programs state the goals and objectives (the operational definitions of the goals) and the activities to be housed and, often, the aesthetic style of a proposed building. It is left to the architect to translate the requirements into spatial and aesthetic characteristics.

Some programs are a mixture of the two. Both types have traditionally focussed on narrowly defined functional concerns—efficiency in movement, the utility of materials and ease of construction. Specifying the general aesthetic qualities a building should have is more problematic. Many architects believe it is their prerogative to prescribe the characteristics of a building's appearance. The need is for a building to fit in with their brand/style of work. Usually, a compatibility of tastes exists between the architect selected for a job and the commissioning client.

2.4.2.2 Designing the Program

The first step in designing the program is usually the definition of the activity patterns to be housed. Often, the patterns are simply implicit in a building type or generic solution studied in thinking about potential designs. Room and space names become surrogates for descriptions of activities that take place in them and the possible furniture arrangements their patterns can accommodate.

One of the major challenges in designing programs is to ascertain what the activity patterns and the aesthetic values would be if those involved understood the range of future possibilities available to them. What are clients' *latent* demands—those that they are unaware of—if the opportunity to fulfil them existed? The same observation holds true for aesthetic possibilities.

A variety of methods are used for obtaining the information required for designing programs (Preiser et al. 1977; Peña and Parshall 2012). One way is simply to look at the layout of precedents. Indeed, such a review is often seen as the first step in architectural programming efforts (Cherry and Petronis 2016). In designing a hospital, the programmer looks at the plans of other hospitals that are deemed to function well and adjusts them to meet the specific demands of the present case.

Interviews and surveys are widely used as the basis for obtaining the information about both the present actions—well- and mal-functioning—of an organization and its aspirations for the future. The question is: Who does the programmer interview? The paying client is an essential interviewee but potential users are also important if a well-functioning future environment is sought. The difficulty is that those being interviewed are usually poor at providing accurate information about the strengths and weaknesses of the present situation let alone about their future demands. Unobtrusive observations (Lee 2019) are widely used in urban design programming but they are more difficult to employ in building design. Careful observations of the present situation, however casually done, have proven to be essential in designing sound programs.

The thoroughness and quality of the program ultimately depends on the questions asked. The quality of questions depends on the quality of the declarative functional theory used as the basis for asking them. Buildings turn out to not be functional enough when the model of human motivations and building needs is a narrow one (Scott Brown 2004; Lang and Moleski 2010); see also Simon (1957). In dealing with wicked problems, it is easiest to reduce the definition of what must be done to what is the easiest to do in the belief or hope that it will be good enough.

The program with which an architect begins seldom remains unchanged during the process of turning it into built form. Conflicts unresolved in the program are reconsidered as the building's design gets defined and refined. New opportunities and constraints arise. Costs are often, if not usually, higher than predicted and changes must be made to the program. Snags arise during implementation.

2.4.2.3 Resource Limitations

Many, although not all, of the most admired, architect-designed buildings in the world have been costly. They were erected by highly skilled workers using contemporaneous advanced construction technologies. Equally admired vernacular and quasi-vernacular architectures draw on relatively inexpensive local materials and are built by workers with important but limited skills. The availability and source of finances and materials constrains what can be built. The question is: When in the design process should resource limitations and the skills of workers be considered? For Laurie Baker (1917–2007), a British-born Indian architect, working with lowly skilled workers it was from the beginning. Material and building techniques choices were made recognizing the strengths and limitations of local construction workers' skills (Bhatia 2003).

In some instances, both the programming and designing of buildings or urban designs proceed without the consideration of resource limitations until possible designs are evaluated. After the evaluation the design that is regarded as best gets whittled away at until it is implementable with the resources available. The assumption is that the consideration of design possibilities is prematurely truncated if economic considerations are introduced prematurely.

2.4.3 The Design Phase and Nature of Designing

Every step of the design process involves the generation and evaluation of ideas—of making and evaluating conjectures. In the design phase, however, architects focus on turning a program into a three-dimensional model of a building that is to be experienced over time in the future. Few thorough building programs, especially performance ones, contain no contradictions in what they specify. The act of designing almost inevitably identifies issues unaddressed in the program. Indeed, anecdotal evidence suggests that the generation of potential designs should play a part in designing the program itself. Designing then becomes an analytical tool.

In professional practice the design algorithm seems to consist of two major steps: schematic design production and then the finalization of a design to be taken into production. The first focuses on the key variables so that an estimation of the cost of a proposed building can be made and the building program evaluated. Its purpose is to enable the commissioning client to decide on what to include in the program. The second involves a more comprehensive look at a design based on the consideration of all the programmatic requirements.

Generating designs is an iterative process of self-argumentation. In arguing, architects draw on their knowledge of the affordances of patterns of built form, design precedents, and structural and constructional techniques. They think both divergently in generating design possibilities, and convergently in synthesizing individual components of the building's design into a unified whole. Both processes include repeated analysis, repeated generation of ideas, and repeated evaluations. Divergent thinking—the production of different ideas—is often considered to be *the* creative design act, but recognizing and judging the utility of ideas is, perhaps, the most demanding creative act of all.

In thinking divergently architects use of variety of idea generation techniques that they find helpful (Koberg and Bagnall 1978; Herring et al. 2009). These techniques include morphological analysis, a method for structuring and identifying the interconnections of a multi-dimensional problem, brainstorming, and the use of analogies and metaphors (Mitzkus 2022). Computer-based form generating algorithms are increasingly used to produce designs. As they become more sophisticated, they will challenge the creative role of the architect and thus the whole nature of the design process.

2.4.3.1 Creative Designs and Creative People

A creative design is often seen as one whose forms depart from the norm. Creative people are those who design them. These definitions are poor. Creativity captures and embraces "ecological and human reality through method, knowledge, and public responsibility" (Tzonis 2014). A creative design is thus best defined as one that resolves problems in a better way than the norm. It is based on an innovative design program that understands and defines problems in a new way.

The research on creative architects (that is, people who are identified as being creative by their peers) has focussed on either their personalities or their ways of thinking (MacKinnon 1962; Barron 1969; Serraino 2016). The former studies offer us little but the latter suggest that people who think creatively have a great *intellectual curiosity* about how the world functions. They are also *constructively* (not cynically) *discontent* with the status quo. They actively examine the world around them and absorbing the functions of a wide variety of precedents.

Creative people are constantly looking at new ways of doing things and can carry different ideas forward without prematurely closing lines of thinking (de Bono 1992, 2015). They are highly productive people who keep on generating ideas. They are, almost universally, good salespeople for their work. Philip Johnson (1906–2005), an architect known for his modern and postmodern designs noted: 'Ninety-five percent of [creative architects] are, of course, businessmen, organizers, salesmen, and hucksters' (cited in Serraino 2016). He later regretted using the term 'huckster'.

2.4.3.2 Designing Styles

Several intellectual activities in creating designs are basic. The ability to develop overall strategies is one. Some architects follow a *depth-first* approach in which a single design is generated as quickly as possible. They then defend it until it is found to be too wanting or too expensive to implement before seeking another. Other architects, or teams, carry several designs forward before evaluating them. They take a *breadth-first* approach in which different lines of thought are pursued simultaneously.

In either case architects often begin with resolving the concern that has the greatest impact on subsequent decisions. The concern may be the overall form of a building into which its components then have to be shoehorned. An alternative is to resolve the easiest components of a problem first and then move on to the trickiest parts or vice versa. Another approach used is to begin with the broadest level concerns and yet others from the details believing that the details are what make a design of high quality. Hill-climbing is one such designing approach. In it, a solution to one part of the program is generated and then another part of the program that is regarded as the best step forward is addressed. Many architects follow a much more haphazard approach. In all these ways, designing involves making detours and retracing one's steps when a dead-end is reached. It is a process that can be both time consuming and frustrating (Zeisel 2006; Makstutis 2018).

Avoiding premature criticism and commitment to lines of thinking are other basic characteristics of creative problem solving. A pattern that is less than desirable in solving one problem may allow other problems to be resolved well. Designers are constantly making choices among such options. How they make them is part of their designing style. Sooner or later, whatever method is used, a decision-maker commits to implement a design or decides to go back to relook at the program or to abandon the whole endeavour.

Universally, architects believe that if they had more time they could come up with a better solution. The desire to create a well-functioning, innovative design, or simply to do a better job, or to receive design awards motivates an architect to keep going. There is no rule for when to stop. The time, or finances, available is the most common criterion for ceasing to generate more design possibilities or improving a design already produced. Sometimes, the architect is simply exhausted and feels that no amount of further effort will lead to a better solution.

2.4.3.3 Design Principles and Standards

Throughout the design phase, architects draw on a declarative knowledge base however well or ill-formed it may be. They also rely on design standards and principles that are recognized as authoritative solutions to recurring problems. Principles are deontic statements in the format *In situation X do Y*. They are value laden prescriptions (Lidwell et al. 2007). The use of such design principles and standards reduces the architect's cognitive load. Their application saves having to reinvent the wheel and to make decisions on matters about which an architect knows little. Most such standards are carefully worked out and are based on observations of what works (Watson et al. 2003; De Chiara and Callendar 2020).

All designers, through their education and experience, have a set of principles in their heads that they draw on consciously or subconsciously all the time. The set defines their brand of work. The nature of design principles took a step forward in the work of Christopher Alexander and his colleagues as described in Chap. 7. Their design principles take the form *In situation X do Y to solve problem* Z because of the following evidence. The empirical basis for each statement is clear (Alexander et al. 1977). The patterns advocated are, nevertheless, value-laden, prescriptive statements based on his and his colleagues' values.

2.4.4 The Choice Phase: Evaluating Potential Designs and Selecting One

In the choice phase of the design process the objective is to reach a consensual agreement to implement one option among design possibilities. Alternatively, it is to decide that none satisfices and that the design program should be amended or another attempt be made to come up with an acceptable design. Often, however, no specific evaluation phase forms part of an architect's design process. The prediction of performance and the evaluation of ongoing design efforts is integrated step-by-step with the generation of ideas and only one design is produced.

Proposed designs are presented for evaluation as drawings, or in three-dimensional models sometimes coloured but often white. They are visual abstractions of the nature of a building or set of buildings and open spaces. The quality and way they are represented is very important because they bias the decisions that clients make (Lockard 2018). Sometimes the models include a proposal's surrounding context but often they are shown as stand-alone objects. Performance characteristics, such as capital investment requirements, operational costs, and return on money invested, are presented in numerical or verbal form. Other variables, such as how well a proposed design will accommodate the everyday activities it is supposed to house, what the spaces will smell or sound like, or the tactile experience that its surfaces offer, or how air will move through it are left to the evaluator to imagine/predict.

Evaluating the performance of proposed buildings accurately is a daunting task. History is replete with examples of once disparaged buildings and urban design proposals that, when implemented, served their instrumental and symbolic aesthetic purposes very well. The opposite is also true. Designs much admired on the drawing board have proven, when built, to be disappointing works in the eyes of the cognoscenti, lay observers, and their inhabitants.

2.4.4.1 Evaluating Proposed Designs: Jury and Systematic Approaches

The evaluation of possible designs, whether they are the work of one or competing firms, can take place informally through general discussions or by formally constituted juries. A jury consists of a group of experts and/or stakeholders (Anthony 1991). Each juror makes a prediction of the performance of the design and evaluates it based on given criteria or, as often, their own criteria that may or may not have been of importance to the designer. In open juries, the discussion among jurors and the arguments they present for their positions can sway each other's views. The power distribution, whether by reputation or by financial position among jury members affects the conclusions drawn and decisions made (Webster 2007).

The systematic approach involves the listing of the decision criteria to be used in the evaluation of the design with weights assigned to the relative importance of each criterion. Jurors then assign scores to the performance of a design on each criterion, based on their predictions of its performance qualities. The results are presented in a goals achievement matrix (Hill and Werczberger 1978). The total score for each competing design is then tabulated and the highest scoring design selected. Often, in practice, the evaluators do not like the result and revert to a more open debate about the qualities of each design based on their own subjective criteria. The listing of criteria to be used in the systematic evaluation of designs, at least, gets the jurors to focus their assessments on similar characteristics.

2.4.4.2 Making Decisions—Seeking Closure

Ultimately one design to be implemented is chosen over the other possibilities at hand. The decision-maker, whether a single person or a committee, must have the legitimacy of the group because the stakeholders need to accept the decision.

Personalities come into play in making decisions. Optimists, or gamblers, choose designs that they predict will function best in the rosiest possible future. Pessimists pick designs that works best in the worst possible future. Realists fall somewhere in between. Once chosen the design then proceeds to be implemented. During implementation all kinds of contingencies usually arise and bits and pieces of the selected design get discarded or altered. In the worst case a scheme gets abandoned during construction. Half constructed examples exist in most cities. The Everglades Jetport in Florida once proposed to be the world's largest in 1968 was one such major project never completed (Prisco 2022). Predictions about the future in which it was supposed to function failed to materialize.

2.4.5 The Implementation Phase

The implementation of a building design takes place in two significant subphases. In the first, the architectural team produces working drawings and specifications in sufficient detail for contractors to begin work on constructing the project. Depending on the society in which a building is being erected, the selected contractors produce refined drawings and then get on with the construction work. Variations to the design often occur as unforeseen problems arise when producing working drawings and specifications. The building program gets amended (Emmitt 2018). In the second subphase, the building is constructed. The process is often supervised by a construction manager working with an architect. Construction management has its own theoretical underpinnings, both declarative and prescriptive. They cover technical, financial, and leadership concerns (Halpin et al. 2017).

In all-of-a-piece urban design, as shown in Fig. 2.3 the enforcement of design guidelines requires the establishment of some authority charged with implementing the scheme. The authority supervises the selection of individual property developers and ensures that they follow the guidelines when components of a project are being designed and built. It is easier to enforce guidelines when there is a demand for development than when the property market is in the doldrums. The whole process of implementation is highly political in nature and many changes to what is proposed can be made during it. (For a description of the political controversies in the development of Battery Park City, New York, see Gordon (1997); for Hudson Yards in the city see Lang (2024).)

2.4.6 Evaluating Completed Designs

From the moment of their completion and first occupancy buildings begin to change. Wear and tear from their occupants' use and from the forces of nature create erosions and adhesions. Sometimes the occupants make radical changes by altering the internal layouts and fenestration patterns to meet the requirements of their ways of life. (For an example see Boudon (1972) on the Le Corbusier-designed Les Quartiers Modernes Frugès.)

Sometimes the intended occupants have simply refused to move in because the design is not what they wanted. This situation usually occurs because potential residents' images of what they need differ from that envisaged by property developers and/or their architects (Weaver 2006). Often what is offered differs radically from what they have experienced. They fail to see the affordances of what, for them, is a new design type or, if they do, it is not what they want.

Completed designs are constantly evaluated by building owners and operators, the people who inhabit them, and by the architectural cognoscenti and critics. Systematic post-occupancy evaluation techniques are available for the multi-dimensional analysis of buildings by impartial researchers (Preiser et al. 2015). The feedback received, when heeded, enhances our understanding of how buildings function over time from different points of view. It thus corroborates or updates declarative functional theory as outlined in Chap. 3. The evaluation techniques are, however, seldom employed. Most evaluations are piecemeal and haphazard. Some architectural critics do more than that.

In the United Sates, critics such as Grady Clay (1916–2013), Ada Louise Huxtable (1921–2013), and Paul Goldberger (b. 1950), have had a profound impact not only on the characteristics of individual buildings and urban designs but on the direction that architectural ideologies—prescriptive theories—have taken. Many societies have such influential observers of the architectural scene. Critics, lay or professional, apply their own, largely subjective, criteria to evaluations. Important though their critiques may be, many buildings highly praised by the cognoscenti have been disliked by those people who work or inhabit them (Michelson 1968; Weaver 2006). Their evaluation criteria differed.

To the cognoscenti, it is the design idea underlying a building's design as a work of art that is important. To the investors it may be the return on capital that is of prime concern. To those who use the building it may be the ambient comfort level and their sense of pride (or not) in the aesthetic qualities of a building that are of consequence. To the neighbours it may be the visual character of the building and the changed exposure to sunlight and breezes that attracts their attention. In the case of the Walt Disney Concert Hall in Los Angeles, it was the glare (Schiler and Valmont, undated). To automobile drivers, rogue reflections may be what matters. To all it is what is wrong that attracts attention. To build declarative theory, what works well needs to be equally noted.

2.5 Three Approaches to Going Through the Phases

All design processes approximate the phases as outlined above but there are different ways of going through them. While each architect approaches the design process in a unique manner some generalizations can be made. A distinction is often drawn between *solution driven* and *problem driven* approaches to design (Rowe 1983; Kruger and Cross 2006). Distinguishing among *design-led*, *type-based*, and *problem-oriented* approaches to design, although the first two are similar, better enhances our understanding. The three are not mutually exclusive in practice. Architects switch from one mode of thought to another when designing.

2.5.1 The Design-Led Approach

The design-led approach to the creation of a building is, perhaps, the preferred way of working for architects (Lawson 2006). Architects start by intuitively developing a design based on their own brand of work. Le Corbusier had his own principles to guide him (Le Corbusier's five points of architecture, undated). Louis I. Kahn began with a geometry and Bernard Tschumi with a philosophical idea in mind. Other architects begin with a structural system or building material leading the way. The program is minimal and unconstrained by resource or other limitations. Nobel Laureate Daniel Kahneman (1934–2024), a psychologist, considered it to be a fast approach to making decisions. Judgments are made speedily based on established paradigms and experience (Kahneman 2011).

The next step is to either fit its instrumental function into the design or rework the form to accommodate that function (Walters and Brown 2004; Frai 2022). Less frequently, architects work from the inside out. They imagine the internal spatial qualities and sequences of the building before developing its overall form.

By minimizing the scope of the program, many issues that are seen as constraints on attaining a creative response are dismissed. In making decisions the architect imposes their image of good architecture on the definition of the problem requiring resolution. Colin Rowe observed that many architects follow a generally accepted contemporary paradigm when designing (Rowe 1983). Each paradigm assumes that some design issues are of prime concern and the design patterns required to resolve them are more important than others. In doing so the paradigm implies that some people or other species are more important than others.

An innate efficiency exists in the design-led approach because it reduces the range of potential solutions that are generated and evaluated. An inherent danger in designing this way is that architects prematurely truncate the search for a good design and the consideration of better, multi-functional solutions. In following a hill-climbing approach of incremental improvements to a design, they are located on the wrong hill.

2.5.2 The Type-Led Approach

Studies of how designers think show that architects rely heavily on existing building types, precedents, and generic solutions as models to imitate (Argan 1963; Colquhoun 1967; Francescato 1994; Calişkan 2012; Graves 2020). They add their own brand elements to it. The type-led approach to design has much in common with the design-led approach. The difference lies in the way a design is first created and thus implicitly the way the program is defined. In the typological approach:

> The design process is a way of bringing the elements of a typology – the idea of a formal structure – into the precise state that characterizes the single work. (Moneo 1978)

The process begins with the selection and imitation of an appropriate plan type, followed by the selection of a structural system to make an exterior shell from a set of possible types and then the selection of decorative elements. It is a combinatory process. J.-N.-L. Durand (1760–1834), an early Beaux Arts architect, developed a generic procedural design method called the *mécanisme*, a step-by-step process of composition based on his typological framework (Hernández 1969). Types are open to endless variation on a central theme. Their use, both shortens the process of defining a building program and finding a solution to it. Giulio Argan (1909–1992), an Italian art historian and politician observed:

> A 'type' as a starting point for the architect's working process does not exhaust his involvement with historical data; it does not stop him from assuming or rejecting definite buildings as models. (Argan 1963)

While the type-based approach to design is most explicitly associated with the educational principles of the École des Beaux Arts (Hernández 1969; Carthian 1979; Salama 1995) it remains a major way of working (Krier 1988; Wozniak 2016). Even avant-garde architects whose work seems to be total break from the past, consciously or subconsciously, draw on the characteristics of precedents (see Rowe 1976 in general and Etlin 1987 on Le Corbusier).

Although each design problem is unique it is almost always a member of a category of design types. Matching the ideals of an antecedent type with present requirements is seen as the heart of the design process. In practice the type is often adapted in a much more parsimonious manner as constraints, often financial, influence outcomes. Housing projects in England based on Le Corbusier's design of the Unité d'habitation in Marseilles are weak imitations of the type (Marmot 1981). The Bauhaus mass housing model is, however, being widely applied with apparent success in East Asia today.

2.5.2.1 Imitation, Copy, and Pastiche

Lucien Steil, a professor of architecture in Luxembourg and the United States, identified three ways of basing new designs on existing types and generic solutions: imitation, copy, and pastiche (Steil 1988). Of the three, imitation is the truly creative act. It is the process of creating something new—not necessarily novel—from the study of the principles underlying a precedent. Imitation requires a thorough understanding of the characteristics of the precedent or building type on which an architect draws. The architectonic and technological mechanisms underlying a type, and what it affords as a set of behaviour settings and aesthetic experiences, must also be clear to the architect referring to it as the basis for a new design. In other words, the affordances and limitations of the type's patterns need to be fully understood. They are often not. It is the geometry that attracts attention.

Antoine-Chrysostôme Quatremére de Quincy (1755–1849), a French archaeologist and architectural theoretician, regarded imitation as the basis for inventing well-functioning new buildings using imagination. A *copy*, in contrast to an imitation, is simply the replication of a precedent while a *pastiche* is based on a mixture or medley of ideas. Copying may seem to be the least creative of the design processes but it often requires great skill particularly in projects requiring high quality craftsmanship to produce them. It can also be the most pragmatic thing to do. *Parody* can be added to the list. It is like a pastiche but it involves exaggerating aspects of a type to make a point.

Imitation, in the sense that the term is used here, is an approach to design that works well when the architect is dealing with generic problems and can creatively reuse design patterns that have worked well in the past. In striving for novelty by departing from well-tried types and generic solutions, architects are not only testing themselves but the marketplace. New design ideas do, however, emerge from such explorations. So do many fiascos.

2.5.3 The Problem/Program-Led Approach

In the problem-oriented approach to design great emphasis is placed on the development of a detailed program specifying the activities to be housed, the levels of privacy associated with each and the level of metabolic comfort required. It also specifies the aesthetic goals, particularly the symbolic ones, to be met by the design. Each phase of the design process is carried out in depth. It is a slow, deliberate decision-making process (Kahneman 2011). In practice, the knowledge of existing design paradigms and building types is part of the essential knowledge base for those professionals developing the program for a project. A multi-functional based program, nevertheless, leads the way (see Chap. 7).

2.6 Commentary

Much is known about the designing process; much remains to be discovered. Studying the design process has not been a topic at the forefront of many architectural theoreticians' interests. Architects, after all, do well enough relying on personal experiences, intuition, and common sense. Continuing research, nevertheless, is improving the quality of declarative procedural design theory. Without a sound understanding of declarative functional theory, all the processes of analysis, design, and evaluation will be limited in their quality. Designing becomes a process guided by habits based on an individual designer's experiences. If architects are happy in generating only satisficing designs that is fine. Artificial Intelligence will replace much of what they do (Toker 2022).

Declarative procedural theory remains largely descriptive. There is very little in the way of explanations of why architects work in particular ways other than much is shaped by the time available to carry out the various tasks demanded of them. We also have limited understanding of the consequences of working one way or another. It must also be recognized that much that is supposedly descriptive in this chapter has a prescriptive overtone to it.

Nowadays another question arises: Where will Artificial Intelligence and computer-based design algorithms take us? Despite the advances in computer-based generative design techniques that promise much, the basic intellectual activities involved in architectural designing remain the same. We rely on the rational model as a point of departure in discussing not only what we are doing but also what we should be doing (as discussed in Chap. 7).

References

Bazjanac V (1974) Architectural design theory: models of the design process. In: Spillers WR (ed) Basic questions of design theory. North Holland, Amsterdam, pp 3–20

Çalişkan O (2012) Design thinking in urbanism: learning from the designers. Urb Des Int 17(4):272–296

Churchman CW (1967) Wicked problems. Manag Sci 14(4):B-141, B-144. Guest editorial: wicked problems. wordpress.com. Accessed 28 Sept 2023

Colquhoun A (1967) Typology and design method. ARENA J Archit Assoc 83(193):11–14

Jones JC (1992, originally 1972) Design methods. Wiley, Hoboken, NJ

Kahneman D (2011) Thinking fast and slow. Farrar, Straus and Giroux, New York

Koberg D, Bagnall J (1978) The universal traveller. William Kaufman, Los Atos, CA

Lang J (1987) Procedural theory. Creating architectural theory: the role of the behavioral sciences in environmental design. Van Nostrand Reinhold, New York, pp 33–72

Lawson B (2006) How designers think: the design process demystified, revised. Architectural Press, Oxford

Makstutis G (2018) Design process in architecture: from concept to completion. Lawrence King, London

Schön DA (1983) The reflective practitioner: how professionals think in action. Basic Books, New York

Simon HA (1969) The sciences of the artificial. MIT Press, Cambridge, MA

Functional Theory

3

Whichever way an architect goes through the process of designing of an environment and/or an object, implicit if not explicit in it, is some theory—some model—of the potential functions of different patterns of built form. *Form follows function* is an enduring aphorism of the Modernist Movement in architecture. It remains a worthwhile prescription guiding architects' work provided *function* is broadly defined. The Modernists considered the function of a building to be its instrumental purpose. All buildings are, indeed, designed to serve an *instrumental function*. Christian Norberg Schulz (1926–2000), a Norwegian architect, referred to that purpose as a building's *task* in his book *Intentions in Architecture* (1965) and Fred Steele, an American professor of psychology, as *task instrumentally* (Steele 1973). Walter Gropius (1833–1969), the eminent Modernist architect and founder of the Bauhaus in 1919, called the purpose simply the *function* of a building (Gropius 1955; see also Behne 1926). Its aesthetic function was something else.

To the Modernists, a functional building was one in which the instrumental tasks of individuals and organizations could be conducted efficiently. Efficiently meant that the movements required to fulfil tasks were kept to a minimum, the illumination sources and levels were appropriate to the task at hand, and the structural and construction methods used for the building were the most suitable for their purpose (Behne 1926). Finnish architect Alvar Aalto (1898–1976) regarded such a view as a vulgar, mechanistic functionalism but offered no explicit alternative model (Wainwright 2021). The Modernists' meaning is still largely with us. It is insufficient (Alexander 1969; Canter 1970; Scott Brown 2004 among others).

The ends and means outlined here are more fully described in *Functional Revisited* written by the author and Walter Moleski and published by Ashgate in 2010.

J. Lang, *The Scope of Architectural Theory*, Design Science and Innovation,
https://doi.org/10.1007/978-981-96-3690-7_3

3.1 Functionalism

Several architects have proposed alternative models of functions that identify the full range of purposes that buildings can serve. Christopher Alexander (1936–2022), an architect and theorist, for instance, considered the work on *essential striving sentiments* of Alexander Leighton (1908–2007) helpful in establishing such a model (Leighton 1959; see Alexander 1969). The model of human motivations proposed by Abraham Maslow, however, explains more.

As shown in Fig. 3.1, people have a hierarchy of needs that they are motivated to fulfil (Maslow 1943, 1987; Huitt 2004). The basic ones—the motivation to survive and to have a sense of security—must be met well enough for people to strive to meet the next higher-level need. In Maslow's hierarchy, that level is one of having a sense of belonging to a group (familial, cultural, socio-economic, or taste), and then to be held in esteem in both their own eyes and the eyes of others. Beyond these are four levels of advanced needs: those for self-actualization, cognitive and aesthetic needs—the desire to understand abstract philosophies including prescriptive architectural positions—and then self-transcendental needs. No need is independent; each is wrapped up in and interacts with the others. Belonging to a group, for instance, is necessary for most people's survival. Individuals are often willing to sacrifice the degree to which they fulfil the very basic of needs to boost their self-esteem. They may even examine the environment, or an object within it, for the enjoyment it gives them rather than for any instrumental purpose it might serve.

The mechanisms used to fulfil people's needs are primarily social but the patterns of built form must afford the required social behaviours to take place. Luckily, the same configuration of the built environment can serve many functions and a function can be served by many different design patterns. This reality makes designing the built world both interesting and demanding.

Maslow's hierarchy of human needs can be configured, as shown in Fig. 3.2, to serve as a framework for asking questions about the ends a design is to serve and then selecting/creating the means to fulfil them. Each need begets a variety of subjects for architects to consider. The concerns that architects choose to address defines their values and brand of work. The selections made must be seen within the standards set by a society's moral code although they are not always heeded.

3.2 The Basic Functions

The fundamental function of buildings and urban designs is indeed to afford and shelter a set of activities. Usually, the design task is to make the conducting of those activities easy, safe, and comfortable. Sometimes, however, the design goal is to make the settings challenging to enable people to maintain or advance their competencies. It is also to provide opportunities for people to participate in and be members of groups. In addition, a design's configuration, illumination level, and pigmentation can be something to contemplate for its composition.

Adapted from Maslow (1987), Huitt (2004), and Lang and Moleski (2010, 56) by Alix Verge: Courtesy of Alix Verge

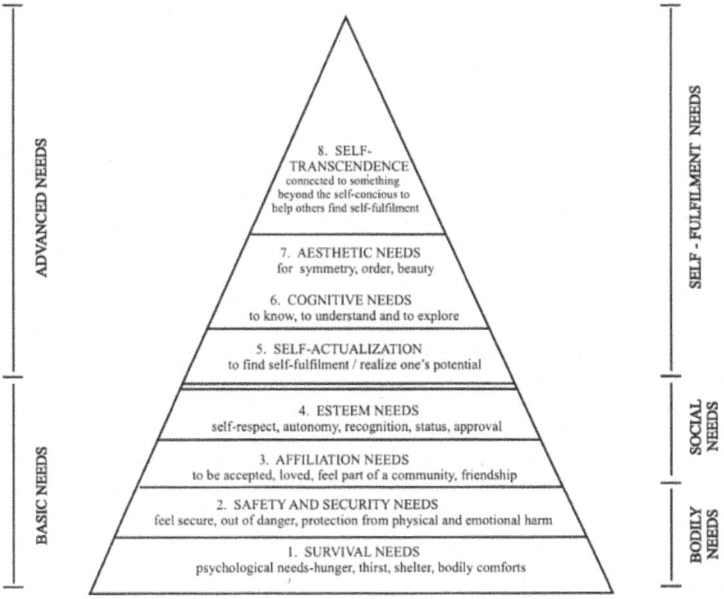

Fig. 3.1 The hierarchy of human needs as seen by Abraham Maslow

3.2.1 The Accommodation of Activities

The design task of architects is to create patterns of built environment that afford the recurrent patterns of behaviour required by their clients to attain both instrumental and symbolic ends. In doing so conflicts between design objectives often arise. Many human activities require the use of machines for them to take place or easier to do. Domestic appliances make household chores less arduous; the use of automobiles and aeroplanes enables movement between places easier; air-conditioning provides comfort in hot and humid weather. In such cases, the patterns of the built environment must enable the machines that we use to function well.

In some places, such as factories, the needs for machines to function well often dominate and their operators must accommodate their behaviour to the way machines operate (Kroemer and Grandjean 1997). Kiyo Izumi (1921–1996), a Canadian architect and educator, called such environments *anthropozemic* in contrast to those that are *anthropophilic* where the function of the environment to meet humanistic values is the primary concern (Izumi 1968) (Fig. 3.3).

Deciding on the *behaviour settings* a building or urban design should afford (that is, in creating the building program) is almost always a subject of debate among those involved. As observed in the previous chapter, analysing and proposing the behaviour and use of facilities by individuals and groups in their various

Adapted from Lewis (1977) and Lang and Moleski (2010) by the author: Courtesy of Nigel Lewis and Taylor and Francis

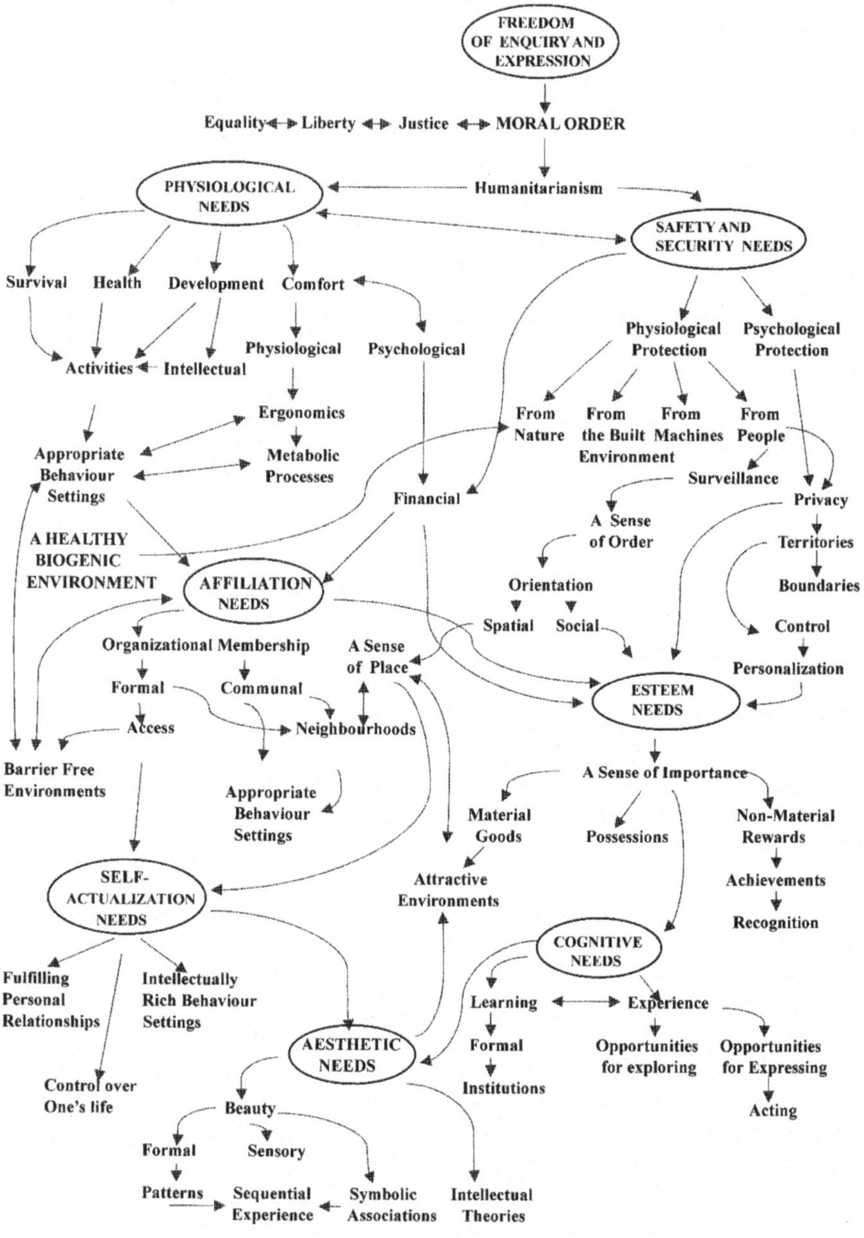

Fig. 3.2 A model of the functions that patterns of buildings and urban areas can serve

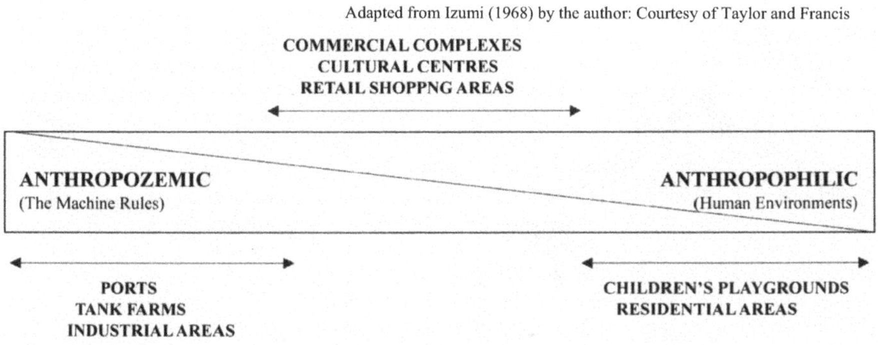

Fig. 3.3 Anthropozemic and anthropophilic environments

capacities is often dealt with indirectly when designing a building program. They are assumed by specifying a program in terms of built form types: places and links (that is, rooms and corridors), parking spaces and other such elements. The goal is for buildings to function well enough to meet some instrumental purpose or purposes efficiently and comfortably.

Anthropometric and *ergonomic* studies have provided architects and industrial designers with the data on human physiological capabilities and what is needed in the design of the fixed features of the built environment and the appliances that ease everyday life (Dreyfuss 2003, originally 1955; Kroemer and Grandjean 1997; Panero 2011). The way daily activities are conducted varies from society to society. The affordances of specific layouts and their fixed features thus need to be understood in their cultural context (Rapoport and Watson 1972; Gattupalli 2023).

How the layout of the environment affords the activities of people of different physiological qualities and competencies, such as those using wheelchairs and the visually impaired, in interior and urban environments has been extensively studied. We know how to design to meet their needs (see, for instance, Goldsmith 1976; Bednar 1977; Barnes 2011). Attitudes to the application of that knowledge differ among societies.

Providing a high quality of shelter for a set of activities may be sacrificed for a building to be in fashion aesthetically. Being in fashion is particularly important for belittled groups and individuals seeking high status. Nevertheless, for people to feel comfortable, the built environment must function well-enough to meet basic metabolic criteria. Our bodies exchange heat with our surroundings. This exchange is affected by what we are doing and wearing, the temperature of the air, radiant heat from surrounding surfaces, air movement, and humidity level. The knowledge base needed to design buildings to be climatically comfortable is well advanced (Givoni 1998; Olgyay 2015). Today, however, most people prefer to rely on mechanical means to attain the comfort that they seek (Barber 2020). Much, however, can be learnt from vernacular designs about how to do so without resorting to mechanical means (Fig. 3.4).

Photograph by the author

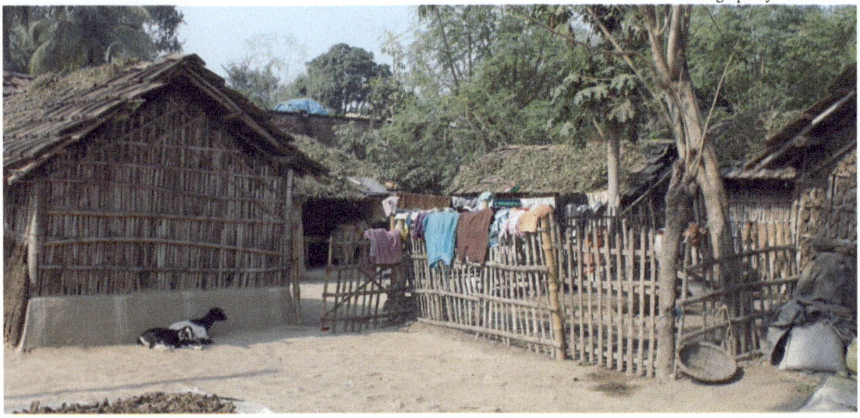

Shelter, Privacy, and the control of space

Photograph by the author

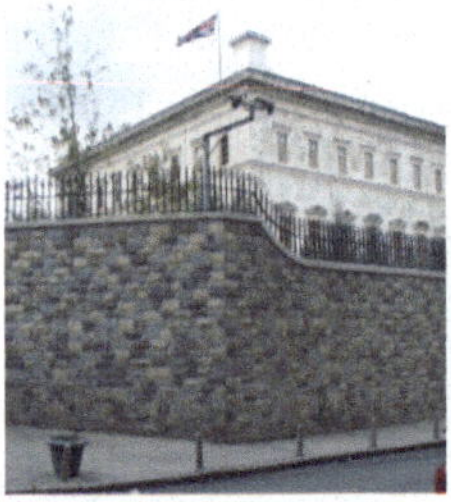

Photograph by the author

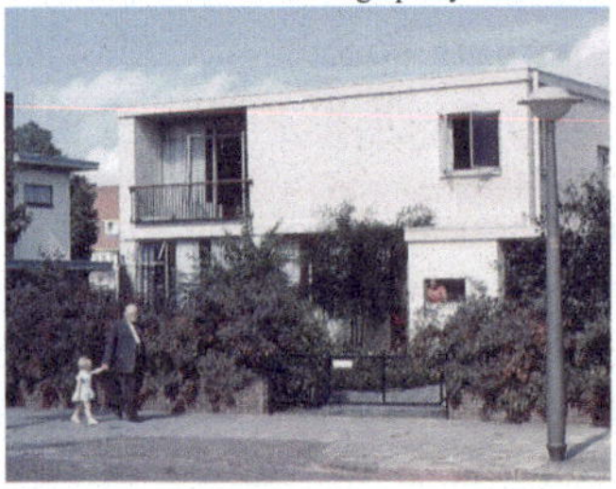

Source: Newman (1975)

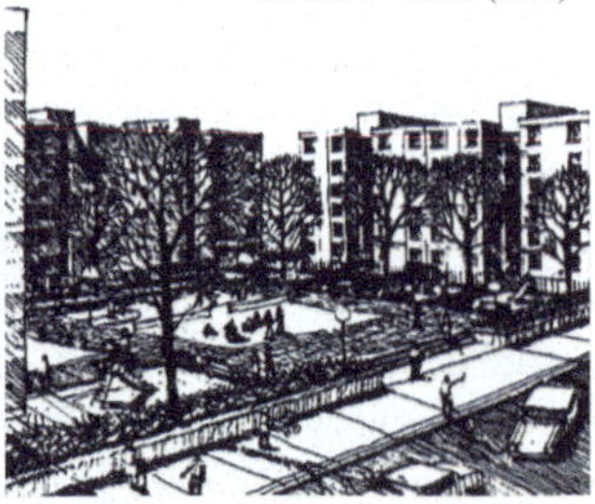

Safety and security using a solid barrier (left), via natural surveillance from sidewalk and front window (centre) and defensible space (right)

Photograph by the author

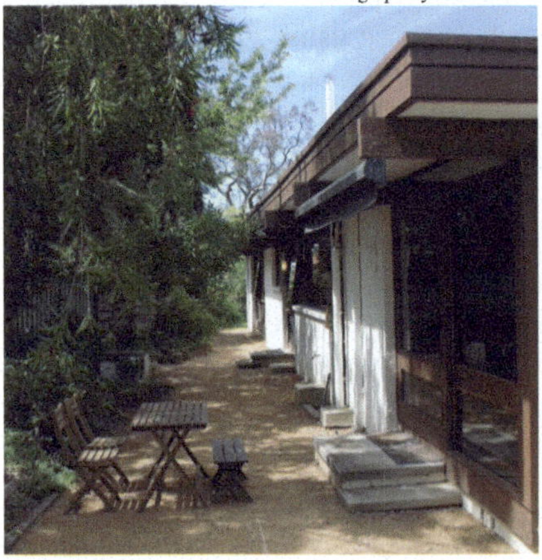

Architecture and identity: An architect's house

Photograph by the author

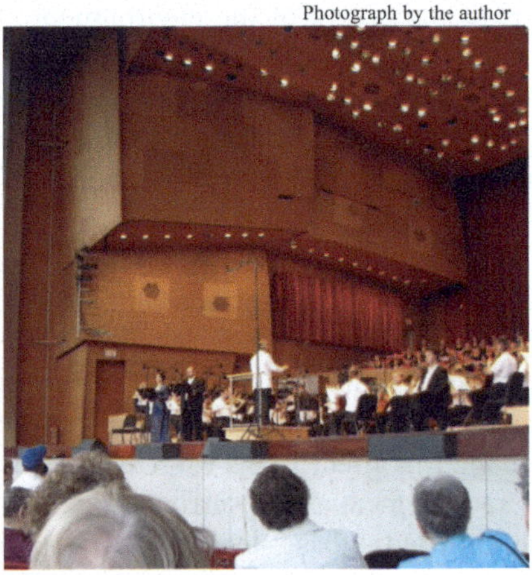

A behaviour setting designed to delight

Fig. 3.4 Basic functions

Vernacular environments that afford their inhabitants tolerable comfort levels are admired for their use of climatic regulating mechanisms that consume limited resources. Schools in Burkina Faso using traditional building approaches stay cool enough for students to study in 40 °C heat (Borràs 2024). Today's demands for comfort are generally much higher than in the past. Some people demand high levels of comfort; others believe that doing so is decadent. Much depends on what people are habituated to accept, and the financial and psychological costs involved in striving for the most comfortable outcome.

Light, natural and/or artificial, entering a behaviour setting is a fundamental requirement for people other than the blind to be able to carry out activities. The location of sunrises and sunsets, the sun's apparent movement through the sky and its height at noon vary by the latitude of a place. The quality of sunlight at a spot is also affected by its topography and adjacent buildings. Sunlight entering a room, city square, or garden is basic to meeting not only comfort and health requirements but to providing feelings of delight (Heschong 1979). The size, placement and reveals of windows function to control the way natural light (and heat) enters a space (Lam 1986; Schielke 2023: see Fig. 4.11). Light needs to have low contrasts and to vary in intensity depending on what one is doing and one's visual capabilities.

3.2.1.1 Coercive Environments: The Non-accommodation of Activities

Some human activities are widely considered to be unacceptable; the appropriateness of others varies from culture to culture or by their location. *Coercive environments* is an umbrella term for patterns of built form that deter or prevent activities from taking place. The terms *hostile* or *unpleasant*, are often used as synonyms. Coercive designs function as substitutes for controlling authorities in preventing behaviours from taking place. They are regarded as silent agents. In public spaces they consist of such items as benches that cannot be lain down on, metal spikes on ledges that discourage people from sitting down on them, metal rings on edges of surfaces to prevent skateboarders damaging them, and such clever devices as spiked aluminium bars that deflect urine flows onto the feet of men urinating in outdoor corners of public places. Speed bumps slow down motorists. Coercive devices are widely used to prevent pigeons from perching, possums or baboons from entering houses, and bears (and people) from climbing fences (Savić and Savićic 2013).

Implicit in the creation of coercive environments is that the desired behaviour of certain people and/or other species is unwelcome. Sometimes they indicate that people such as vagrants are unwelcome in a place. Coercive environments challenge the idea of inclusivity in designs (Fig. 3.5).

More generally, environmental layouts, whether within buildings or outdoors, are coercive if desired behaviours cannot take place. For instance, if there is no exterior place for children to play, many traditional childhood games cannot take place. (In this case the availability of information technology has led to major changes in the way children play.) At other times if an activity is tiresome people

Photograph by Nancy Marshall Photograph by the author Photograph by the author

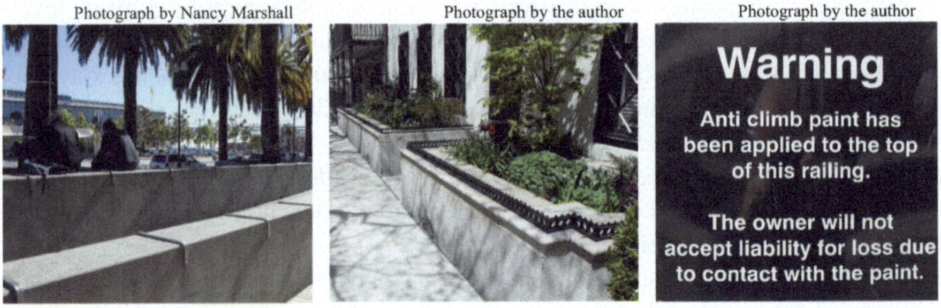

Fig. 3.5 Coercive design: 'You can sit but not skateboard here' (left), 'Do not sit down here' (centre) and 'Do not climb here' (right)

will forgo it. If, for instance, the time required to walk to a desired shopping location takes more than ten minutes, car owners will usually drive rather than walk there. Similarly, highways that cut through neighbourhoods making it a long way to go from one side to the other can disrupt traditional neighbourhood patterns of life.

3.2.2 Salubrious Environments

Designing salubrious environments was a preoccupation of the architects of the Modern Movements in architecture. The deadly impact of the Great Influenza Pandemic (Spanish Flu) was fresh in their minds. Their concern was with creating patterns of the artificial environment that afford the leading of healthy daily lives (Beatley et al. 2018). Many of their observations and designs have been reinforced by scientific and quasi-scientific research.

First and perhaps foremost, urban design and building layouts and fenestration patterns must afford the penetration of sunlight into inhabited spaces to promote both physiological and psychological health. Access to fresh air, opportunities for exercise in the spaces between buildings and surrounding areas away from pollutants is, as the Modernists argued, fundamental to a good life.

At the building level, the location of windows to allow sunlight to penetrate rooms as much as possible is important in temperate places. The requirements in hot and arid places differ. Cross ventilation of rooms reduces internal temperatures. Reticulated air should, ideally, be filtered to remove particles where artificial heating and cooling equipment is used to maintain desired comfort levels (Beatley et al. 2018; McManus 2018).

3.2.2.1 A Note on Healing Functions

Healing environments are those that promote the peace of mind of people recovering from illnesses, physiological or mental. Such environments have both social and physical dimensions. The most important social factor is for individuals to have some control over the decisions that are made about them. Healing interior and exterior spaces possess face-to-face laid-out furniture arrangements where people can choose to interact with others such as family members and fellow patients. Building interiors consist of familiar, home-like patterns. Such environments have been shown to reduce mental anxiety and blood pressure levels and speed post-operative recovery (Sternberg 2009; Beatley et al. 2018).

Healing is aided by shaping buildings and interiors to eliminate environmental stressors such as noise pollution, the lack of privacy, and poor air quality. The configuration of rooms and the way they are coloured affect moods that, in turn, affect health outcomes (Ulrich 1991). Building patterns that connect internal behaviour settings to the outdoors through views of gardens, atria and courtyards have also been shown to aid the healing process (Van den Berg 2005).

3.2.3 Safety and Security Functions

Two sets of mechanisms—physiological and psychological—provide people with a sense of safety and security. A building's function as a haven in the former case is to protect people from the potentially harmful effects of natural events such as earthquakes, fires, and storms. It is also, sadly, to protect them from man-made events such as home invasions, riots, and wars. Buildings must, at least, be structurally sound in the climatic and geological conditions of their terrestrial context.

To meet psychological security needs the built environment must be structured to enable people to have the culturally appropriate level of privacy for the activity in which they are engaged as shown in Fig. 3.6. Privacy, according to *privacy regulation theory*, involves the control of information flows—visual, sonic, and olfactory—from within a setting to outsiders and from the unwanted intrusion of sights, sounds, and smells from the outside into it (Altman 1975; Madanipour 2003). Levels of tolerance of departures from norms vary from society to society and among members of a society. The expectations are not static; they change as cultures evolve. When our competencies are low, we often must sacrifice privacy levels to get the assistance we need.

Some of the ways that individuals attain privacy, such as turning one's back on other people, are behavioural but others are architectural. The latter include separating one space from another by screens and walls that control the flow of information between settings. The degree of seclusion they provide depends on whether the elements are opaque, translucent, or transparent, flimsy or substantial. Other boundary elements, such as changing levels of illumination or floor textures, are symbolic. Such territorial demarcations do not provide barriers to physical transgression. Walled compounds with guarded gates do (Low 2003).

Gated housing areas surrounded by walls provide a sense of security to their inhabitants. Their function is to reduce the possibility of criminal and other anti-social behaviours penetrating a locale from the outside. Such communities exist across the globe. They not only provide their residents with security but also offer

Adapted from Altman (1975) and Lang and Moleski (2010, 146) by the author. Courtesy of Taylor and Francis

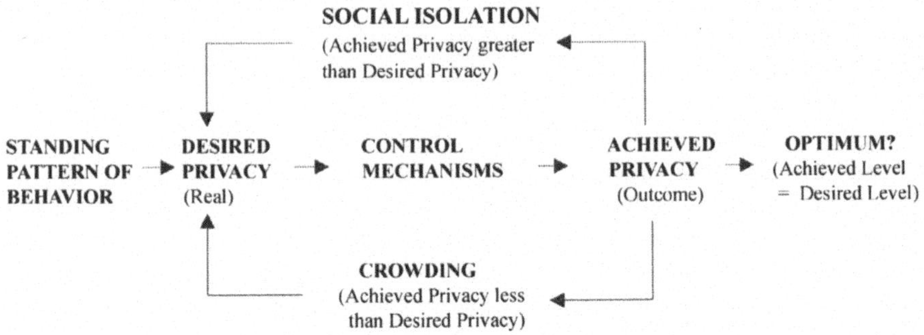

Fig. 3.6 A dynamic model of privacy

a sense of belonging to a socio-economically similar group that possesses some shared values.

Oscar Newman (1935–2004), an American architect, suggested that less conspicuous means function in a similar manner (Newman 1972; Colquhoun 2003; Cozens and Love 2015). Newman introduced the term *defensible space* to the architectural vocabulary although similar concepts had been presented earlier (for example, in Jacobs 1961). His observations have been challenged but have not been dismissed. Applying them, at least, provides visually rich places.

Newman argued that organizing the layout of neighbourhoods and buildings into a hierarchy of physically or symbolically defined territories, from public, to semi-public, to semi-private, to private, encourages the inhabitants of places to take control over and, essentially, guard them. Open spaces, he argued, need to be territorially under the guardianship of a specific group of people and not be no-man's land. In addition, natural surveillance opportunities—those that occur as part of the activities of everyday life—reduce opportunistic crime. Newman also believed that places that are poor in appearance and maintenance are seen as easy targets for anti-social behaviour. Security is, however, more often sought by strategically placed cameras relaying images of activities in a setting to a control point.

Within settings, the way furniture is arranged affects the interactions among people. Humphrey Osmond (1817–2004), a British psychiatrist, distinguished *sociofugal* from *sociopetal* seating layouts (Osmond 1966). The former are grid-like with seating facing outwards making face-to-face interactions difficult while the latter make them easy. The back-to-back seating arrangement in airport lounges is an example of the former and conversation pits the latter. Sociopetal seatings tends to be radial with seating facing inwards. The shape of rooms, the way they are entered and fenestrated dictates what they can offer for comfortable furniture arrangements. Public spaces with both sociopetal and sociofugal spaces give both habitués and strangers the opportunity to engage with others or not (Meagher and Marsh 2017).

In some cultures, complying with sacred geometries gives people who inhabit buildings a feeling of security from evil forces. In India, the application of the *Vastu-Purusha* mandala to the plans and elevations of buildings, is important to many people (Chakrabarti 1998; Bubbar 2005; Sachdev 2011). Such compliance also provides them with a sense that this architecture "belongs to us" in the face of secular, globalizing architectural design forces. In China, *Feng-Shui* design principles function to serve the same end (Smith and Stewart 2006; Anderson 2014). Islam too has its sacred geometries based on cosmological interpretations (Critchlow 1976). Migrants from these societies often take the rules with them to their new countries of residence and want their houses to follow the rules. The sustained psychological impact of not doing so can be severe.

To feel secure people also need to have control over where they are in space and time and to be able find their way around buildings and cities. The level of need varies depending on the situation and the competence level of the person involved. The ease with which we find our way around buildings, neighbourhoods, and cities depends not only on signage (Gibson 2009), but also on their internal structure. Those layouts that enable people to construct a clear spatial image—a cognitive map—of their surroundings enable them to orient themselves (Lynch 1960; Heft 1996; Lang and Moleski 2010; Kitchin and Freundschuh 2018). The formation of such maps has its theoretical explanation in the Gestalt psychology laws of visual organization.

Clarity in the way paths (or corridors) meet at nodes, and that landmarks (significant objects that are visually distinctive in their context) are organized gives people clues as to where they are in geographic space (Lynch 1960). They can also be organized to make way-finding difficult. Sometimes that is done for fun, as in labyrinths, and sometimes for defence by making it easy for outsiders to get disoriented, as in Venice.

Cognitive maps are built up over time through our movement among places. We soon learn to find our way around but for strangers and for people with low competencies, such as those suffering from dementia, orienting oneself in the environment can be difficult. They scan the environment in different ways to highly competent habitués. We do, however, know how to design the layout of the environment at both the building and the neighbourhood level to enhance the abilities of those less competent to find their way (Mitchell et al. 2004).

Providing for financial security is a very different function of buildings. The ownership of buildings or control of the occupancy of rented apartments or buildings is a major source of financial and psychological security. All buildings function as financial investments and their owners seek a return, financial or psychological, on the capital spent. A fundamental function of some buildings is, indeed, to simply maximize financial returns. Meeting this end can clash with how well the basic functions of a building are addressed (Willis 1995; Wainwright 2019).

Often it is the front-end capital costs of a building that are deemed to be the most important by property developers and financiers. Operating and maintenance costs are then regarded as the concern of future property owners. Fulfiling the instrumental function of a buildings is perceived to be the major concern but the aesthetic quality of a building is also proving to a have a monetary value (Clarke 2004). Being able to say that a building has been designed by a renowned architect provides a marketing advantage and is thus a financial asset.

3.2.4 Community and Social Status

Everyone has a unique identity but each person also belongs to a set of people—a community. In combination, the two give individuals a social status. Some communities are locally based but others are of widely dispersed groups of like-minded people who never meet face-to-face. The patterns of the built environment function in two ways to afford people the opportunity to develop a sense of belonging to a local community. The first is through the way the layout of the milieu affords the seeing and meeting of people and the communal facilities provided. The facilities required depend on whether one is dealing with a building or urban design. The second way they can occur is through the meanings people read into the appearance of buildings.

Sociological communities in which people interact with each other can be *formal* organizations in which bonds are established by contract as in commercial firms, or they can be *communal* ones that grow from the common interests of people. Formal organizations can be designed from the top down; the latter grow from the grassroots up based on the opportunistic meetings of people. Specific settings for the former can be designed, while only places allowing for the latter to develop can be provided. These generalizations apply at the building, neighbourhood, and metropolitan scale of design.

Privacy and community go hand in hand in the growth of territorially based (that is, local) communities whether in commercial organizations, institutions, or neighbourhoods (Alexander and Chermayeff 1964; Greenberg 1995; US Dept of Housing 2000). The likelihood of interactions occurring increases if the layout of the environment clearly affords people places for exposure to and for withdrawal from dealing with others. Bedok Court in Singapore is an example where the inclusion of places to see one's neighbours come and go and to gather fostered a sense of community in a large housing development (Bay and Ong 2006; Lang and Moleski 2010).

Dreams of creating traditional neighbourhood communities must be tempered by realities. The strength of local communities formed of many formal and communal organizations, such as those described by sociologist Herbert Gans (b. 1927) in *The Urban Villagers* (1962), has been eroded in many cultures. Increases in physical and social mobility have led to people leading metropolitan and digital rather than parochial lives. Most neighbourhoods have become areas where residents have limited obligations to each other. Nevertheless, designing facilities for common usage can still foster a sense of belonging to an area. The idea of a clearly bounded precinct with strong edges and a bold centre—a heart, and buildings of a similar visual character—a district—will, at least, give its resident a place with which to identify.

The size, massing, structural qualities and materials and the way they are patterned not only give a building an identity, they also are cues of the social status of their users and inhabitants (Butina-Watson and Bentley 2007). Outsiders predict, correctly or incorrectly, the characteristics of the inhabitants—whether they would be helpful or not—from the appearance of buildings (Nasar 1988). Whether we choose to inhabit a place or are obligated to do so because we have no choice, the qualities of buildings establish an image of ourselves to ourselves and who we are to outsiders (Cooper 1974).

3.2.4.1 Buildings as Repositories of Memories and Their Function in Reinforcing Identities

Buildings can function as the repositories of histories that reinforce the self-images of groups of people. The structures may represent good examples of a style or period of architecture that are of importance. Buildings can also carry meanings that have little to do with their intrinsic nature but because of who lived there or what occurred there (Lynch 1972; Yarrow 2018).

Many buildings around the world have become museums because they were the homes of famous politicians, authors, or scientists. They honour an individual of note and boost the esteem of the various communities of which they were members. Similarly, buildings may be associated with groups of people such as trade unions, military squadrons, or professions and be of importance to their members. Other buildings are regarded as important because of the events, such as celebrations, revolutionary meetings, or assassinations, that took place in them. These buildings may be of local, national, or international importance to one group of people but not others.

Individuals and organizations vie to have buildings preserved, restored, or reused to keep a memory alive because of their importance in giving them an identity and sense of pride. If the associations are negative for a group of people, the group may seek to have a building demolished even if it is of architectural importance. This situation often arises in countries formerly colonized. What should be done with the fine buildings of an era that a nation wants to forget?

3.2.5 Delight and Experiential Aesthetics

Experiential aesthetics deals with the pleasure we receive from subconsciously perceiving, via all our sensory systems, the world around us as we inhabit and move from one behaviour setting to another (Rasmusson 1959; Chelkoff 2003). Experiential aesthetics stands in strong contrast to intellectual aesthetics which involves the self-conscious examination and analysis of patterns of built form based on some artificial mental construct.

3.2.5.1 Sensory Aesthetics

The world around us is the source of rich sensory experiences. Such experiences, positive, neutral, or negative, arise from the arousal of our perceptual systems: visual (except for the blind), auditory (except for the deaf), olfactory (except for the anosomic) and haptic (except for the hypoesthesic) (Diaconu 2006). We pay little attention to our sensory experiences except when they depart from the norm but, subconsciously, they affect our perceptions of the quality of places (Rasmussen 1959; Heschong 1979; Pallasmaa 2012, originally 1994; Lewis 2020). We pay attention more to the source of the experience than the experience itself.

In fields such as textile design, the tactile quality of materials is an important consideration in the design of a product. Architects, however, seldom pay attention to such sensory experiences in their designs. Those experiences, nevertheless, contribute to the appreciation of places (Gallace and Spence 2011; Faucheu et al. 2019). In contrast, the formal organization of the patterns of built environment is central to architects' hearts and work.

3.2.5.2 Formal Aesthetics

All patterns of built form have a composition of parts in relationship to the whole. Compositions exist at various levels of order and complexity and are examined with various levels of attention as an observer views them from a station point (Kepes 1944; Arnheim 1977; Alexander 2000–2005; Lewis 2020) or moving around or through them (Cullen 1962). Formal aesthetics is primarily concerned with the appreciation of the visual character of objects by sighted people.

The research shows that regular forms are easiest to see but complex forms attract greater attention and are more interesting. Interestingness and pleasurableness are correlated positively up to the point when patterns become chaotic (Berlyne 1970; Weber 1995; Van Geert and Wagemans 2020; see also the essays in Nadal and Vartanian 2019). Such observations have their theoretical base in Gestalt psychology's concept of field forces.

Field forces are analogous to the concept of equilibrium in mechanics. Static compositions are simpler than the more complex forms that deviate from a state of visual equilibrium. Such designs are said to be dynamic (Levi 1974; Arnheim 1977). The horizontal plane acts as the base for the vertical axis. A symmetrical scheme in which parts are related to the whole based on a common proportional system, is seen as static. Forms are seen as increasingly dynamic depending on their degree of departure from the basic static pattern (Kepes 1944; Lang and Moleski 2010).

Debates remain as to whether such perceptions are innate or result from learning to associate specific patterns with specific feelings. Many architects learnt these associations from what is stated in invented visual languages such as that of the Basic Design course developed by artists at the Bauhaus and promoted by Johannes Itten (1888–1967), an expressionist painter (Itten 1964). What is important for declarative theory is that forms do communicate meanings. It is also clear that individual's and culture's subconscious preferences for types of composition vary over time.

There appears to be a sequence of societal preferences for increasingly complex forms over time until they reach very high levels of complexity. Tastes then return to preferences for simpler forms followed by the incremental increase in preference for more complex forms until they are seen as too complex (Tyng 1975). Individual tastes go through this sequence more rapidly than those of societies, and more rapidly for easy to change items than for those, such as buildings, where changes are more difficult to implement.

Except for the blind and the colour blind, the world is pigmented. The effort to understand the effect of colours on the appreciation of the built environment goes back to the days of Aristotle. It is generally agreed that certain colours shape our impressions of the size of places. Some colours are said to make spaces seem smaller; others make them appear to be larger (Birren 1965; De Grandis 1986). Colours are also said to arouse specific emotions.

Warm colours such as reds, oranges, and yellows are claimed to evoke energy and excitement while blues and greens, the cooler colours, feelings of calmness. These observations seem to be culturally specific and to depend on the observer's feelings at a particular time. The effect of the colours of settings on the performance and productivity of workers has been much studied (see Elliott and Maier 2013; Shandrow 2015).

Much of the experimental and empirical research on formal aesthetics has focussed on the perception of two-dimensional patterns. That on three-dimensional objects has viewed buildings and cityscapes from a stationary point as in a photograph. Accurate perception of the world around us depends on our moving around objects or through environments (Gibson 1979). Our experiencing and appreciation of the built environment is a function of the way vistas open up from behind occluding surfaces as we move past them (Cullen 1962; Appleyard et al. 1965; Lewis 2020). We attend to different elements of the environment when we are walking from one space in a building to another than when we are driving through places. The sequential experience that we have largely dictates our feelings of liking or disliking places. In addition, the activities that occur in a setting and the people involved shape our feelings about it. We respond to the totality of the experiencing of a behaviour setting not simply the formal structure of its milieu. We can, however, treat the milieu as an object and examine its character, but that requires a special mode of attention.

Symbolic meanings are derived from what an observer imputes, consciously or subconsciously, to an object or a scene. The association (that is, the meaning of an object) maybe widely understood within a society or be highly individualistic. Carl Jung (1875–1961), a Swiss psychiatrist, believed that some meanings are part of the collective unconscious of everybody and are shared as innate archetypes inherited from our ancestors (Jung 1968; Cherry 2023a) (Fig. 3.7).

3.2.5.3 Symbolic Aesthetics

The relationship between an object, or an element of the built environment, and its meaning has been explained using the semiological triangle linking symbol,

Photograph by Kirill Neiezmaklov/Shutterstock.com

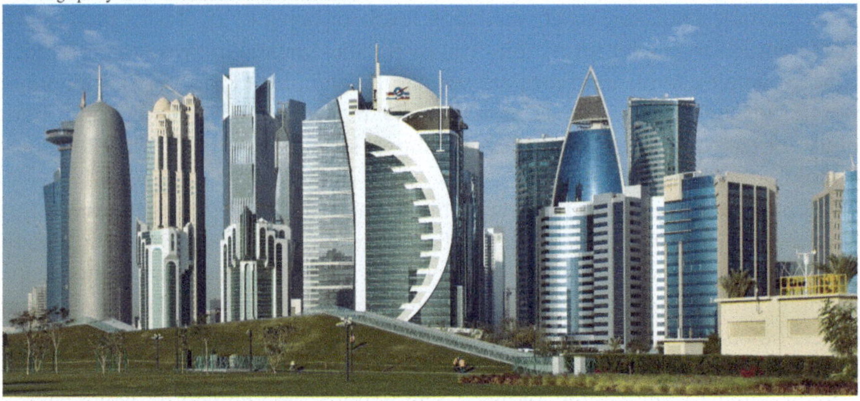

Unity through diversity: Doha skyline in 2020, the work of many designers

Photograph by High Contrast/Wikimedia Commons

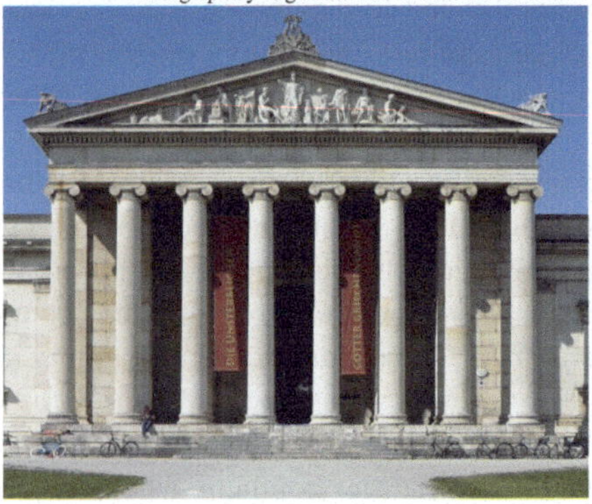

Photograph by Bill Price III/Wikimedia Commons

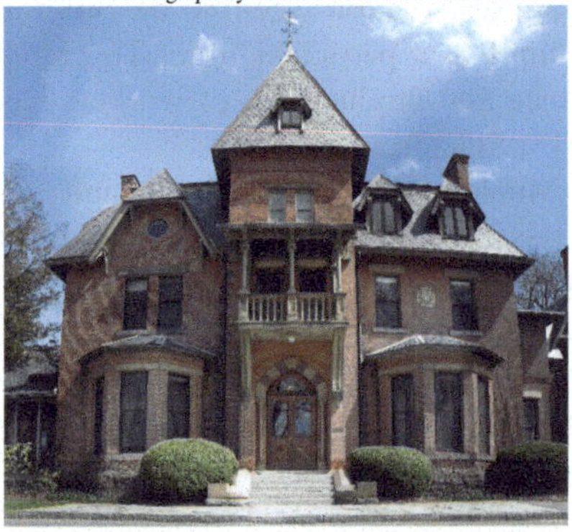

Static compositions: Glyptothek, Munich (1815) (left) and a more dynamic one Andrew D White House (1871) (right)

Photograph by Trungydang /Wikimedia Commons

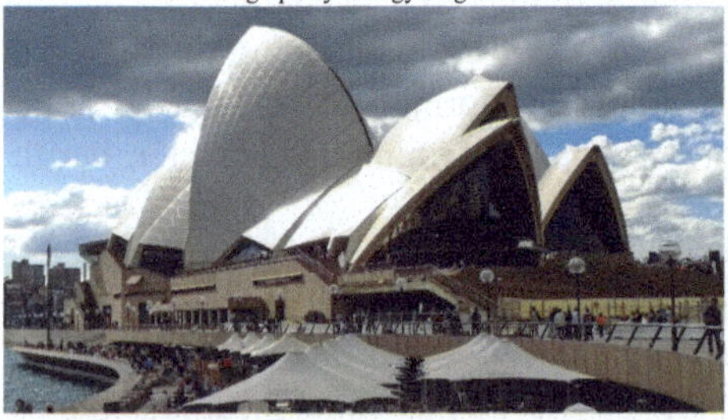

Photograph By Carol Highsmith/Wikimedia Commons

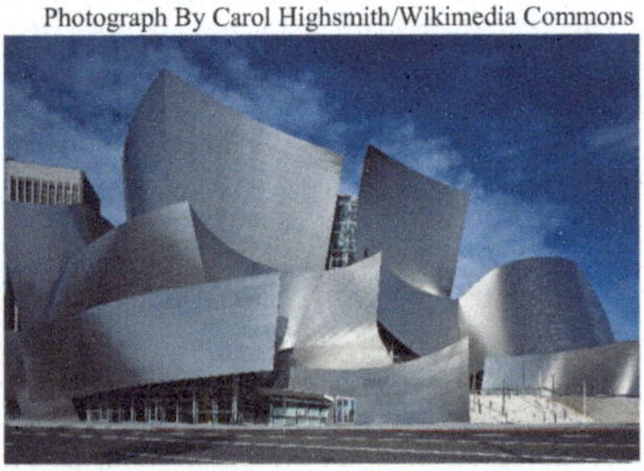

Increasingly dynamic compositions: Sydney Opera House (1956-1973) (left) Walt Disney Concert Hall, Los Angeles (1995-2003) (right)

Fig. 3.7 Formal aesthetics

Source: Lang and Moleski (2010, 42). Courtesy of Taylor and Francis

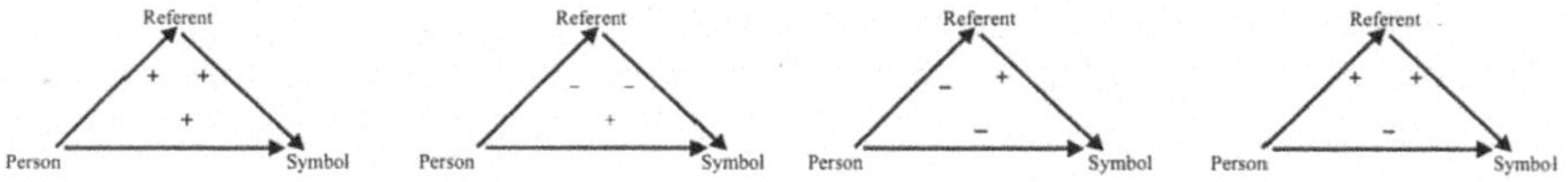

Fig. 3.8 Cognitive consistency and balance theory: three congruent relationships (left) and one incongruent (far right)

referent, and thought as originally developed by Charles Sanders Peirce (1839–1914), an American philosopher (Hardwick 1977). The pattern of built form and/or its materials, pigmentation, and/or the nature of illumination is the symbol, or signifier. Its attributes depend on what one associates with it. That referent may be an idea, a memory, a place, or another object. The signified is the thought evoked from making these connections, consciously or subconsciously (Broadbent et al. 1980; Short 2009).

Whether one regards specific meanings as positive or negative can be explained by Balance Theory as developed by Fritz Heider (1896–1988), an Austrian psychologist (Heider 1958). Heider suggested that the relationship among the three elements in the semiological triangle are seen as positive (good, liked, favourable) or negative (bad, disliked, unfavourable). The basic tenet is that people strive to keep the relationship between the elements congruent and mutually supportive (Festinger 1962; Crandall et al. 2007).

If one person has a positive attitude towards another person, people or set of ideas then the first person will have a positive attitude towards the symbol (such as the architectural style) associated with that second person, people, or ideas (see on the left in Fig. 3.8). Similarly, if the first person has negative associations with the second then, to keep the system in balance, the first person would have to have negative feelings about the symbols associated with the second person. When the relationships are incongruent people strive to make them so by changing their attitudes—their beliefs and the values associated with them.

Building forms can mean different things to different people depending on the associations that they make between patterns and referents. The Taj Mahal in Agra is presented to international tourists as a symbol of love and romance. It can also represent the exploitation of labourers, the subjugation of Hindu India to Moghul rule, or the repressed role of women in society (Gupta et al. 2011). Meanings can change as new associations are formed.

Façades have been the most important variable in communicating a building's inhabitants' values and status. More subtle variables also communicate meanings. The internal organization and sequence of spaces, the proportions of enclosed to open space, and the degree of penetration that an outsider is allowed to enter the interior spaces of a building are all meaningful. Colours are important signifiers and their use in various contexts is related to the norms of different cultures (Wu et al. 2018). The use and nature of decorations has been equally important. Referents can be varied too. To architects, the referent is usually other designs and/

or architectural ideas. Indeed, to some of the cognoscenti such associations is the sole variable of concern in evaluating buildings.

Attitudes change over time. Influential tastemakers attempt to make a society develop new relationships among objects and what they signify. They strive to change the referents associated with specific objects and events (Lynes 1983). Sometimes they succeed in changing people's tastes and sometimes they fail. Much depends on the esteem in which the influencers are held. The policy of Sinicization of mosques in China has led to 'Arabic' domes being demolished and minarets taking on pagoda forms. The actions reflect the iconoclasm of the Chinese Communist Party. It seeks a unified culture as displayed in architectural tastes.

3.2.6 Taste Cultures

Pierre Bourdieu (1930–2002), a French sociologist, argued that there is no such thing as 'good taste'. He believed that our tastes and the way we define them identify who we are (Bourdieu 1984). Tastes are formed through the way we are socialized and our formal education (Imani and Zafarmandi 2017). The places where we live, the schools and universities that we attend, and the clothes we choose to wear all reflect our tastes and affect how we are perceived by others. The aesthetic quality of a place is based on the values of the beholder; it is not a property of the object viewed although we may consider it to be (Table 3.1).

Herbert Gans suggested that in late twentieth century United States, at least, five overlapping taste cultures existed (Gans 1999, originally 1975). His analysis has held up over time although we might use different labels to his. Gans identified: (1) A high taste culture of intellectual, academic, and avant-garde concerns, (2) An upper-middle taste culture whose concerns are self-consciously about current fashions, (3) A lower-middle taste culture whose concerns are eclectic, democratic, and popular, and (4) A traditional taste culture, whose concerns are largely unselfconscious and anonymous. Finally, he suggested, America has: (5) Quasi-folk taste cultures whose concerns are communal and ad hoc. Another way to identify differing tastes is to distinguish among academic, self-conscious popularist, and unselfconscious taste cultures (Mann 1979). Seeing them in a hierarchy or of equal value depends on the analyst, but each group tends to hold itself in high-esteem and to be dismissive of the tastes of the others.

The legitimate taste of a society is considered to be that of the ruling class and the cognoscenti (Bourdieu 1984). Although societies consist of more than they, they and the people with upper middle-class tastes form the base group of architects' clients (Jenkins 1961; Eaton 1969). Building designers do the bulk of work for the remainder of the population except where being a registered architect is legally required to design buildings beyond a certain size and/or monetary value.

Table 3.1 Taste cultures, aesthetic, standards, and patterns of built form

Focus	Taste publics (Gans 1975)	Aesthetic standards	Architecture			Contextual relationship
			Form	Context	Composition	
Academic	High	Intellectual and academic	Abstract and idealistic	Implicit and symbolic	Canonic and coherent	Heroic, original and monumental
		Avant-garde	Theoretical		Unified	Unique
Self-conscious	Upper-middle	Self-conscious current	Conceptual	Sophisticated cosmopolitan	Classical refined	Foreground
		Fashionable		Associative	Ambiguous	Pretentious
Popularist	Lower-middle	Eclectic democratic popular	Informal additive arbitrary	Manneristic stylistic referential: tudor, Californian, etc.	Contradictory and complex	Homogeneous eccentric ubiquitous
		Traditional	Pragmatic	Ordinary	Repetitive	Uniform
Unselfconscious	Low	Anonymous	Concrete	Austere		Obscure
		Unselfconscious	Logical	Explicit	Open-ended	Background
	Folk	Communal adhocist	Direct	Picturesque	Chaotic	Quaint

Source Lang and Moleski (2010, 216). Adapted from Mann (1979) by the author: Courtesy of Taylor and Francis

3.3 The Advanced Functions

Although considered to be advanced functions, the need to be continually learning and for understanding the surrounding world parallels our motivations to get basic needs met. In addition, we can examine the patterns of the world around us that have been self-consciously designed to carry specific meanings. Enriching people's experiences is a potential function of architecture.

3.3.1 The Cognitive Function of Environments and Objects

Formal institutions such as schools socialize us into ways of thinking and provide us with the knowledge base that enables us to perform as members of a society. Architectural schools serve this role for the architectural profession. Much learning, however, occurs vicariously when people interact with others and participate in, or observe what is occurring in the behaviour settings that form part of our lives. George Santayana (1863–1952), a Spanish-American philosopher, once observed: "A child only educated at school is an uneducated child" (Santayana 1935). Fulfiling cognitive needs is basic to life and to the good life. Architectural forms, per se, offer little, in comparison to the social environment in fulfiling this need but they do reflect the culture of which they are a part.

The cognitive function of behaviour settings is more an issue in forming policies and programs for city planning and urban designs than for interior design. The layout of the built environment can assist or detract from the opportunity to get exposure to the behaviour settings that enrich people's learning about life and about oneself. Many Modernist urban designs, although much admired for their intellectual formal aesthetic qualities, are boring places. They contain a paucity of behaviour settings. This end has been achieved by focussing on creating sanitized, salubrious places. It is also a by-product of the change in scale of economic activities and the substitution of direct participation in life with the mediated mechanisms afforded by advances in information technology.

3.3.2 Intellectual Aesthetics and the Cognoscenti

To many architects the reasoning supporting the sculptural forms of buildings or urban designs is the central concern of architectural theory. The focus is on the intellectual aesthetic function of an artwork (Santayana 1896). Buildings can be analysed as works of art by focussing on their exterior three-dimensional character and their interior spatial configurations for the pleasure, or not, derived from understanding their creator's logic in organizing them into specific patterns. Buildings thus function as objects for the contemplative analysis of a person as an art critic.

The vehicle for studying a building as a work of high art is the *parti*. A parti consists of the design principles that give a composition its structure. These principles may not be evident to casual observers but must be explained to them by the architect (Johnson 1994). This is particularly true of much Modernist and Post-modernist architecture, but less so of Classical and Neoclassical works that are more easily open to interpretation.

One group of the cognoscenti are architects themselves. They are interested in the logic of how their designs' patterns relate to the sculptural qualities of other architects' works and the way the forms they employ establish their own identity and brand of work. Such architects have an *extrinsic* concern with establishing their reputations and careers. The other two groups of the cognoscenti consist of the broader art community and the potential patrons of architecture. Theirs is an *intrinsic* concern with buildings as public art. They promote the field of architecture as an intellectual discipline (Allport 1954). They may be simultaneously promoting their own social standing as being 'in the know' as part of the high-order taste culture identified by Gans.

All the buildings illustrated in Fig. 3.9 reflect the ideas of their architects although the designers of Akbar's tomb in Agra are unknown. The tomb reflects the intricate work and craftsmanship of seventeenth century Indian architect-builders influenced by Persian aesthetic ideals (Asher 1992). All the other buildings in the figure are clearly the work of their authors and are celebrated for the intellectual logic of their designs. The Palazzo Nuevo on the Capitoline Hill in Rome may not depart from contemporary norms as much as the other buildings shown in the figure. It is a background building to the life in the square that it fronts.

The other buildings in the figure, except for the Richmond waterfront, are clearly foreground buildings and objects in space open to analysis by the cognoscenti and lay people alike. The former are more likely to celebrate the buildings' intellectual nature and avant-garde designs than the latter. The cognoscenti feel uncomfortable with the retrogressive aesthetic qualities of the Richmond Waterfront however aesthetically precise it is or how well it meets the needs of its clients, or how it frames an active outdoor behaviour setting (Watkin 1996).

3.4 The Catalytic Function of Buildings: Multiplier and Side Effects

All structures have an impact on their surroundings. They affect the ecological nature and the financial value of adjacent areas and the character of the social environment. Sometimes buildings are erected to purposefully lead to a change in their surroundings but it is seldom a fundamental concern of architects. It may be of municipal governments. The positive aspects of these impacts are their *multiplier effects* and the negative their *side effects*.

Public policy investments in museums and other cultural facilities, parking garages, and public spaces that have been designed to make locales more attractive for property developers to build on adjacent sites have had some success (Attoe and

Photograph by the author

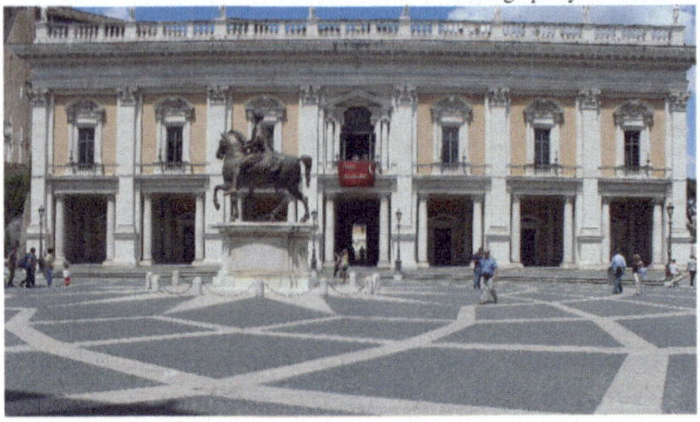

Palazzo Nuevo, Capitoline Hill, Rome (1603)
Michelangelo Buonarroti (1536-1546), architect

Photograph by the author

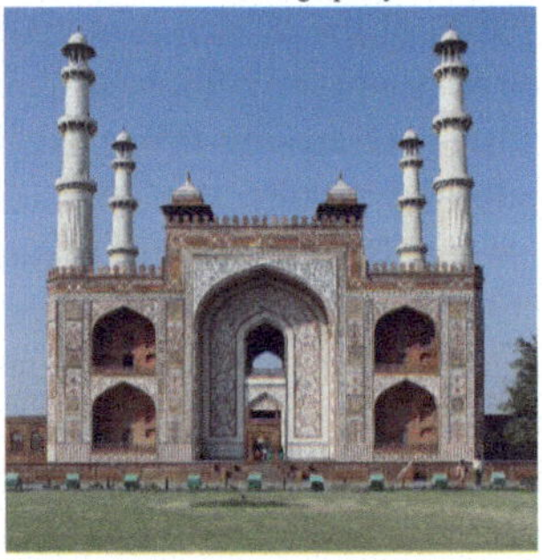

Akbar's Tomb, Agra (1605)
Anonymous Indian designer/builders

Photograph by the author

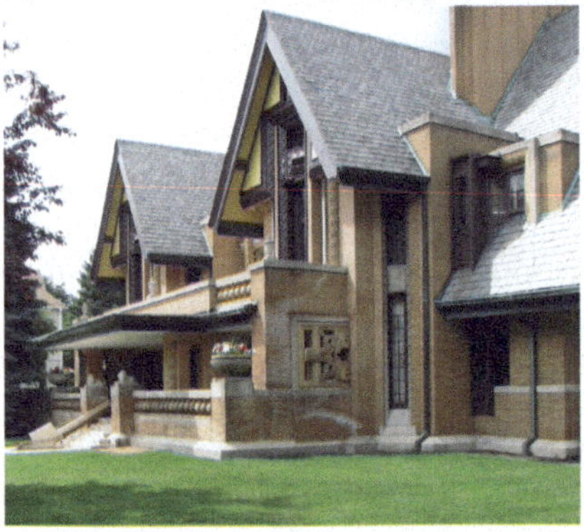

Nathan G. Moore House, Oak Park (1895)
Frank Lloyd Wright, architect

Photograph by the author

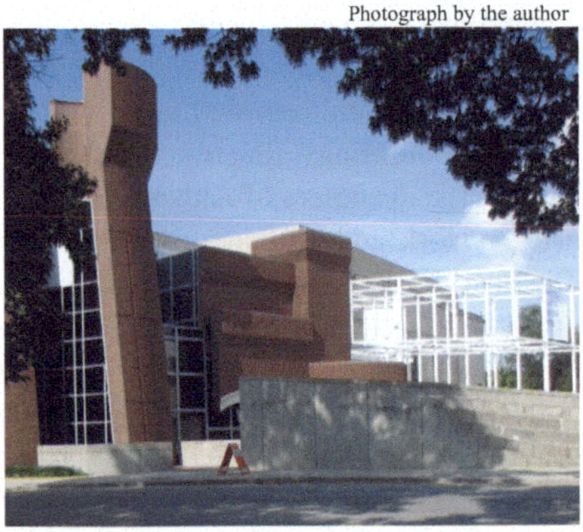

Wexner Center for the Arts, Cincinnati, Ohio (1989)
Peter Eisenman (b. 1932) Associates, architects

Photograph by Alix Verge: Courtesy of Alix Verge

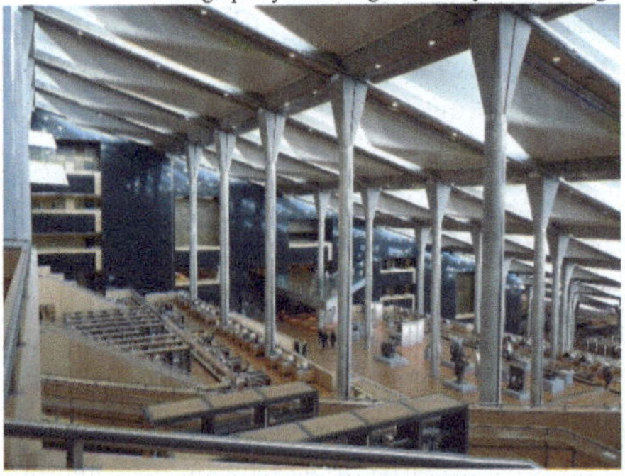

Bibliotheca Alexandrina, Alexandria (1999)
Snøhetta Architects (founded 1988), architects

Photograph by the author

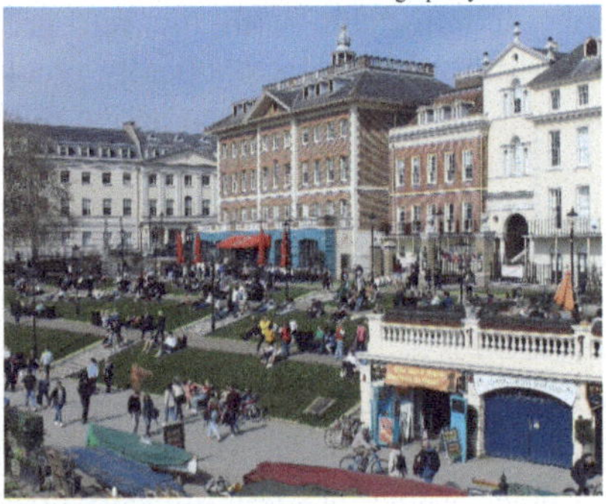

Richmond Waterfront, London (1989)
Quinlan Terry (b. 1937) Architects, architects

Fig. 3.9 Sources for intellectual analysis: buildings representing their designers' prescriptive theories

Logan 1989; Sternberg 2002; Domanski and Gwosdz 2010; Kristo and Dhiamandi 2016). The investment in the Guggenheim Museum in Bilbao is an often-cited example. Many such buildings designed by starchitects simultaneously function as tourist destinations; they enhance the local economy. New buildings in low-income areas can, however, lead to the gentrification of neighbourhoods and displacement of local people and businesses. Sometimes that end has been the purposeful public policy goal.

The social impacts of new buildings vary. They, at least, change pedestrian traffic flows. They may affect vehicular traffic movements. That can lead to the deterioration of neighbourhoods and consequently an increase in opportunities for anti-social behaviour (Newman 1975; Skogan 1990; Colquhoun 2004). The way automobile parking is managed affects the experience of pedestrians as does the character of the ground floor of buildings and the way the two relate to streets. Such design variables and use characteristics of places can enhance or reduce the interestingness of the environment for passers-by and, indirectly, their willingness to walk between destinations. Tall buildings can affect the privacy of the residents of adjacent low-rise properties that they overlook. Invasion of privacy problems arise in hot-arid areas of the world where flat roof tops are used for family life when evening temperatures drop (Hakim 1986).

All new buildings change the *edaphic* and *biotic* environments—the physiology and the fauna and flora of a location. Buildings and plantings inevitably change the microclimate of their surroundings. They change air movement patterns. Buildings, particularly high-rise structures, cast extensive shadows especially when located at high latitudes. Paved surfaces of building roofs and urban open areas add to the heat island effect of cities. Buildings, themselves, can be the sources of glare and may shed heat onto their neighbours as the Walt Disney Concert Hall in Los Angeles did (Schiler and Valmont undated). Some buildings even cause noises as winds whistle through protrusions. Some of the problems get fixed after the buildings have been in operation. Others are simply tolerated.

3.5 The Concept of Fit

The cockpits of racing cars are often custom-made for a specific driver. For architects, the question is: 'How congruent should the patterns of the environment be with the patterns of behaviour predicted to take place within them and the taste culture of the people who will be using or observing them?' As Maslow noted, once people's basic needs are met, they are motivated to achieve other ends. Designing solely to meet the immediate shelter needs of people in mass housing schemes has shown that higher order needs should be considered when establishing building programs. As noted in the previous chapter on procedural theory, predicting future behaviour patterns is fraught with difficulties.

One argument is that because the instrumental functions of buildings keep changing and consequently the behaviour patterns and tastes of their users change, architects need not worry about how well the environment that they are creating fits specific tasks. People adapt their behaviour or adapt the environment to meet their needs. The alternative argument is that the environment should be closely designed for the specific needs of the original uses and be designed to be easily adaptable to meet the requirements of changes as they occur. Certainly, some buildings have had as many lives as a cat is reputed to have while retaining their essential architectural integrity (that is, the central idea, or brand, shaping them).

3.5.1 Fixed-Feature Spaces

Some features of buildings are fixed in place. An empty room affords many different configurations of furniture layouts but what can be done well is, nevertheless, constrained by the location of doors and windows and elements such as fireplaces. Many buildings have staircases and/or lifts that are fixed in places. Bathrooms and kitchens have elements whose location and characters are fixed by the demands of plumbing. Sometimes seating is built in.

The shell of some buildings affords many options. Row houses have been converted into offices and back to houses with their fixed features remaining largely intact. Office buildings have been converted into apartments. Abandoned suburban shopping malls have been converted into colleges and churches; more have been demolished. Some churches, such as the Catholic Cathedral in Tripoli in Libya, have been converted into mosques; others with internal adaptations have become concert halls, houses, architectural studios, and commercial spaces. Their internal fixed features have been changed to accommodate the new instrumental functions they house.

3.6 The Functions of Buildings for Architects

Architects are businesspeople who run or participate in practices that range in size from those of single practitioners to multi-national corporations that have hundreds of employees. Designing buildings and urban designs and seeing them implemented is their source of income. To stay in business, they must have a steady flow of commissions. To obtain commissions architects are advocates for their own work. They must convince potential paying clients that they will not only get a good design but also their money's worth. Major architects have been good salespeople in obtaining work. A few have been successful in finding patrons who give them a free hand to explore innovative design ideas.

The character of the buildings that they design gives architects an identity, a niche in the marketplace, and a social status. Well-received projects reinforce the self-image and feelings of self-worth of an architect; they also serve as an advertisement for their work. The function of buildings as a vehicle for architects to communicate their identity through the central message they offer the world is an important means of establishing and developing a professional practice. Architects are motivated, as Steven Izenour (1961–2001) of Venturi and Rauch and co-author of *Learning from Las Vegas* used to say, get their own "jollies" out of their work (Fig. 3.10).

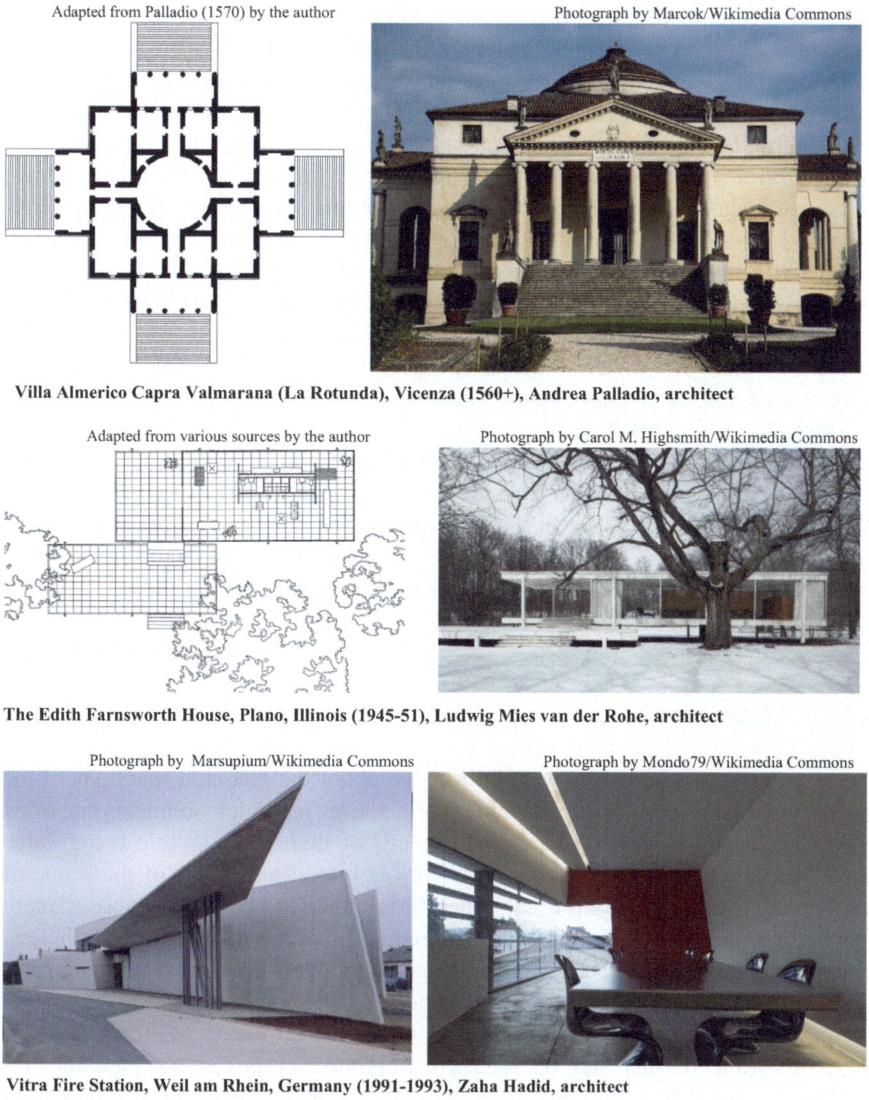

Fig. 3.10 Three buildings based on strong central geometric ideas

3.6.1 Central Ideas and Brands

Many architects downgrade the relative importance of certain functional requirements of a design in order to impose a controlling idea on its character. The central idea may be unique to a scheme to make it a cohesive whole or it may be a general image that reinforces the public's view of an architect's brand of work.

The controlling idea may be a formal aesthetic style, or a preferred structural system, and/or a construction material. The Villa Almerico Capra Valmarana, designed by Renaissance architect Adrea Palladio (1508–1580) for a retired priest, is based on a strong geometric idea. The plan is symmetrical. The façades have the same design despite their differing orientations. The design represents Palladio's brand of work. The designs of Spanish architect and engineer Santiago Calatrava (b. 1951) are known for their structural ingenuity. Louis Kahn, Ulrich Franzen (1921–2012) in the United States, and Laurie Baker (1917–2007) in India creatively used brick in their work. For Khan and Franzen it was very much an aesthetic choice; for Baker it was a cheaper, readily available material and the workmen employed on constructing his designs knew how to lay bricks.

The Villa Savoye, the Farnsworth House and the Vitra Fire Station demonstrates strong guiding ideas. None of the three functioned well in fulfiling its instrumental purpose. They are creative works of art. Some critics believe architects' foci on illuminating central ideas in their designs shows that they are detached from everyday life (Weaver 2006). At the same time, having strong central ideas gives a powerful image to a building and an identity to its architect in the eyes of both other architects and the public.

3.7 Commentary

The goal in establishing a body of knowledge about the potential functions of the built environment is for it to be a value free resource on which architects can draw in asking questions about the ends and means of their work. While being value free is an honourable goal, functional theory is biased by a view of the scope of architecture's concerns. As presented here it is inevitably based on an assessment of what is important for architects to understand. Functional theory, nevertheless, aims to be value free. What has been presented here has only been a structure for thinking about the ends that architectural design can serve.

Many architects will regard the outline of functional theory as structured here to be irrelevant to what they do. To them it assumes, a procedural paradigm for architecture that differs from the one that they rely on. Architects instead base their designs on existing building types and generic designs and do not ask questions about the range of functions a building could serve. In working this way they need to ask questions about how each type functions in the cultural context of its formation. Otherwise, they may be using a type outside its realm of applicability. That has often happened.

References

Behne A (1926) The modern functional building. Translated from the German by Robinson M. The Getty Research Institute for History of Arts and Humanities, Santa Monica, CA, 1996, republished

Gropius W (1955, originally 1953) The scope of total architecture. Collier, New York

Lang J, Moleski W (2010) Functionalism revisited. In: Architectural theory and practice and the behavioral sciences. Ashgate, Farnham. Republished by Routledge in 2016

Lewis NC (2020) Design and order: perceptual experience of built form: principles in the planning and making of place. Wiley Blackwell, Oxford

Maslow A (1987) Motivation and personality, 3rd edn. Revised by Fraeger R, Fadiman J, McReynolds C, Cox R. Harper and Row, New York, NY

Scott Brown D (2004) The redefinition of functionalism. In: Venturi R, Scott Brown D (eds) Architecture as signs and systems for a mannerist time. Belknap Press, Cambridge, MA, pp 142–174

Types and Typologies

4

Languages are full of common nouns that cluster objects and processes into groups based on their similar characteristics. 'Sofa' refers to a *type* of seating that has a common geometrical form. The type includes cabrioles, chesterfields, and camelbacks. Each of these types consists of a subgroup of unique sofas. An architectural type is simply a category of buildings that have something in common. *Typology* is, strictly speaking, the study of types. The term is, however, widely used to mean a category of types. A *generic solution* is a type that is regarded by its protagonists as a *paradigm*, an exemplar of good practice, a model to be imitated in resolving a standard problem type. Romanesque church forms, for example, share general plan patterns (see Fig. 4.1 left). Each of the church plans in the figure is unique but clearly of a type that affords the activities required by a common liturgy.

The analysis of a building type is the point of departure for many architects when approaching a design task. As noted in Chap. 2 on procedural theory, architectural typologies provide a vocabulary of patterns for an architect to copy or imitate in devising a new building that fits within a general category. The evolution of the design of the Papal Basilica of St Peter in Vatican City shows how different architects, all considered to be highly creative, dealt with a common, generic geometrical plan type in a unique manner (Fig. 4.1 right). They were imitating not copying a basic type.

4.1 Architectural Typologies

Designers' use of types of building plans, façades, and structural and construction techniques as the basis for new designs goes back to prehistory. The architects of ancient Greece had an image of the commonality of temples plans, elevations, and decoration in mind when siting and building new temples (see Fig. 6.2). Many

J. Lang, *The Scope of Architectural Theory*, Design Science and Innovation,
https://doi.org/10.1007/978-981-96-3690-7_4

Collection of the author: Various sources

Collection of the author: Various sources

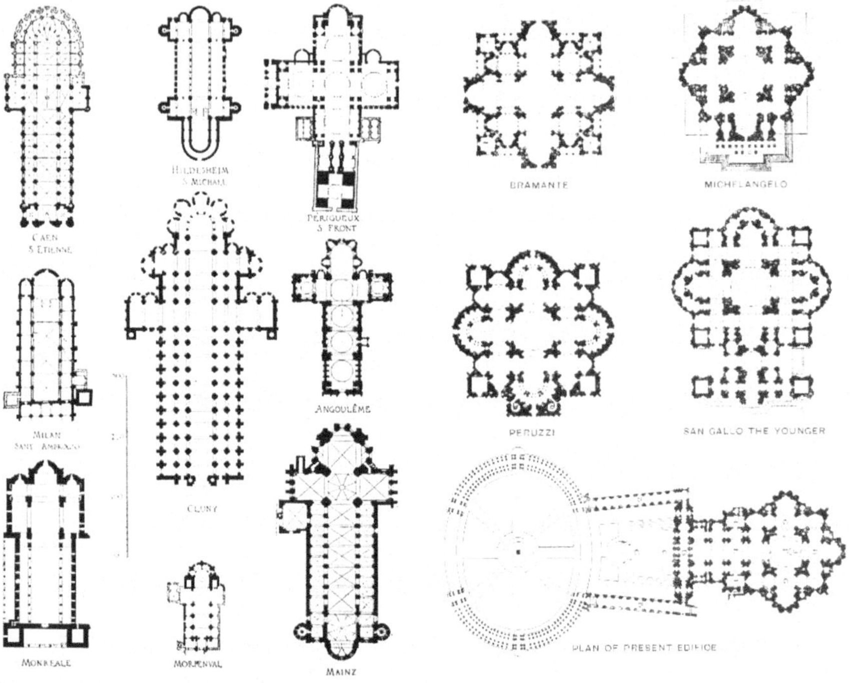

Fig. 4.1 Romanesque church plan types (left) and a sequence of designs based on a generic solution for St Peter's, Vatican City (right)

organizational rules were implicit in the temple types as were the activities that they were expected to house. The design of the Parthenon is an example of the selection of parts from a set of component types.

The systematic categorizing of the plans and cross-sections of buildings as an aid to architects began with the efforts of Abbé Marc-Antoine Laugier (1713–69), a Jesuit priest. Laugier, in his *Essai sur l'Architecture* (1755), argued for a columnar and trabeated architecture and showed plans and details of such buildings for architects to imitate. A contemporary, Pierre Patte (1723–1814), an architect, presented a comparative study of theatre plans in his *Essai sur l'Architecture Théâtrale* (1782) to assist architects when designing theatres. Jean-Nicolas-Louis Durand (1760–1830), an influential neoclassical architect and forerunner of modern rationalist design, took the study of building typologies a step forward.

Early in the nineteenth century Durand developed a systematic way of illustrating typical plans, elevations, and building components. In his *Précis des Leçons d'Architecture à l'École Royale Polytechnique* (1809) he showed plans and elevations of buildings based on their geometric commonalities in overall form and internal organization (Villari 1990). Durand advocated a system of design based on the assembly of types of building modules. These modules ranged from plan forms to decorative elements. Durand's contemporary, Quatremère de Quincy (1755–1849), differentiated between a *type* and a *model* (Ruprecht 2021). A type is derived from several existing buildings that have geometrical patterns serving instrumental functions in common. A model, to him, is a type that is worth imitating. A model, in this sense, if not perfect is close to it. The criteria for making that judgement may be explicitly stated but they are often simply implicit in the model. Today we do use the term 'model' in the same way although in everyday usage the term 'model', as already used in this book, is often simply any abstraction of an artifact or process.

The design of buildings in the eighteenth century was based on the widespread use of pattern books that present models—ideal types—for less accomplished architects and engineers to emulate in their designs. *La Science Des Ingénieurs: Dans la Conduite des Travaux de Fortifications et d'Architecture Civile* (1719) by Bernard Forest de Bélidore (1698–1761), a French hydraulics engineer, was widely referred to in Europe and in Europe's Asian colonies by army engineers when designing buildings. *A Book of Architecture* (1728) by James Gibbs (1682–1754), an architect trained in Rome, followed, but it was *The Architecture of Andrea Palladio with Notes by I. Jones* (1742) written by Giacomo Leoni (1886–1746), a devotee of the Renaissance architect and polymath, Leon Battista Alberti (1404–1472), that established an influential neo-Palladian architectural design paradigm.

The theme of design as a combinatorial process—the composition of the elements of proven types—was furthered by architects trained within the tradition of École des Beaux Arts in Paris. William Ware (1832–1915), founder of the architectural programs at Massachusetts Institute of Technology and at Columbia University, produced *The American Vignola* (1904) that contained illustrations of types of building components—arches, vaults, roofs, etc.—and rules for amalgamating them into a design. The title shows the author's debt to Giacomo Barozzi da Vignola (1507–73), a highly influential Italian Mannerist architect, famous for his documentation of the classical laws of proportion. In a similar vein, John Theodore Haneman (1880–1960) produced *A Manual of Architectural Compositions* (1923) in which he presented many building and building component types for architects to combine in their compositions (Figs. 4.2 and 4.3).

To inspire architects' urban design work, Werner Hegemann (1881–1936), an architectural critic and city planner, and Elbert Peets (1886–1968), a landscape architect, produced *The American Vitruvius: An Architect's Handbook of Civic Art* (1922). It was the urban design equivalent of Ware's and of Haneman's architectural treatises. The monumental tome catalogues plazas, groups of buildings, streets, and city plans to provide architects with a compendium of good urban designs to imitate. The book remains a source of ideas for architects when dealing with cities.

Contemporaneously, the Rationalists among Modernists sought a functional architecture devoid of historical precedents. They disparaged the consideration of designing as a combinatorial process of composition. Yet they too drew on precedents and they produced ideal urban and building types for others to imitate. Pattern books continued to be produced to provide property developers and

Source: Patti (1782, plate 3)

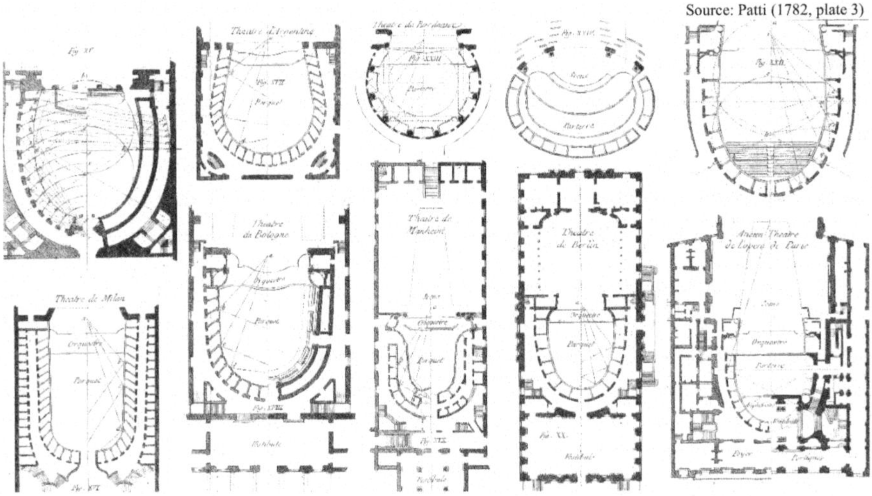

Fig. 4.2 Pierre Patte's comparative study of theatre plans

Source: Durand (1808, plate 2)

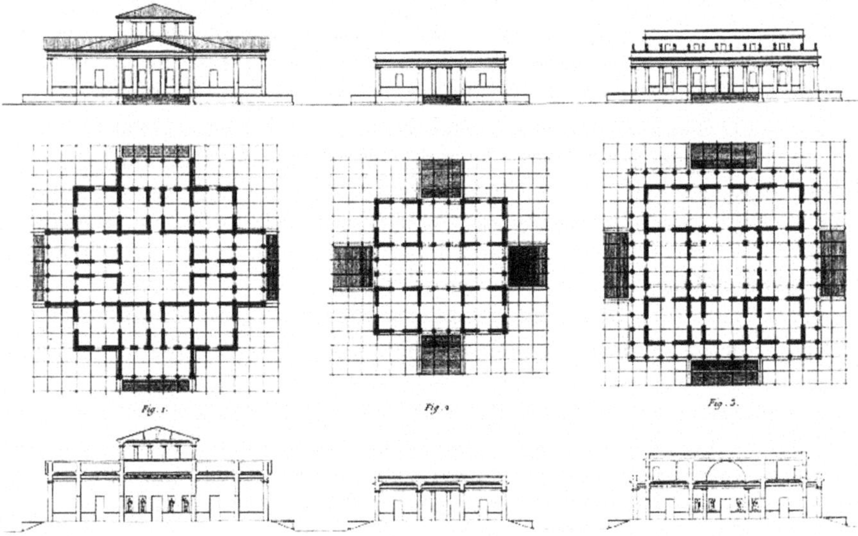

Fig. 4.3 An illustration of horizontal composition types in elevation, plan and section from Durand's *Précis des Leçons d'Architecture données à l'École Royale Polytechnique*

architects with models to follow when proposing new developments. For instance, a pattern book for US Army recreation centres comes with guidelines on how to localize facilities (Department of the Army 1976). In 2024 the Government Architect's office in New South Wales, Australia is producing a pattern book that provides designs and guidelines for sustainable, high quality housing types. Developers who follow the models will get speedy approval of their proposals from consent authorities. The necessity for the creation of the pattern book reflects the perception that the standard models guiding development are not good enough.

4.1.1 Building Typologies and the Potential Functions of Architecture

Implicit in each building type is a program that serves both basic and advanced functions. Each type implicitly provides shelter and a sense of safety and security to its inhabitants. Implicit in the type's spatial character is that it provides the metabolic comfort level required by its users and the level of privacy associated with the activities that can take place within it. The experiencing of a sequence of spaces as an observer moves through them is implicit in their plans. Some building types provide food for thoughts of the cognoscenti.

Few building type studies explicitly specify the cultural basis of their layouts or their formal and symbolic aesthetic characteristics. To the extent the implicit set of activities required to meet instrumental ends are universal, the type is universal. Most types are, however, culturally specific although this reality is not always obvious. Often building types are latitude specific. They implicitly cater to the height of the sun and its apparent movement through the sky over the course of the seasons in a particular location. Often types are imitated without an understanding of their cultural and geographic implications.

4.2 Typology and Types Today

The idea of types remains constant but the utility of types in design is regularly being reconsidered (Colquhoun 1967; Moneo 1978; Franck and Schneekloth 1994; Heyda 2019). Giulio Argan recognized the utility of building types:

> A type is an answer to a complex of ideological, religious, or practical demands which arise in a given historical condition of whatever culture.
>
> In the process of comparing and superimposing forms to determine the type, particular [and peculiar] aspects of each individual building are eliminated and only those remain which are common to every unit of the series. (Argan 1963)

While the design process is typically based on the manipulation of standard types to meet a current situation, a problem-solving approach to design is needed in dealing with a building that has no precedent to imitate. A program driven approach rather than a solution-drive approach to design is needed. Views of how we *should* conduct the design process are presented in Chap. 7 (Fig. 4.4).

4.2.1 Urban Form Types

Sociologists have categorized types of city forms based on the distribution of land uses and modes of transportation. In North America, the types include the concentric, star, and multi-nodal forms that have resulted from a laissez faire approach to city development. Each type explains much about how metropolitan areas function, but they are two-dimensional representations of urban forms. It is the three-dimensional world that is of interest to architects. Settlement types serve as a shorthand reference for the description of places. The patterns of the medieval city have been something that architects have decried and lauded since the days of Alberti. We immediately have an image of what they raged for or against. The same can be said of modern suburbia and the early twenty-first century city. Different suburban types such as the streetcar suburb and the motor car suburb reflect contemporary modes of transportation (Stern 1981). Other categorizations of urban types are based on different criteria. For instance, Anslem Strauss (1916–96), a sociologist, in his book *Images of the American City* (1961), distinguished between types of cities based on their distinguishing instrumental function—the industrial city, the historic city, the recreation city, etc.

Scholars have studied the difference in patterns of typical North American, Latin American and Asian cities. Their research has been of greater interest to city planners than architects and has had little impact on city form policies. The Transect Model of DPZ CoDesign is an exception. Assuming a concentric ring model of the city it shows the types of development that take place in the zones of a cross-section of the twentieth century North American city (Fig. 4.5). It is used as the basis for the contextual design proposals of the firm (Duany and Talen 2002; Bohl and Plater-Zyberk 2006). The firm's prescription is that new buildings should follow the building types of the zone in which they are located. The position can be disputed. Many clients and their architects want designs that stand as unique figures against a background of the existing.

To architects concerned with urban design it is the types of elements of urban form that have more immediate applicability in structuring development proposals that are of greater interest.

4.2.2 Elements of Urban Form

Architectural scholars have had different foci when identifying the elements of urban form that are helpful in giving order to the design of new towns, urban districts, and suburbs. Some have looked at street pattern types, others at land use areas—commercial, residential, industrial—and their subtypes that make up the

Photograph by the author

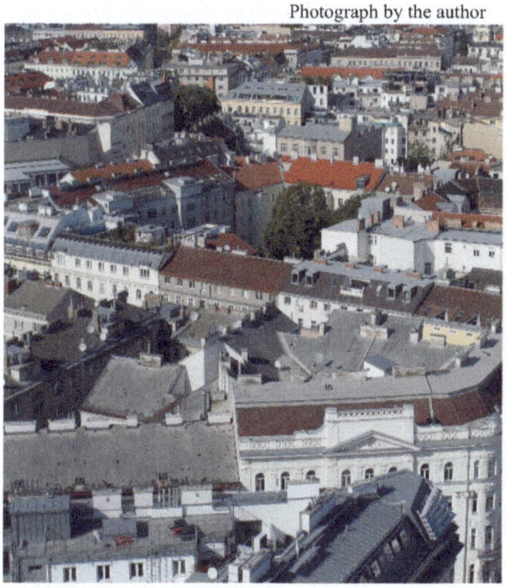

A medieval city: Vienna, Austria

Photograph by M. Kebra/Shutterstock

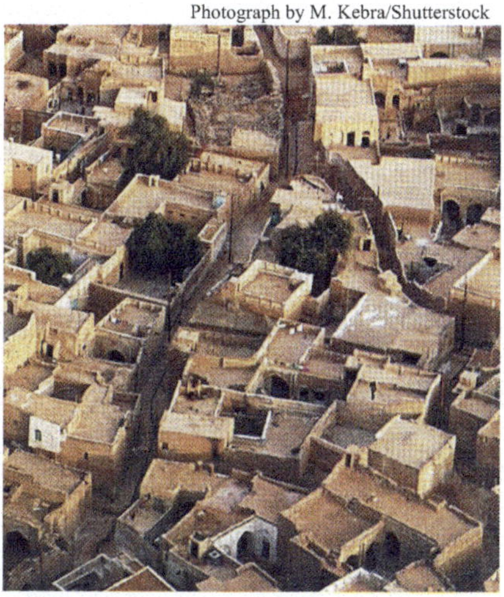

A medieval street: Shustar, Iran

Photograph by Hansueli Krapf/Wikimedia Commons

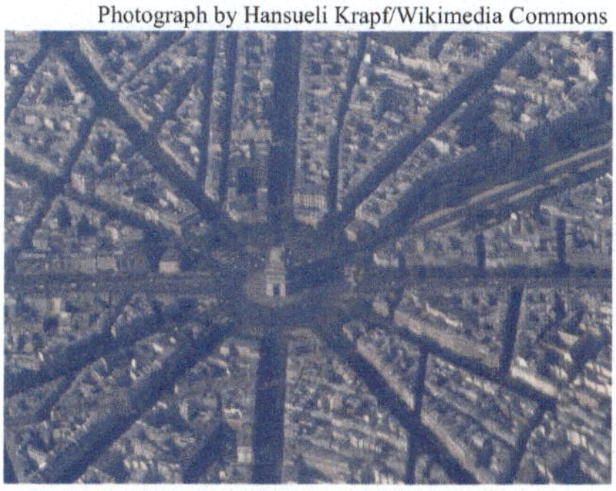

A nineteenth century city, Paris, France

Photograph by iofoto/Shutterstock

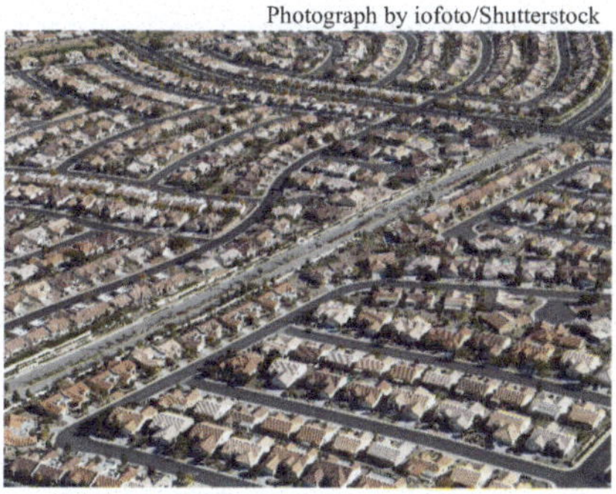

A mid-twentieth century automobile suburb, Homewood, Illinois, USA

Photograph by Ahmed Hussain/Wikimedia Commons

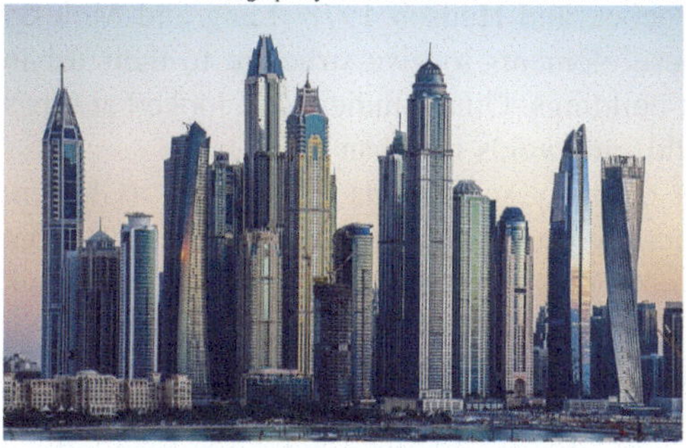

Photograph by the author

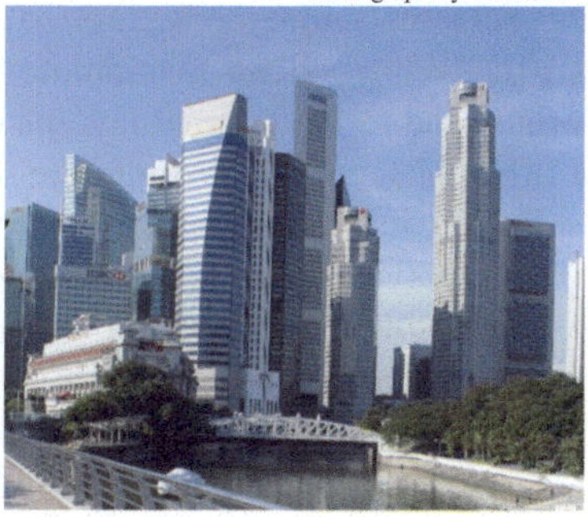

The early twenty-first century city: Dubai in 2023 (left) and Singapore in 2018 (right)

Fig. 4.4 Urban form types

Source: Bohl and Plater-Zyberk (2008): Courtesy of DPZCoDesign

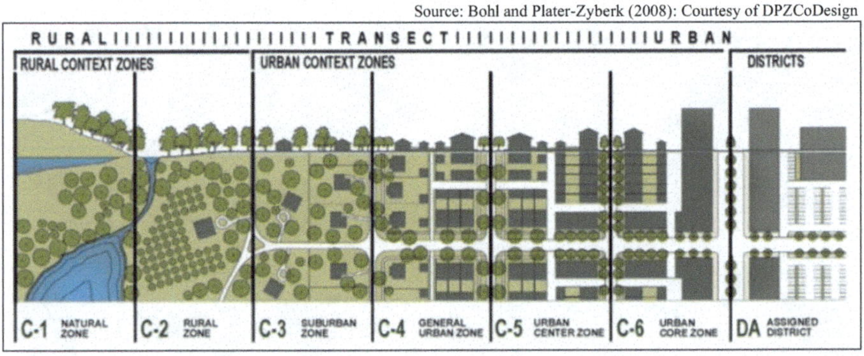

Fig. 4.5 The transect model: types of North American urban districts

city. Others have considered the city in terms of neighbourhood types (Trudeau 2013) and yet others in terms of building styles.

In *The Architecture of the City* (1982), Aldo Rossi (1931–1997), a Neo-Rationalist Italian architect and scholar, focussed on the role of buildings in the city, their enduring qualities, the memories embedded in their architecture and those elements that define a city's particular culture. He was interested in how these elements change over time reflecting social class struggles among financial forces (Rossi 1982). Other ways of looking at cities have, however, had a greater impact on how architects think.

One of the best-known urban form typologies was introduced in Chap. 3. Kevin Lynch (1918–1984), a professor at the Massachusetts Institute of Technology, identified five elements that shape people's cognitive maps of cities. If clearly organized, they enable people to easily navigate their way around cities (Lynch 1960). His findings also apply to the internal organization of buildings. The typology consists of *paths*, or corridors, which meet at *nodes* of activities, clear sub-areas, or *districts*, and significant objects, or *landmarks. Edges* or defining borders are also important but, apparently, less so (Pocock and Hudson 1978; Lang and Moleski 2010). Architects can manipulate these elements to give structure to their urban designs and to the internal layouts of buildings. Other studies have looked at types of urban space that constitute the solids and voids of urban forms.

The basic morphological element of many settlements is the block (Ghisleni 2023; Shakya 2023). A block is generally seen as the space between streets that has been subdivided into plots that contain buildings. More importantly a block is the space spanning a street bounded by opposing buildings. The architecture of the block is composed of the street itself and the façades facing it. The street acts as a seam making the block a unit. If traffic volumes on the street are not high, this type of block forms the basic social unit of a city. The ribbon shopping street is such a type.

Several block types are well known. Shustar's historic pattern as shown in Fig. 4.4 is an unselfconsciously designed layout that follows generally understood design rules. Hausmann's Paris's design is, in contrast, a highly self-conscious act of will. The gridiron plan is a pragmatic urban type. The Barcelona block with its chamfered corners, designed by Ildefons Cerdà i Sunyer (1815–1876), has become a type on its own. In many modern cities the block is an island and the roads surrounding it do not act as seams. They are dividers. The superblock where vehicular traffic is kept on the outside and the interior pedestrianized is a type that has strong advocates.

Several major studies of types of urban spaces and their functions at both an instrumental and symbolic level were conducted during the late nineteenth and early twentieth centuries. The most influential are those of Camillo Sitte (1889), Hegemann and Peets (1922), as already mentioned, and Rob Krier's study of squares (Krier 2005, originally 1979). Sitte (1843–1903), an Austrian architect, catalogued existing European city squares in plan form drawn at the same scale. The various types provide urban designers with a vocabulary to draw on when crafting a new urban space. Sitte says little about how the squares function other than as a setting for the display of a foreground building. Krier's typology is more thorough (Shane 1976).

Krier was concerned with squares' roles in establishing a sense of place, as locations for communal public life, and as settings for formal (that is, planned) events. His typology begins with the identification of the basic platonic shapes of square, circle, and triangle and how they are angled, segmented, added to, merged, overlapped, and distorted (see Fig. 4.6). He then distinguishes between those that are regularly or irregularly framed and how they are entered from adjacent streets. Another study of public open spaces analysed urban squares as places, links, and aesthetic displays and categorized them according to their sizes, configurations, and other attributes (Lang and Marshall 2016). The goal of that typology was to explain which types are effective and which are not, and for whom in specific locational contexts.

Typologies of streets, street hierarchies, and patterns based on the instrumental functions that they serve inform designers of the options available to them (Watson et al. 2003). The primary function of some street types is to speed traffic flows. Other types, such as the *woonerf*, slow vehicular traffic down to create spaces shared by drivers and pedestrians. Standards specify the patterns that are required for other elements of the city such as parking lots and details such as storm water drains. They provide easy sources of patterns for architect to copy, hence their name *Time Saver Standards.*

4.2.3 Changing Technologies and the Emergence of New Types

As technologies develop and patterns of life change so new types emerge either from practical experience or rational thought. Artificial Intelligence and communications technology will no doubt reshape many existing urban types and introduce new ones in the coming years. The industrial revolution certainly launched many

Source: Krier (2005, 29): Collection of the author

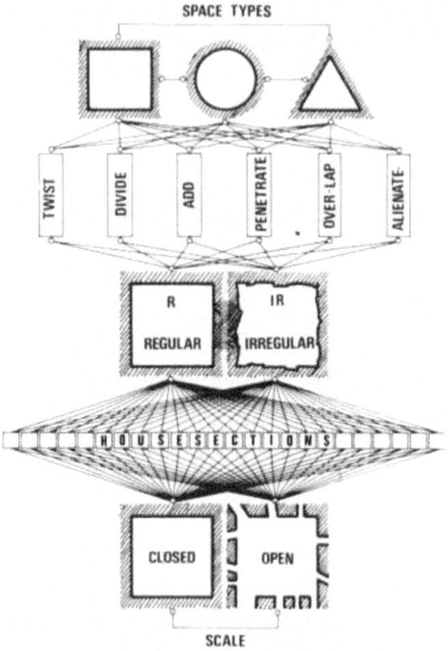

Krier's categorization system

Source: Krier (2005, 33): Collection of the author

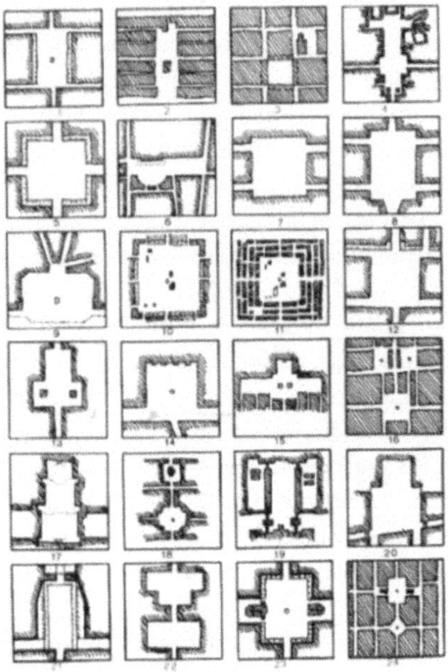

Orthogonal plans for squares

Fig. 4.6 Rob Krier's typology of urban spaces

new types of buildings. The safety elevator and new structural processes made high-rise buildings possible. New types of single-family detached house plans emerged in response to changing gender roles, parent–child relationships, and ways of entertaining. In addition, architects have produced new types of building forms simply as a reaction to traditional types.

Developments in technologies have led to some places becoming obsolete. The abandonment of railway lines in the centre of cities provided an opportunity for a new type of linear, urban park to evolve. The Promenade Plantée (or Coulée Verte René-Dumont, 1980–1993 and 1999–2015) was once the Vincennes railway viaduct. Its conversion to an almost five-kilometre-long linear park became a precedent for others to imitate. High Line Park in New York was built in the first decade of this century and its success as a walkway, as an avant-garde design, and as a catalyst for adjacent development has spurred projects in other places based on the generic idea. De Hofbogen in Rotterdam and Seoul's Seoullo are variations of the type (Fig. 4.7).

Photograph by the author

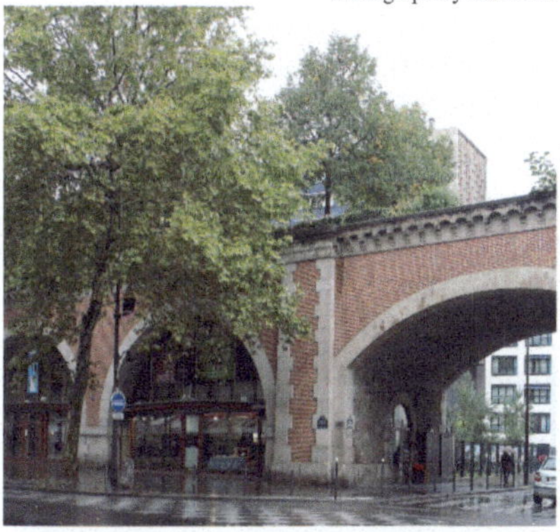

Photograph by the author

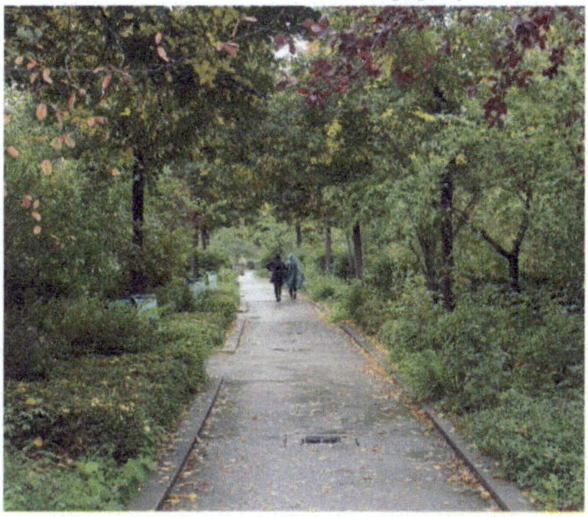

Promenade Plantée on the former Vincennes railway viaduct, Paris

Photograph by the author

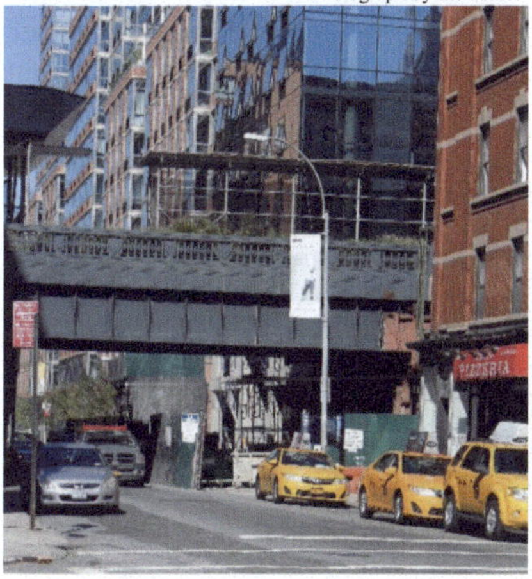

Photograph by the author

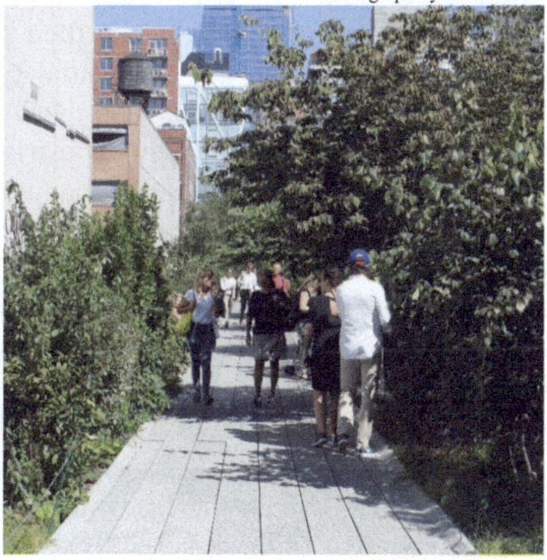

The High Line and High Line Park, New York

Photograph by Mw0007/Wikimedia Commons

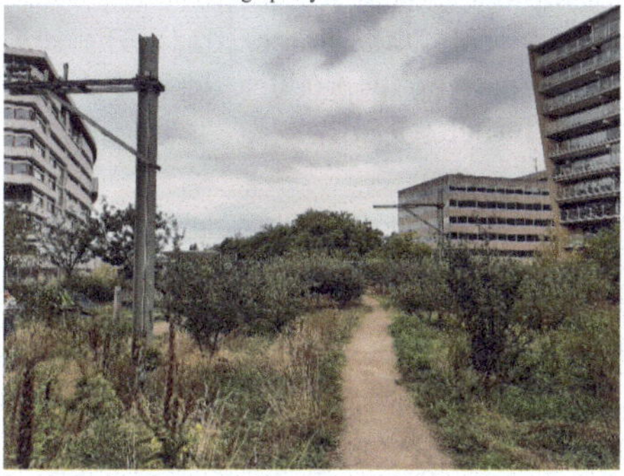

Luchtpark, Hofbogen (formerly Hofplein Viaduct) , Rotterdam

Photograph by Brit in Seoul/Wikimedia Commons

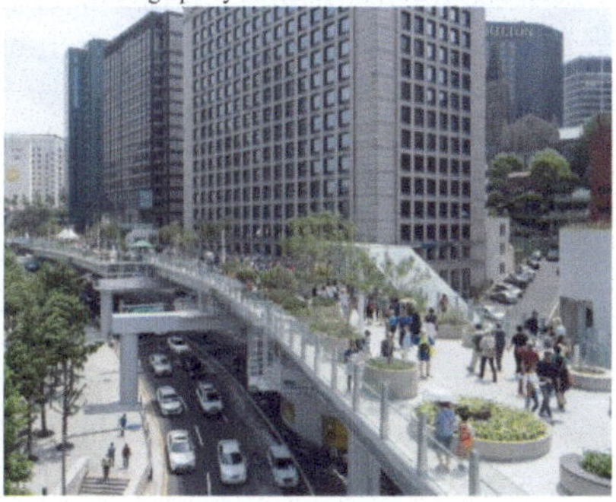

Seoullo (Skygarden), Seoul

Fig. 4.7 A new type of urban park: the linear corridor

4.3 Urban Design Types

Urban design project types have been categorized based on modes of implementation, their instrumental function, and the design paradigm on which they are based. They can be *total* urban design schemes where one architectural firm has control over the whole project from beginning to conclusion and from the large scale to the details. *All-of-a-piece* urban design describes projects that have a master design but are divided into smaller components that are implemented by several property developers and designed by their architects according to the design guidelines presented in the master design (see Fig. 2.3). *Plug-in* urban design consists of placing new buildings, often cultural facilities, into a district to function as a catalyst to spur new developments in their surroundings (Lang 2017, 2021).

The largest scale of urban design types are new towns of which there are three basic types. The Garden City and the Neo-traditional city are empiricist or contextual types and then there is the Rationalist City (see Fig. 4.12). Town types are divided into districts based on their primary instrumental function: business, industrial, institutional, and residential types among them. Commercial districts can consist of offices, retail shops and/or entertainment uses although the last is often a category on its own. Industrial types range from tank farms to large scale factories and smaller scale manufacturing operations. Residential districts may consist of houses and/or townhouses, or a mixture of medium density and high-rise buildings.

The designs follow various urban design paradigms. I have distinguished among Modernist, Postmodernist, and Neo-classical paradigms. Each can be subdivided into Neo-rationalist, Neo-empiricist types (Lang 2005, 2017). Harrison Fraker, a professor at the University of California, Berkeley, has provided a more descriptive categorization: Everyday Urbanism, Generic-Urbanism/Hypermodernity, Hybrid Urbanism, New Urbanism, Transformative Urban Morphology, and Urban Ecological Reconstruction (Fraker 2007).

Urban design, not as project design, but as public policy formation, also comes in various forms. Zoning ordinances and building codes, where enforced, shape the three-dimensional qualities of cities (Ben-Joseph 2005; Talen 2012). The zoning ordinances may be restrictive or offer incentives to produce desired design outcomes as in New York's Theatre District (Barnett 1974).

4.4 Building Types

Three basic building form-giving types serve as the potential declarative theory core for architectural practice. The first categorizes the configuration of buildings. The second is concerned with the similarities in structural elements and the third with types of decoration. Each consists of subtypes. Giulio Carlo Argan identified examples of each:

> [In] the first category are centrally or longitudinally planned buildings; of the second, flat or domed roofs trabeated or arcuated systems; and of the third, orders of columns, ornamental details, etc. (Argan 1963)

Argan suggests, as noted in Chap. 2, that this categorization follows the way architects go about, consciously or sub-consciously, turning a building program into a design. They begin by formulating the plan and then go on to selecting its structural systems and end up by considering the surface treatment of the building they are designing. Construction details follow.

While Argan is drawing on the European experience, the categorization of building types has been conducted elsewhere although a disproportionate number have been the work of European scholars. Hindu temples, or *mandirs*, for instance, have been classified into (1) overall form types—rectangular, octagonal semicircular, (2) structural types—domes or steeples, and (3) by their components—inner chambers and halls, porches, but also (4) how they are circumambulated (Michell 1977). Few new Hindu temples deviate far from idealized models.

4.4.1 Form Types

Buildings come in a variety of forms although the rectangular is basic. Others may be circular, pentagonal, octagonal or in free forms. François Blanciak, a professor at the National University of Singapore, has drawn a thousand possible forms (Blanciak 2008). How they would be constructed and what instrumental functions they might serve are open to conjecture.

Argan's categorization of building form types is one way of identifying commonalties in building form. Another common way is by their style—their mode of expression. Style becomes a catch-all rubric for a cluster of design variables: mass, structure, façade organization and decoration. The evolution of styles reflects changing attitudes to the visual presentations of buildings to the world. At any given time, several styles may be popular. They attract different clients. During the first decades of the twentieth century Neoclassical houses were being designed along with Arts and Crafts, Art Deco and Prairie style ones. The use of all four overlapped the acceptance of the International Style for domestic architecture. Postmodernism was a later reaction to the strictures of the International Style (see Fig. 4.8). Most house designs are based pragmatically on types that meet the demands of the contemporary marketplace.

4.4.2 Structural Types and Construction Types

Argan gave "flat or domed roofs, trabeated or arcuated systems" as examples of structural types (Argan 1963). Domes can be categorized in several ways. One way is by their structural nature: cross-arch, cable net, geodesic, etc. Another way is by their shape: beehive, hemispherical, onion, etc. Still another is by the material

Photograph by Wdzinc/Wikimedia Commons

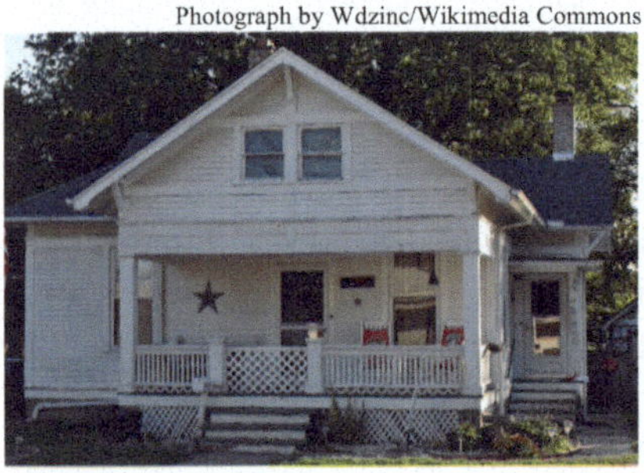

Californian bungalow

Photograph by Michael Reed/ Wikimedia Commons

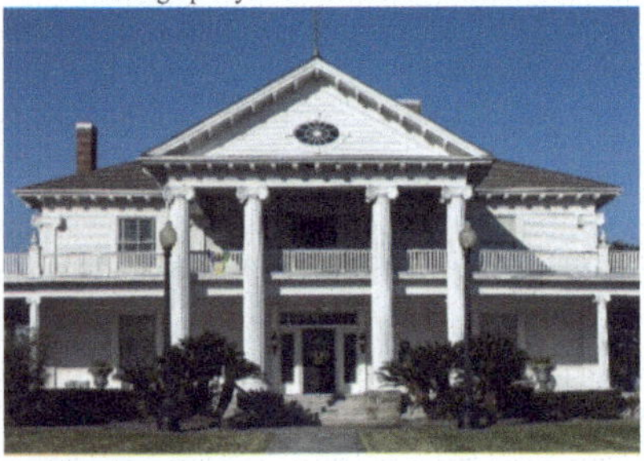

Neo-classical

Photograph by Alexander Cuthbert: Courtesy of Alexander Cuthbert

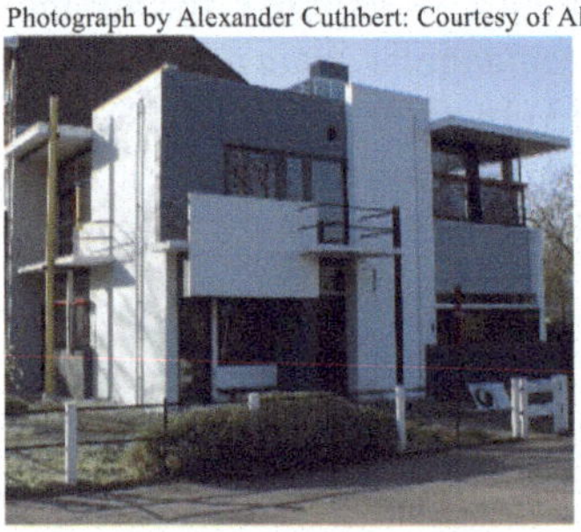

De Stijl

Photograph by the author

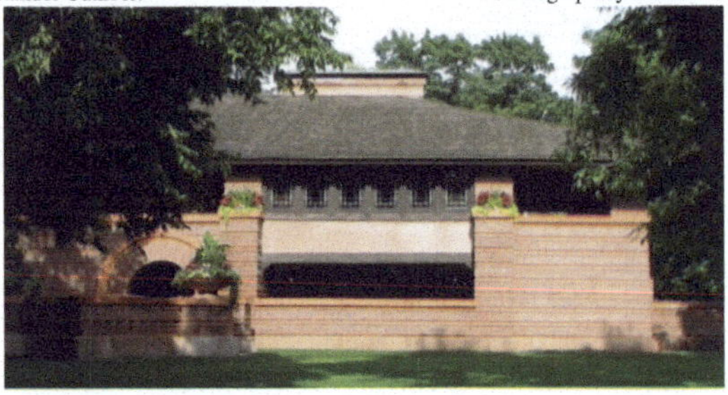

Prairie

Photograph by Sardaka/Wikimedia Commons

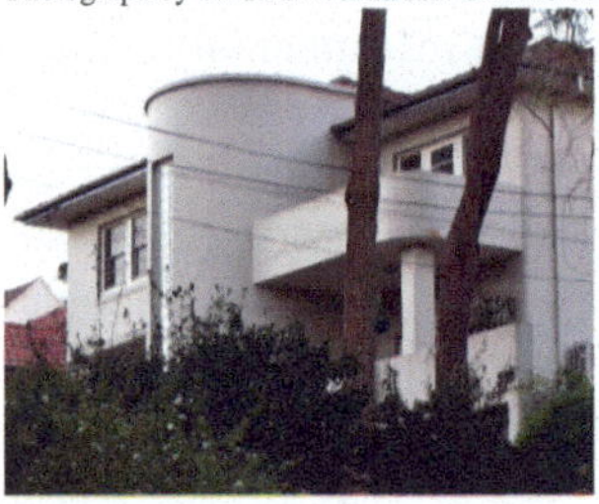

Art Deco

Photograph by Romany V. Sansky/Wikimedia Commons

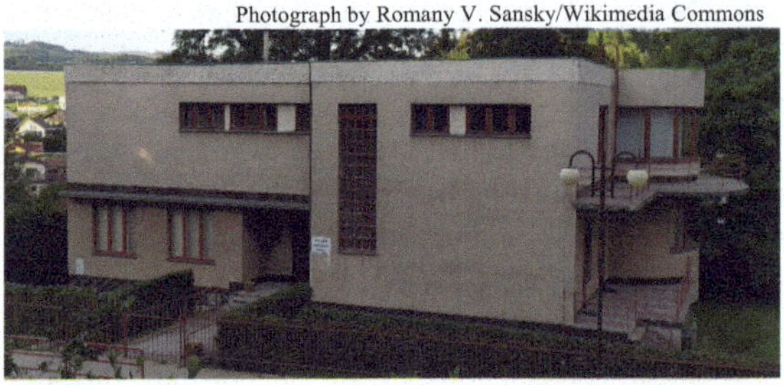

Modern functional

Photograph by No Swan so Fine/Wikimedia Commons

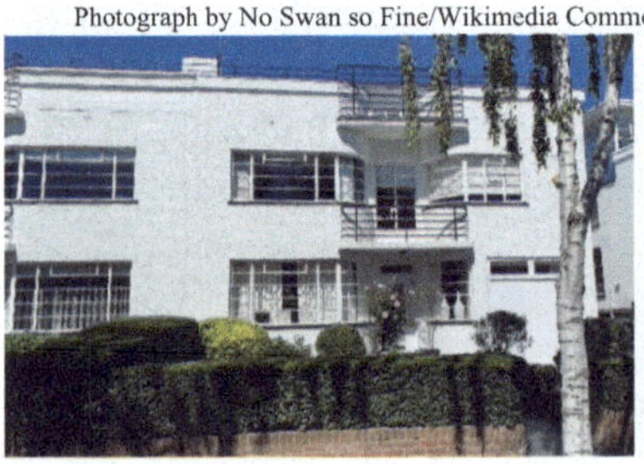

International Moderne

Photographer unknown/Wikimedia Commons

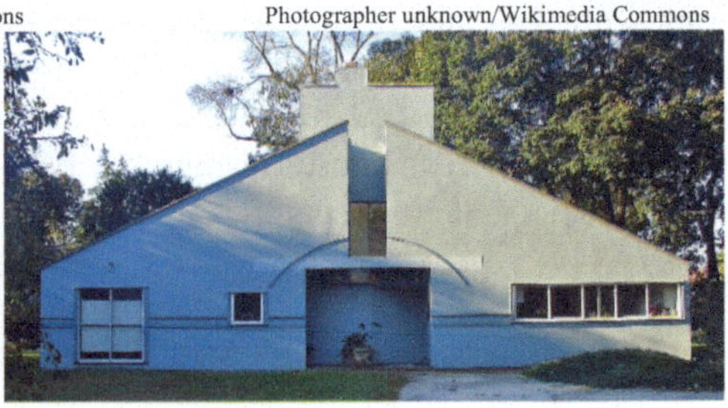

Postmodernist

Fig. 4.8 A sample of twentieth century house style types

of which they are constructed: concrete, mudbrick, timber, etc. Each type affords a range of instrumental ends and symbolic interpretations. Domes, for instance provide interior spaces serving a variety of purposes and during the early twentieth century they were considered to be a sign of modernism in many parts of the world (Koenigsberger 1994).

Roof configurations can be A-frame, flat, gable, jerkinhead, etc. What is important is how well each type functions in shedding rainwater or snow within its climatic context. The best design in serving an instrumental purpose often gives way to its function as a symbol—the type's association with some referent that the building's designers and/or the clients hold in their minds. Flat roofs are associated with Modernist buildings in all climatic zones. The same observation can be made about construction types. Some have high utility in creating sound buildings but have negative connotations because of their associative meanings. The opposite also holds true.

Several approaches are used to categorize construction types. A simple one is to distinguish among wood framed, steel framed (both light and heavy gauge), reinforced concrete framed, and load bearing masonry (stone, brick, or concrete block). Wood framed is the most widely used world-wide to construct buildings from squatters' shacks to multi-storey commercial buildings although steel framed and reinforced concrete framed are much more usually applied in tall building design. Each type has its advantages and disadvantages in serving different functions, instrumental and symbolic.

Materials can be categorized in different ways. Their utility depends on their strength, durability, appearance, and cost. Some are regarded as luxurious because they are not readily available; others are dismissed as cheap. Each material has a specific structural and insulating quality. Each affords a unique tactile experience for the person walking or touching it depending on its flexibility and texture(s). The possibilities are vast. The range of brick types alone is substantial.

4.4.3 Instrumental Functional Types

The categorization of buildings by their instrumental function goes back to prehistoric times. Today standard layout patterns for hospitals, hotels, schools and other institutional buildings, commercial offices, factories, warehouses, and housing abound (De Chiara and Callander 2020). Each has subtypes. Hotels, for instance, consist of full-service, extended stay, boutique, resort, and motel types. Each of these types can be further subcategorized. The list of housing types is vast. That for single-family detached homes alone is extensive. Houses come in detached and attached forms. They can have courtyard, corridor, and 'shotgun' plans. Aesthetically, they fall into various style types too as already shown.

An example of the detailed study of instrumental types is the work of Bernard (Bernd) Becher (1931–2007) and Hilla Becher (1934–2015), German conceptual artists and photographers. In Europe they established typologies of timber framed houses and industrial buildings and in the United States they photographed such

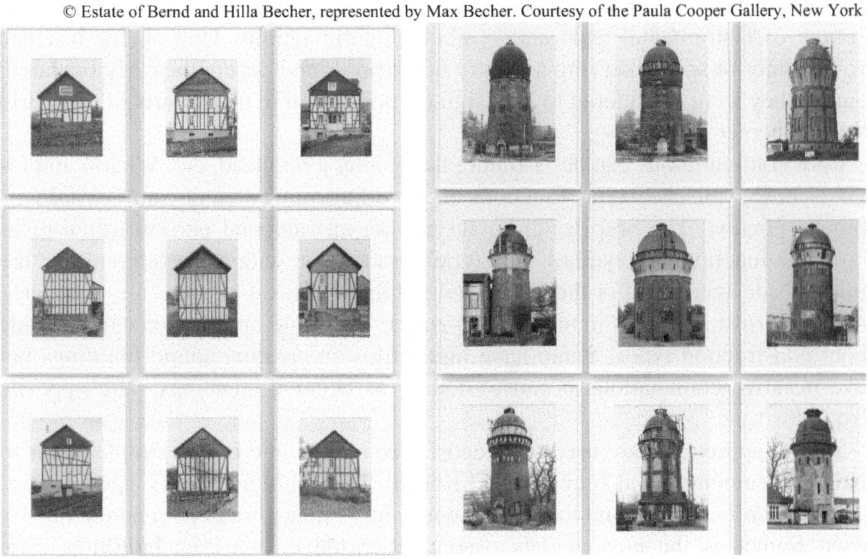

Framework houses (left) and Water Towers (right)

Fig. 4.9 The typographic studies of Bernd and Hilla Becher

places as water towers and coal tipples (Becher and Becher 1991). They were not interested in producing a prompt to architects but in recording buildings for their cultural value. The Bechers presented their photographs in grids so that rows and columns showed the variations on the basic type (Rosenheim 2022). Their photographs were taken on cloudy days to avoid differences in shadow lines on the images. Over the course of their lives their output of typologies was prodigious (Fig. 4.9).

The layouts of buildings serving the same instrumental function are similar in plan and geometric form. Basic platonic forms—rectangular box, prism, pyramid, and sphere—are common. One specific type of rectangular box is a 'shoebox' where the length is at least double its width.

Almost all concert halls with sound acoustic properties have a shoebox form. They possess a greater dynamic acoustical range (Pätynen et al. 2014). Their maximum seating capacity is about two thousand. The world has hundreds of such halls. Among the oldest existing are the Aberdeen Music Hall (1858) in Scotland and the Wilton Music Hall (1859) in London. Shoebox concert halls continue to be built. The Konsertsaal in Lucerne, designed by Jean Nouvel (b. 1945), was completed in 2000, the Schermerhorn Symphony Center in Nashville, designed by David M. Schwarz, Architects, opened in 2006 (see Fig. 6.12). The 2017 proposal for the Munich Concert Hall by Zaha Hadid Architects and Mecanoo is an almost pure shoebox. The Berliner Philharmonie is an exception. Designed by Hans Scharoun (1893–1972) it has a vineyard type seating arrangement. The building is praised for successfully breaking away from the shoebox mould while providing sound acoustics for both performers and audience.

Many shoebox music halls are not easy to recognize from the outside. Carnal Hall at Institut Le Rosey was designed by Bernard Tschumi Associates, Casa de Musica in Porto is the work of OMA (Office for Metropolitan Architecture), and Philharmonic Hall in Szczecin, Poland was designed by Barozzi Veiga architects. None of them looks like a shoebox on the exterior.

Other instrumental form types have been as persistent as the shoebox concert hall. Many Christian churches being built today are variants of the standard basilica or cruciform model. The change in the liturgy of the Catholic church following the Vatican Council held between 1962 and 1965, resulted in a new generic plan form that affords a congregation's participation in the service better than the basilica form that has been the standard type for centuries (Hackett 2012).

4.4.4 Building Component Types

Typologies exist for the detail of buildings from the grand to the most basic. Books specifying the utilitarian types of layouts of buildings and rooms within them are useful to architects.

> [Standards] give basic design criteria for each major type of building. It will give those unfamiliar with a specific building type a talking and working knowledge of its functions, organization, and major components. (Watson 2005)

Some of the patterns they espouse are international in character but other are culture specific. This observation applies particularly to bathrooms and kitchens and other fixed feature interior and exterior spaces that architects design.

The purpose of standards is to truncate the search for and the testing of potential design solutions. Applying them saves an architect's time in having to deduce patterns from basic performance requirements (Crosbie and Watson 2005). The difficulty is to keep the specifications up to date in a changing world. Luckily, many types are persistent.

4.4.5 Lighting Types

The quality of a behaviour setting is determined by how well its configuration affords the activities taking place within it, its acoustic and tactile qualities, and thus the quality of the materials of which it is made. A major criterion in establishing the utility and aesthetic character of a setting is the quality of light, natural or artificial, that illuminates it.

One typology of daylighting for internal spaces categorizes lighting patterns into ten types. They range from high contrast and variability types to low contrast and variability types. The categorization was produced by Siobhan Rockcastle and

Marilyne Anderson (Schielke 2023). It is shown in Fig. 4.11. Each category is shaped by the nature of windows and their reveals. One type consists of light from spatially indented openings through roofs and walls. Its use is exemplified by the designs of architectural luminaries such as Americans Louis I. Khan and Steven Hol (b. 1947) (Fig. 4.10).

Diagram by the author

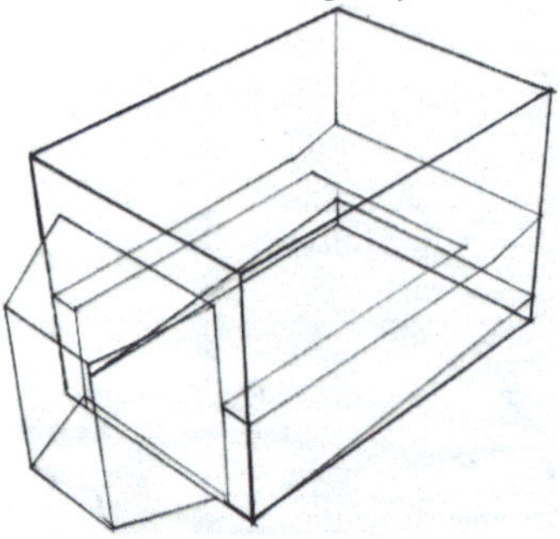

The generic shoebox shape

Photograph by Ian Watt/Wikimedia Commons

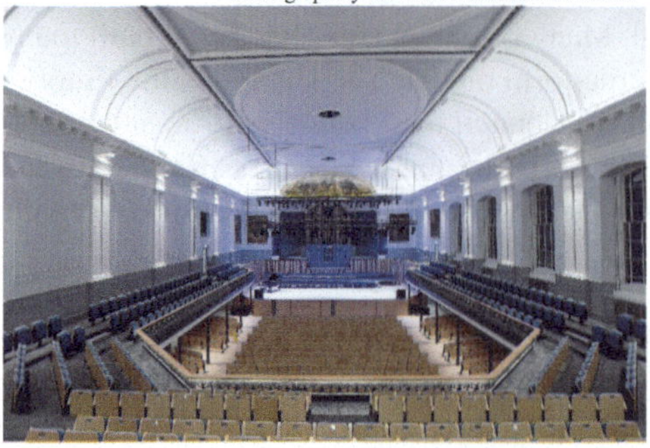

Aberdeen Music Hall (1858) as refurbished in 2018

Photograph by Rovaniemi taidemuseo/Wikimedia Commons

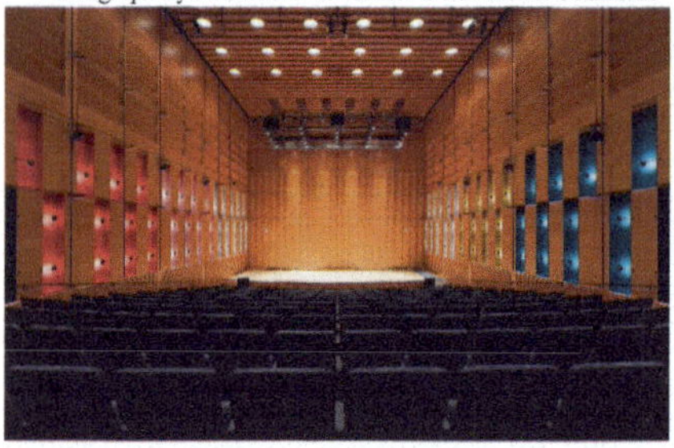

Culture house Korundi Rovaniemi, Finland (1993)

Photograph by Clew2807/Wikimedia Commons

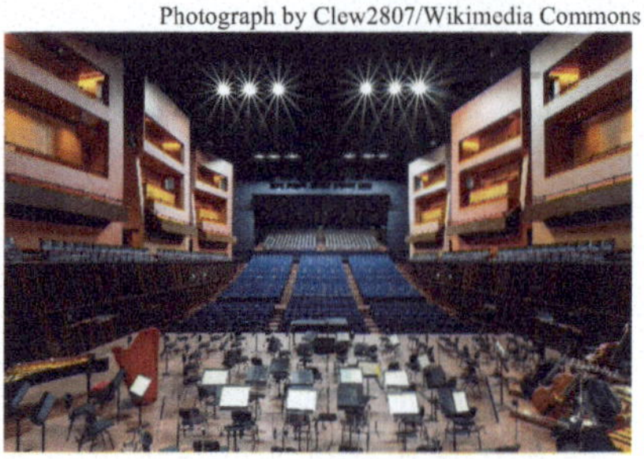

Philharmonie Großer Konsertsaaal, Luxembourg (2005)

Photograph © Iwan Baan: Courtesy of the photographer

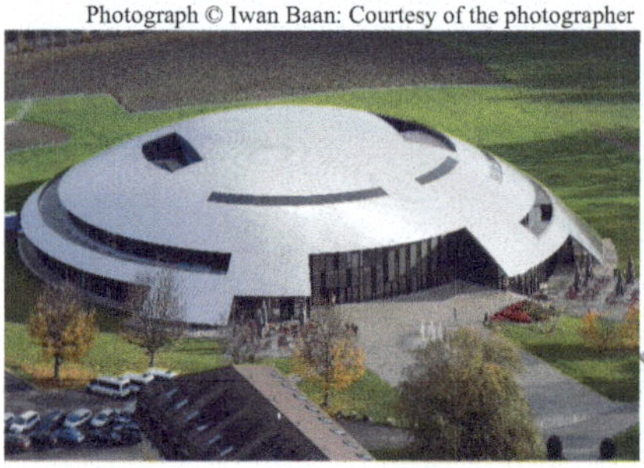

Courtesy of Bernard Tschumi Architects

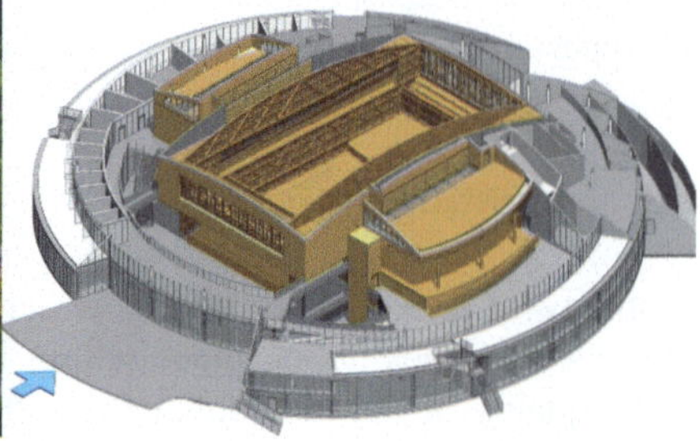

Paul and Henri Carnal Hall, Institut Le Rosey (2014): bird's eye view (left), isometric of the interior of the dome (right)

Fig. 4.10 Shoebox music halls

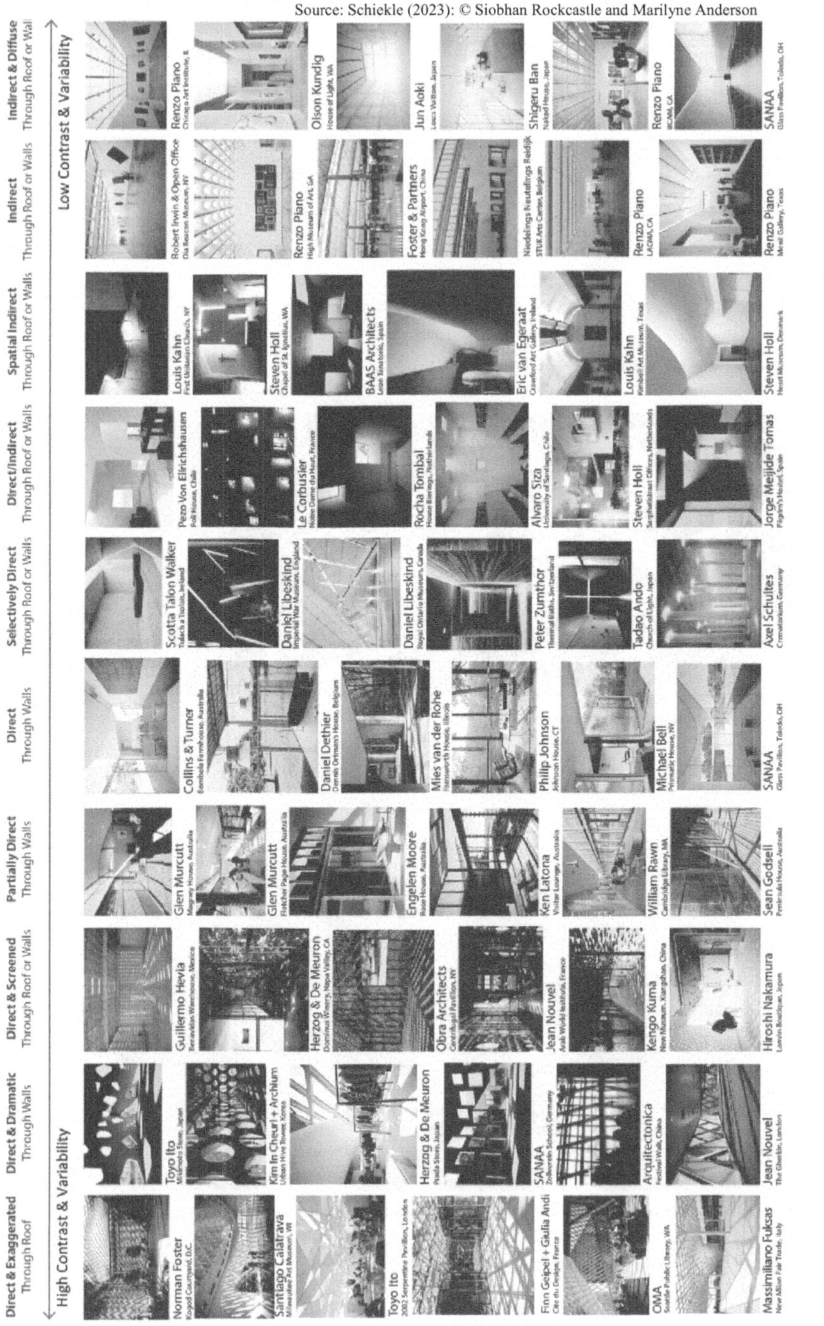

Source: Schiekle (2023): © Siobhan Rockcastle and Marilyne Anderson

Fig. 4.11 A typology of daylighting types

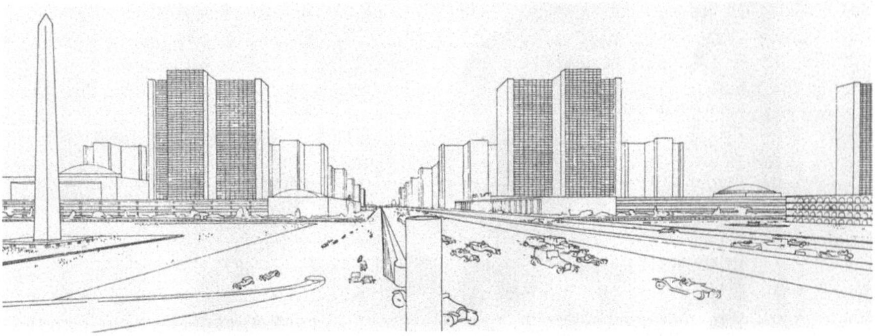

A generic design for a city of three million on a flat site

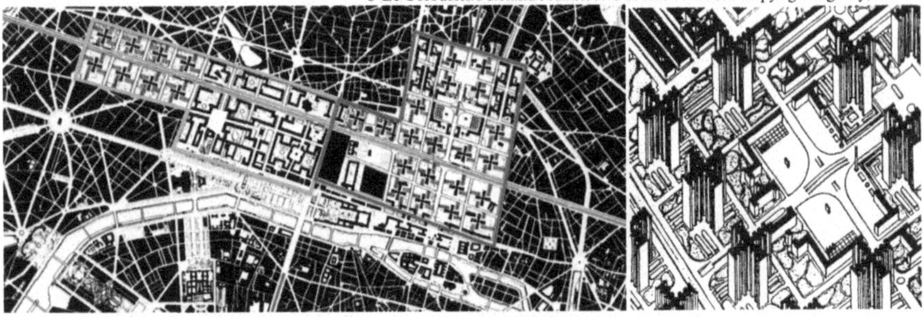

A speculative application: The Plan Voisin proposal for Paris (1925)

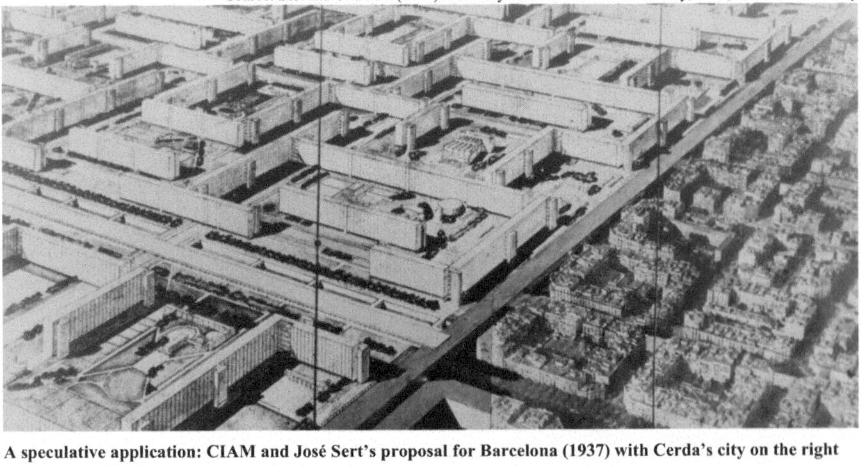

A speculative application: CIAM and José Sert's proposal for Barcelona (1937) with Cerda's city on the right

Fig. 4.12 The generic rationalist city

Such typologies bring architect's attention to the functions that light can serve and the ways daylighting ends can be met. It provides a toolbox of mechanisms for lighting spaces that extends beyond a concern for the energy consumption level of artificial lighting. The concern is an aesthetic as well as a practical one.

4.5 Invented Ideal Types and Generic Solutions

Generic solutions are a category of types that are presented as ideal models to be imitated. Prescriptive in nature, they are conceptual examples of the patterns that resolve repetitive, or generic, problems such as the provision of housing for low-income groups or the speeding, or slowing, of vehicular traffic along city streets.

Invented generic solutions may result from rationalist thinking, empirical observations or emerge from the pragmatic requirements of the capitalist marketplace. One that was widely used for structuring new towns is the decomposition model. Each town would have a centre with a set of districts surrounding it each with its own centre. Each district would be a village divided into several neighbourhoods with their own centres. It was the model used for organizing the overall layout of capital cities such as Canberra, Brasília, Chandigarh, and Abuja and the first generation of post-war British new towns and for many other new towns around the world.

Some generic solutions are based on explorations of problems of interest to individual architects. The best known of these generic ideas are Le Corbusier's designs for a 'city of three million' and the 'radiant city'. Never applied in their pure form to specific situations, their qualities have formed the basis of exploratory urban designs by Le Corbusier himself in Paris (Le Corbusier 1925a, 1960) and by José Luis Sert (Josep Lluís Sert i López, 1902–1983), working with a group of other Catalan architects, in Barcelona (Sert and C.I.A.M. 1944) as shown in Fig. 4.12.

Several other types of urban forms have attracted the interest of architects. The linear city is one (Lang 2021, 140–141). Originally proposed by a Spanish tramway operator, Arturo Soria y Mata (1844–1920), the linear city model captured the imagination of architects such as the Russian Nikolay Alexandrovich Milyutin (1889–1942), and Le Corbusier. The design was seen as an efficient layout for new towns by minimizing travel times between work and living places. Edgar Chambliss (1870–1926) took the model into the third dimension. American Modernists Paul Rudolph (1918–1997), and Michael Graves (1934–2015) and Peter Eisenman (b. 1932) produced linear development proposals for New York City and New Jersey respectively. A 105 mile (168 km) long linear city without roads was proposed for Saudi Arabia in 2021 (Alter 2022).

The mass housing model developed by the Bauhaus masters and C.I.A.M. in the 1920s and 1930s has been applied across the world in very different cultural and geographic contexts. Its application has been a response to the pragmatic need to provide shelter for swelling numbers of families and individuals. In some cultural contexts the model has functioned well in doing just that. Its quality as a place to live a full life has varied depending on whether its affordances fit the ways of life and aesthetic values of the people who live in them and, more importantly, if they have chosen to live in them or have been forced to do so because no alternative is possible.

Streets are for people and not just vehicular traffic (Rudofsky 1969). The recognition of the importance of streets as seams rather than boundaries of blocks has resulted in a change in the generic solutions for residential neighbourhoods. A model for neighbourhood layouts for the United States was proposed by Clarence Perry (1872–1944), a sociologist and city planner, in the 1920s. It was empirically based on his perceptions of Forest Hill Gardens in New York as a fine place to live. His generic design diagram had facilities clustered at the unit's centre. The revised plan has them along a street. The principle that local facilities should be located within an easy walkable distance of homes remains one of its key elements (Perry 1929) (Fig. 4.13).

Le Corbusier developed and promoted the Unité d'habitation as a vertical neighbourhood. It has been copied in a watered-down manner in the United Kingdom and the United States. Influenced by Le Corbusier, British architects Peter (1923–2003) and Alison Smithson (1928–1993) proposed the streets-in-the-sky model as an alternative generic mass housing solution. Based on their observations of urban life, it was presented as:

Source: Lang (2021, 216):.Originally (Perry 1929)

Adapted from Lang (2021, 216) by the author: Courtesy of Taylor and Francis

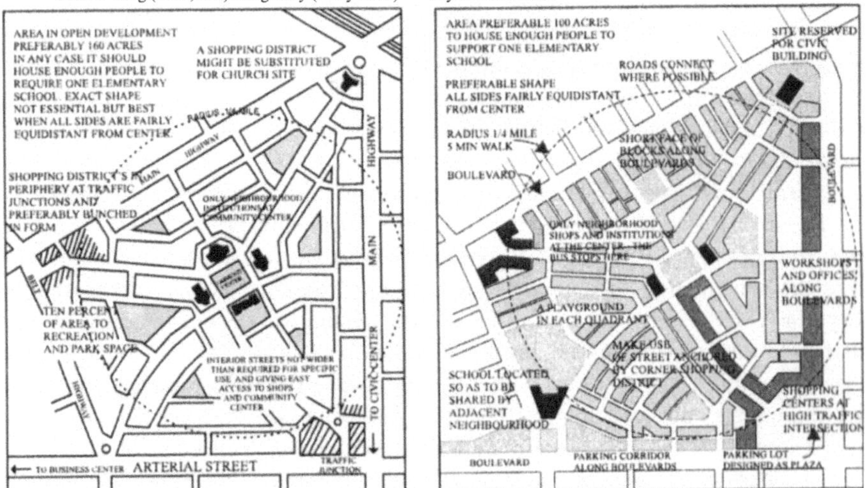

Fig. 4.13 Perry's generic plan (left) and DPZ CoDesign's update (right)

> ... an exemplar – a demonstration of a more enjoyable way of living in an old industrial part of the city. It is a model of urban organization which can show what life can be like. (*The Smithsons on Housing* cited in Powers 2011)

They overestimated the model's efficacy. The utility of streets in the sky differed from streets on the ground.

Generic designs exist for almost all instrumental building types. For hotels there are a dozen basic types and a generic solution for each type (Penner et al. 2013). The standard floor plan of a central corridor and bathrooms and sleeping areas aligned on each side can be twisted into many configurations. Another type is based on a single-loaded corridor around a central atrium. Both types often sit on a podium that accommodates communal facilities and meeting spaces. This format has become a generic type for hotels in central business districts across the globe. Indigenous decorative features are used to locate hotels in their geographic setting. Designing with the climate in mind has seldom been a major form-giving criterion.

Recent commercial office buildings around the world have a variety of external shapes and are clad with different materials. Despite their visual dissimilarity most have a similar plan (Hascher et al. 2002). The plan is a square with services (elevator, escape stairs, rest rooms and kitchens) in a core at its centre. The distance between core and periphery was once dictated by the distance natural light could penetrate the interior. This criterion is now often waived and floor plates are larger. Another, but less common, generic office building plan is to have two cores attached to the outside freeing up space for many interior design configurations (Fig. 4.14).

The generic North American suburban shopping centre design has a dumbbell layout with 'big box stores' as attractors that lure shoppers at both ends. They are linked by a longitudinal covered hall lined by shops. It is a type that has been applied in several other countries and to a university in Bielefeld in Germany (Universität Bielefeld 2015). Very few shopping centres of this type have been built in the United States during the past two decades. Hundreds of them have ceased to operate and have been converted to other uses (Lutz 2016). People purchase goods online and/or use traditional local ribbon retail streets. Few such streets are, however, as economically healthy as they were before the automobile came into common usage. Now the economic viability of such streets in threatened, although not universally, by internet shopping. The strip of shops with parking in front is a common type in many places where the usage of automobiles for transportation dominates. It is a type generally deplored by architects.

4.6 Commentary

Studies of types remain at a descriptive level. The way each type functions for whom in what circumstances is seldom fully explained. In imitating a type, it is important to first understand what its patterns afford. The experience with a common housing type in Philadelphia shows the importance of matching types

Adapted from Richards (1962) and Lang (2021, 178) by the author

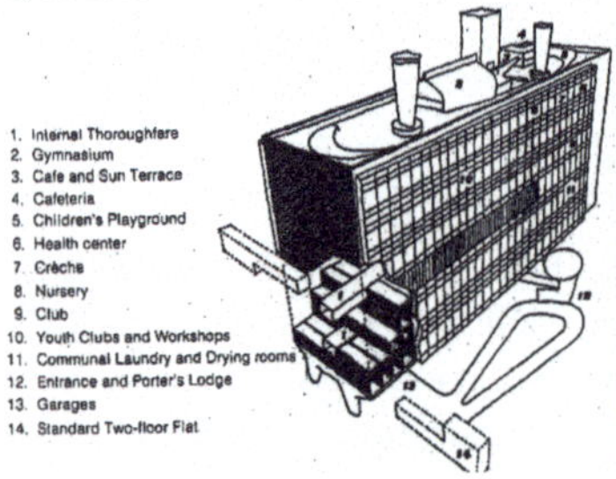

Photograph by the author

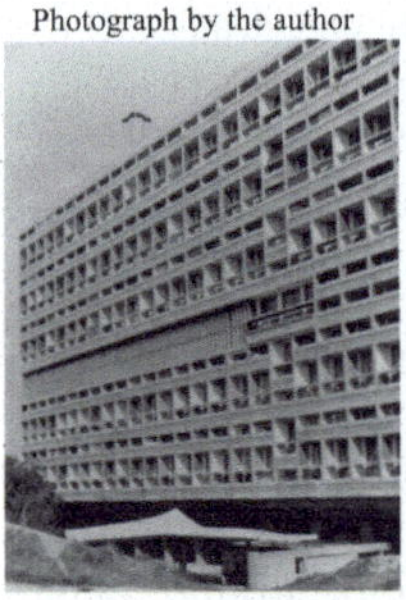

Source: A Savin/Wikiphotospace

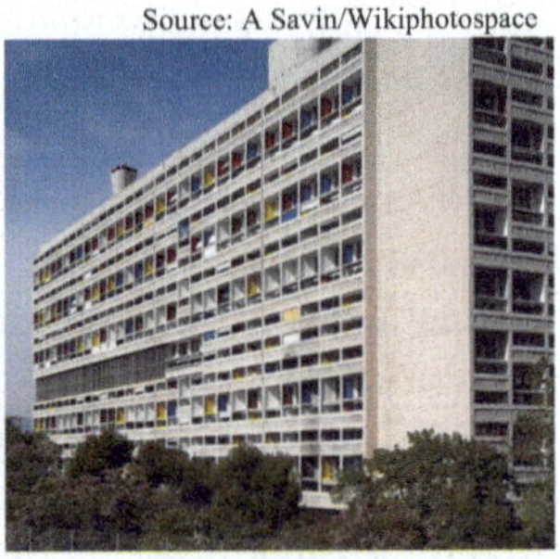

The Unité d'habitation: The generic idea (left), the Unité, Marseilles (centre), the Unité, Berlin (right)

Adapted from Hilberseimer (1927) by the author

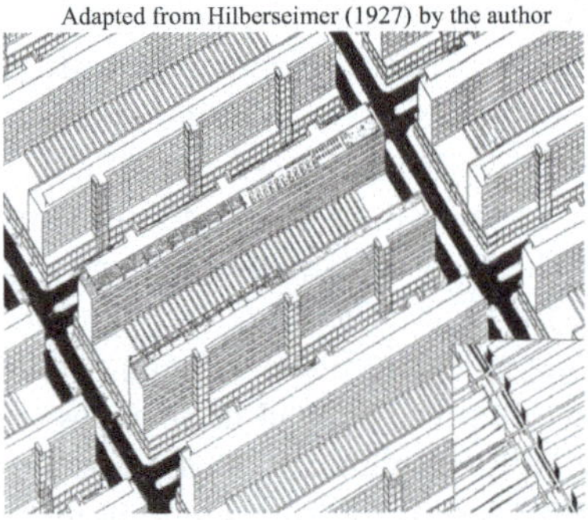

Unknown photographer/Shutterstock.com

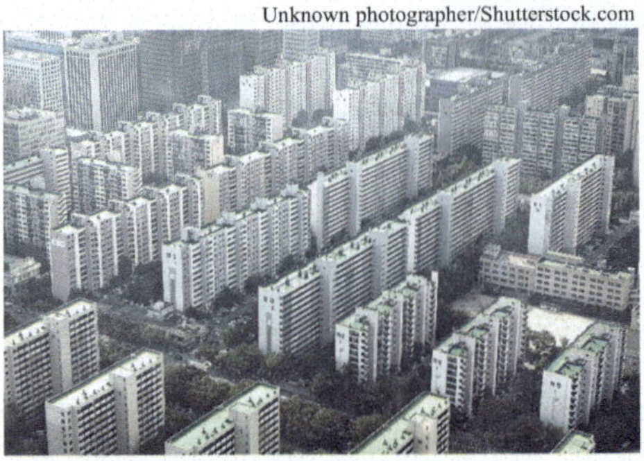

Generic housing design by Ludwig Hilberseimer (1927) (left) and housing in Yeouido, Yeongdeu, Korea (2010)

Drawing by the author

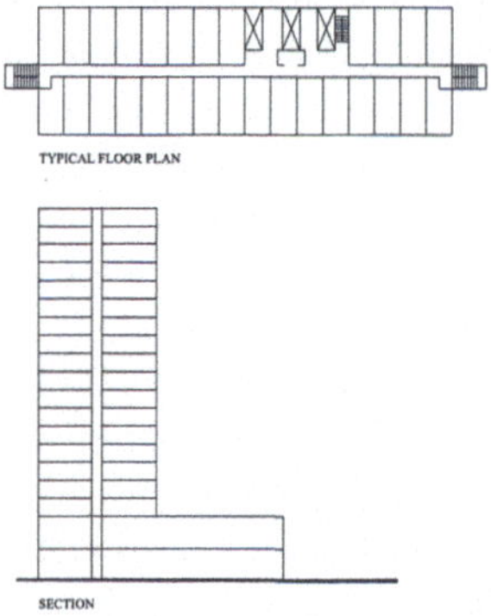

Photograph by Erhard Pfeiffer: Courtesy of Erhard Pfeiffer

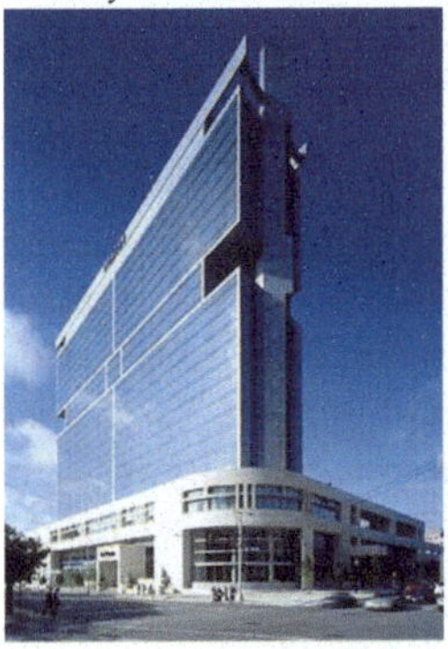

Photograph by Abdullah/Wikimedia Commons

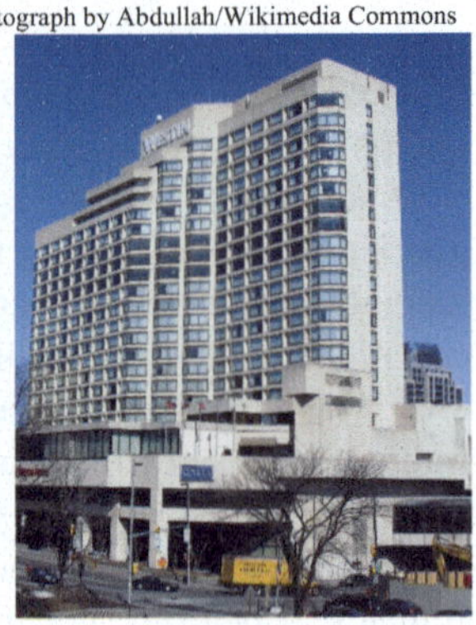

A generic hotel design (left) and two examples of specific hotels: Westins in Charlotte, NC and Ottawa

Fig. 4.14 Generic types and particular buildings

with problems they can resolve. The type consists of three high rise towers surrounded by low rise row housing or apartment blocks located within a pedestrian superblock. It was widely applied in the United States and United Kingdom as a model for mass housing projects. The goal was for new developments to simultaneously modernize cities and blend into existing low-rise surroundings.

Three examples from Philadelphia exemplify the use of the type. All were designed by internationally renowned architects. The first is Mill Creek Homes designed by Louis I. Kahn of Kahn Stonorov in association with Kenneth Day and Louis McAlister and later by Kahn on his own, possibly with Anne Tyng (1920–2011) in 1959 (Hartman 2022). The second, Southwark Plaza, was designed by Oscar Stonorov (1905–1970) of Stonorov and Haws, and the third is the Society Hill Housing designed by I. M. Pei (1917–2019) and Associates. The client for the first two was the Philadelphia Housing Authority and for the last a private developer, Webb and Knapp.

The type, as designed in Southwark, was described in the *Philadelphia Inquirer* as follows:

> When you enter the plaza Southwark surprises with the makings of a nice community. The towers look into a community centre, open square and trees and from these extend little streets of rowhouses with hedges and yards. It is a campus like setting full of potential. (cited in Finkel 2020)

Twenty years later the situation at Mill Creek Homes and Southwark was very different. They were demolished early in this century.

Mill Creek and Southwark did not work out well for low-income people with few resources, Society Hill Towers flourishes. It is inhabited by a very different socio-economic group of people. It was originally designed for rental accommodation but soon switched to condominium ownership. It remains a popular place to live and it has had a major catalytic effect on its surroundings. The Society Hill neighbourhood has been turned from being a semi-abandoned, dilapidated locale associated with the failing dockyards on the Delaware River into a sought-after place to live. The housing stock afforded easy gentrification.

Mill Creek Homes and Southwark Plaza, in contrast to Society Hill Towers, had low-income single-parent, female-headed households beset with social problems. Society Hill Towers was not inhabited by people with few resources. In addition, it does not have the image of being inhabited by vulnerable people. The other two did; they were cheaply built and inadequately maintained and looked like public housing for the poor.

There is nothing inherently wrong with the basic type. In Philadelphia, the mismatch between the design type, the quality of construction and maintenance and the needs of residents was exposed. Given the wisdom of hindsight, a recognition of the functions that the type as a generic solution does and does not afford needed to be recognized before it was selected to be the basis of a design. That observation applies to all urban and building types and generic solutions (Rapoport

1990). Following the accept paradigms and the types that leading architects have promoted has often led architects astray (see Jenkins 2024 on Britain).

References

Argan GC (1963) On the typology of architecture. Translated from the Italian by Joseph Rykwert. Archit Des 33:564–561. scribd.com. Accessed 2 Nov 2022

Colquhoun A (1967) Typology and design method. ARENA J Archit Assoc 83(913):11–14

Franck KA, Schneekloth LH (eds) (1994) Ordering space: types in architecture and design. Van Nostrand Reinhold, New York, NY

Lang J (2017) Urban design: a typology of procedures and products, 2nd edn. Routledge, London and New York

Moneo R (1978) On typology. Oppositions 18:23–45. moneo_on-typology_oppositions.pdf. wordpress.com. Accessed 1 July 2023

Rapoport A (1990) History and precedent in environmental design. Plenum Press, New York and London

Ray N (2023) The typological critique. In: Thinking through twentieth century architecture. Routledge, London

Villari S (1990) JNL Durand (1760–1845) art and science of architecture. Rizzoli, New York, NY. Originally published as Arte e Sciencia dell'architettura in 1987

Part II

Prescriptive Theories: Views of the Way Matters Ought to Be

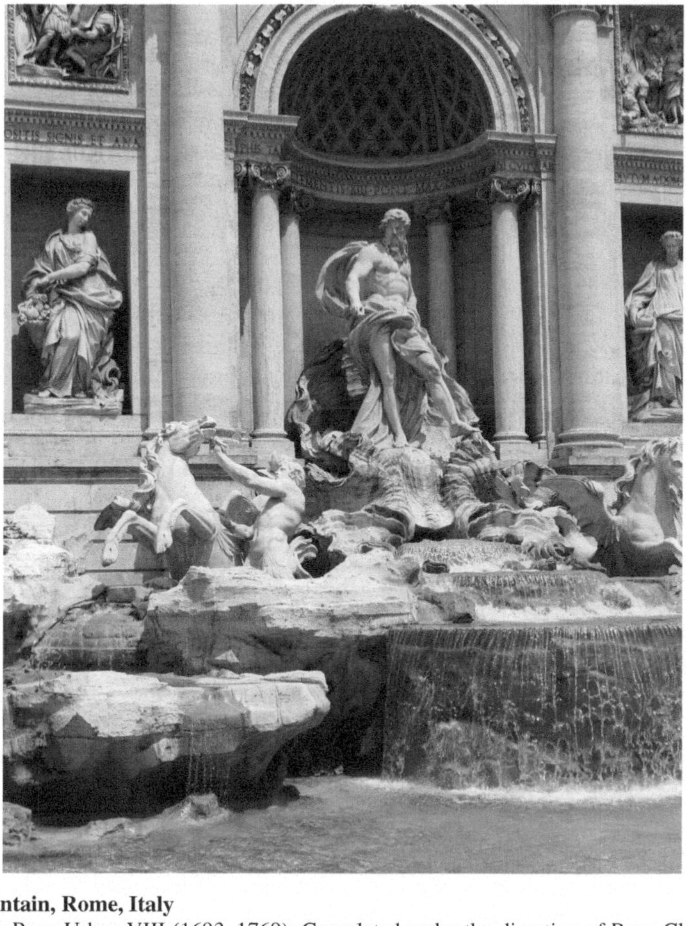

Trevi Fountain, Rome, Italy
Initiated by Pope Urban VIII (1693–1769); Completed under the direction of Pope Clement XIII (1693–1769)
Design initiated by Nicolai Salvi (1692–1751), architect, completed by Giuseppe Pannani (1718–1805), architect
Sculptures by Pietro Bracci, Filippo Valle, Giovani Grossi, and Andrea Bergondi
Photograph by the author

Part II deals with the images held by architects of good worlds, good buildings and urban environments, and good ways of designing. It begins by describing the importance of recognizing clients' aspirations as well as the socio-economic and technological tenor of the times in order to understand architectural products. It then goes on to outline the historic development of the characteristics of architects, their clients and work before discussing different views of how clients and architects *should* produce designs together.

Every design reflects the distribution of knowledge and interests among clients, sponsors, and potential users involved in its production. Individual architects have, as businesspeople, their own interests at heart too. For them to continue practicing, their works must appeal to (or be sold to) some sector of the marketplace seeking services. The characteristics of clients and the demands they make on architects are reflected in the designs produced. Clients' attitudes are embedded in the description of the work and ways of working of architects.

The focus of attention in the architecture promoted by individual architects reflects the functions—the ends—that they prioritize when designing and the patterns of built form—the means—that they apply in fulfiling those functions. Some architects are highly articulate about their attitudes; others are reticent. The former include such early twentieth century luminaries as Frank Lloyd Wright (originally Frank Wright, 1867–1959), Walter Gropius and Le Corbusier. Later in the century we have the views of those architects who rejected the Modernists' manifestoes. Robert Venturi (1925–2018) and Zaha Hadid (1950–2016) were among them. Today, architects such as Frank Gehry (né Frank Goldberg, b. 1929) and Norman Foster (b. 1935), and those associated with the Office for Metropolitan Architecture (founded in the 1970s) are leaders in establishing what architects should be doing. Lesser-known architects also promote the attainment of specific functional ends and/or the types of patterns required to fulfil them. Prescriptive theory thus consists of the descriptions and explanations of the ends and means, and of the design processes that establish an architectural movement's and/or an individual architect's brand of architecture.

Meeting a broad range of human needs is not easy. Identifying the ends that a design is to meet and trading off one set for another is a highly political act. Louis Kahn recognized that people have many latent needs that they do not recognize:

> The creation of art is not the fulfilment of a need but the creation of a need. The world never needed Beethoven's Fifth Symphony until he created it (cited in Latour 1991).

Many people have no need for Beethoven's work nor enjoy having it thrust on them but the world or, at least, the Euro-centric part of it, would be a lesser place without it. If architects consider their task is solely to create the equivalent of the Fifth Symphony the world would also be a lesser place.

Organizing different architects' thoughts on what the ends of architecture should be and how to go about designing to achieve them can be done in various ways. The method chosen here explains much. The first chapter *Ends and Means* reviews the similarity of advocacies of groups of architects based on their philosophical stances. Architects are primarily Pragmatists, Empiricists, or Rationalists although their approaches often overlap. The second chapter, *Architectural Prescriptions: A History*, outlines the unfolding story of the nature of buildings and building types, the functions they serve and the patterns used to fulfil them. Implicit in the discussion is the emergence of architects as a distinct body of professionals who design buildings better than non-architects do. The history of architecture is thus a story of evolving social systems, the development of new technologies and aesthetic tastes and the attitudes of architects towards their work and ways of working. The third chapter, *Prescriptions for the Design Process*, considers different advocacies for how the design process should be conducted.

None of the three chapters provides a comprehensive coverage of the topic it addresses. Each simply brings attention to the basic characteristics of prescriptive architectural theory. Little of the contents of the three chapters is independent of the other two. They are intertwined as are the relationships among society, professionals and clients. That topic is more explicitly discussed in *Part III: Architects and their Prescriptive Theories, and Their Designs* of this book.

Ends and Means: A Prolegomenon to the Study of Prescriptive Theories of Architecture

5

Architects have produced many manifestos on architecture. A handful have been particularly influential. Vitruvius's *De architectura: Libre decem* was the first. The second was *De re aedificatoria: Libre decem* (1452) written by Leon Batista Alberti. In it, Alberti rejected existing design paradigms and promoted a new architecture leading to the High Renaissance in Italy. The next manifesto heralding a new architecture was Adolf Behne's *The Modern Functional Building* (1926) although the Modernist position is more clearly stated in *The Scope of Total Architecture* (1955) written by Walter Gropius. Gropius prescribed, although he denied it, what is generally called the International Style. The challenge to Modernist orthodoxy came from Robert Venturi. His *Complexity and Contradiction in Modern Architecture* (1963) set the foundation for a postmodern era. A reaction to these lines of thought came from the architects proposing a neotraditional architecture as presented by Andres Duany (b. 1949), in 'A general agreement for the practice of contemporary architecture' (2021, originally 1991).

Along the way were major statements on architecture by Le Corbusier in *Vers Une Architecture* (1923) although it is more of an attack on the past than a statement about the future. It furthered the arguments that Henry Provensal (1868–1934), an artist and architect, presented in *L'arte de demain: vers l'harmonie intégrale* (1904) (Hewitt 2023). Le Corbusier's ideas on architecture and urban design, along with the members of the Congrès Internationaux d'Architecture Moderne (CIAM), were propounded in a series of declarations. Cumulatively, they had a profound impact on architecture and architectural education. The clients for architectural services swayed by these ideas are seldom identified. Without them there would be no buildings.

J. Lang, *The Scope of Architectural Theory*, Design Science and Innovation,
https://doi.org/10.1007/978-981-96-3690-7_5

5.1 Clients and Their Ends

'The client has a role to play [in the design of a building] which from one point of view is more important than the architect's own' (Van Rensselaer 1893). Every building is designed for a client or set of clients who, along with architects and contractors, are the major stakeholders in bringing a project to fruition. The relationship between clients and architects and the way they exchange ideas about ends and means determines much in the outcome of a design (Jenkins 1961; Eaton 1969; Soth 1983; Brown 1996; Cooper 1996).

Edwin Lutyens (1869–1944), an English architect known for his adaptations of traditional architectural forms to meet contemporary needs, believed that "there will never be great architects or great architecture without great clients" (cited in Brown 1996). Dutch architect Nanne de Ru (b. 1976) goes even further. "All the beautiful, most iconic buildings in the world are conceived by brilliant clients" (cited in Thakar 2021). For architects the ideal client is a patron. Patrons allow their architects to define the ends—the program—and the means—of the design sought. Many were wealthy people who solicited buildings that they, themselves, disliked to boost their self-esteem (Wolfe 1981).

The architectural transition from revivalism to modernism and from arts and craftsmanship to industrialization had notable patrons. In Barcelona, Josep Batlló, a textile magnate, sought a house that was lively and imaginative and like no other. In hiring Antoni Gaudi i Cornet (1852–1928) to refurbish the Sánchez house, Batlló not only got an inspired work but established Gaudi's professional career and led to the Association of Worshippers of Saint Joseph entrusting him with the designing of Basílica i Temple Expiatori de la Sagrada Família. Le Corbusier had similar support (Fig. 5.1).

The clients for Le Corbusier's early houses in France allowed him to dictate almost all the qualities that a house should possess (Eaton 1969; Soth 1983). The Villa Savoye was a work of fine art with Pierre and Eugène Savoye as patrons. A luxury automobile manufacturer (Gabriel Voisin of Avions Voisin) sponsored Le Corbusier's early urban design explorations. Le Corbusier himself owned a series of Voisin cars. It is not surprising that highways along which cars could speed played such an important role in his designs.

The early clients of the highly prolific Frank Lloyd Wright in Oak Park, Illinois were self-made people full of ideas, some of which Wright absorbed into his work. Nancy Willey, the daughter of a painter, an avant-garde woman, worked closely with Wright on the Malcolm Willey House. She taught Wright much about designing for the middle class. James Rogers who hired Wright to design a house for his daughter and son-in-law was a successful grain merchant. He is typical of Wright's clients. The contemporary Chicago clients of Howard Van Dorn Shaw (1869–1926) were different. They were or aspired to be members of the established, conservative elite (Eaton 1969; Cohen 2015). The differences are clear in the brands of houses the two architects designed (Fig. 5.2).

Other clients were patrons of rationalist buildings. In Argentina, for instance, Corina Kavanagh (1890–1984), a wealthy socialite, was the client for the first

Photograph by Anaïs Napoleon/Wikimedia Commons

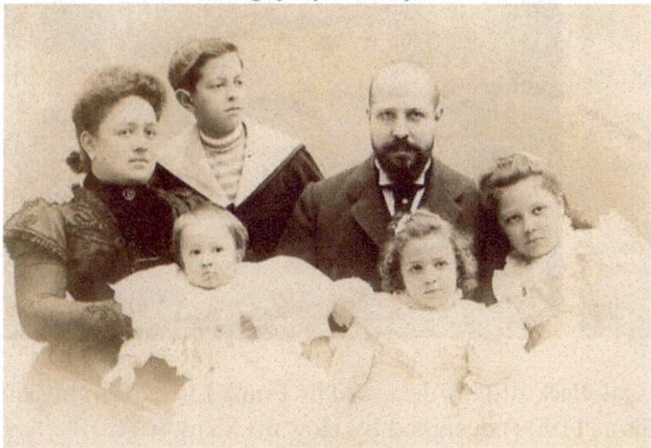

The Josep Batlló i Casanovas (d. 1934) Family

Adapted from US Library of Congress/Wikimedia Commons

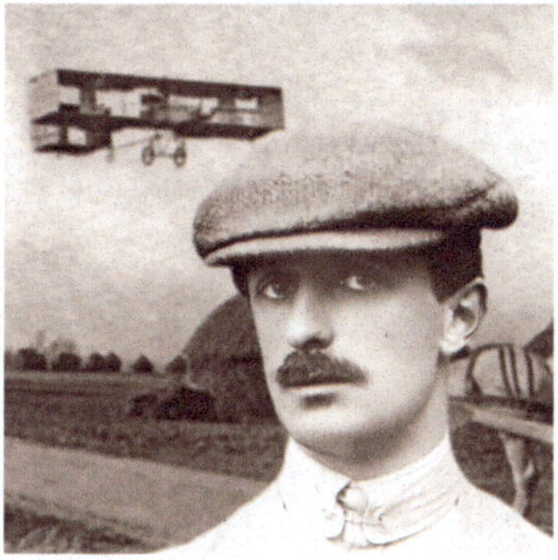

Gabriel Voisin (1880-1973)

Shawn Beckwith collection: Collection of the Author

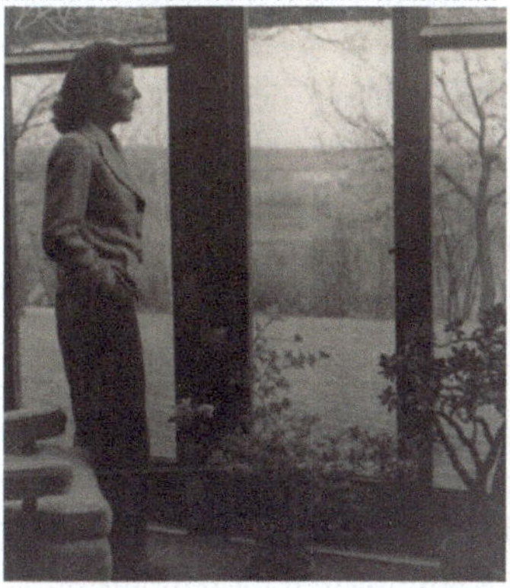

Photograph by William E. Dunlap/Edith Farnsworth House Historical Site

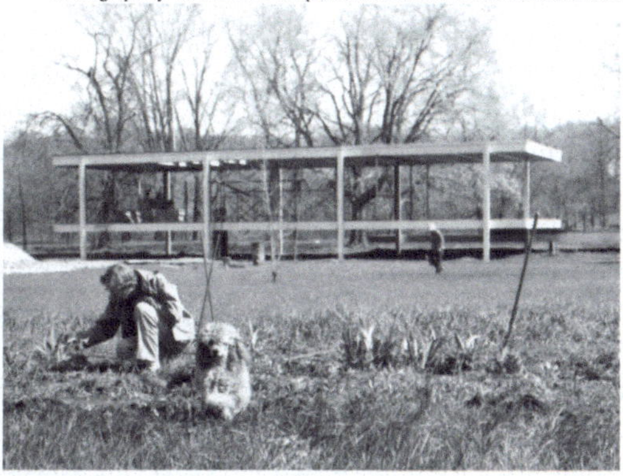

Nancy (Boyd) Willey (1902-1998) in the Malcom Willey House (1933) (left) and Dr. Edith Farnsworth (1906-1971) and her dog at the Farnsworth House during construction (1950)

Michiel Hendryckx/Wikimedia Commons

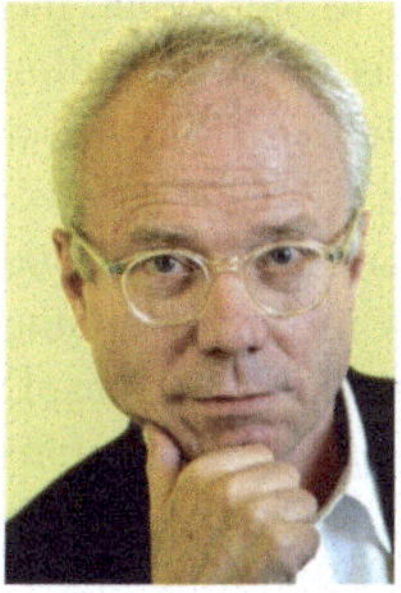

Photograph by Kathy Hutchins/ Shutterstock.com

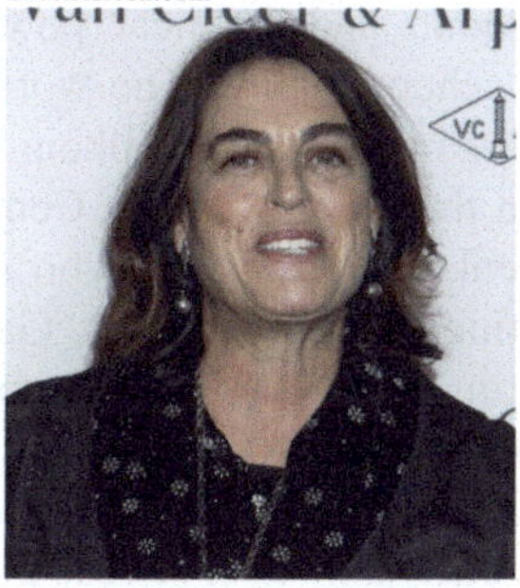

Photograph by President.az/Wikimedia Commons

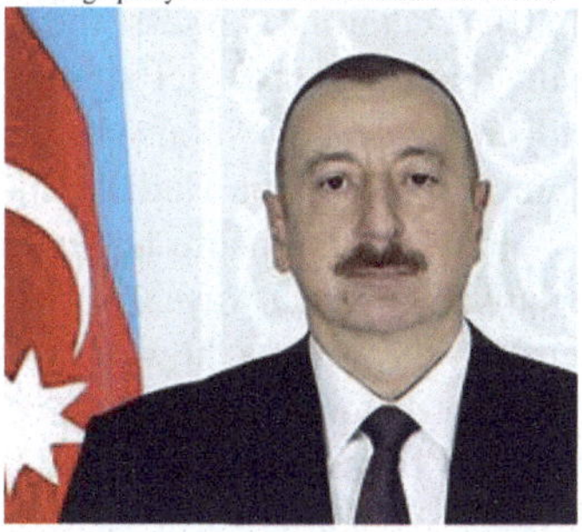

Three recent patrons: Rolf Fehlbaum (b. 1941), Maja Hoffmann (b. 1956) and President Ilham Aliyev (b. 1961)

Fig. 5.1 Patrons

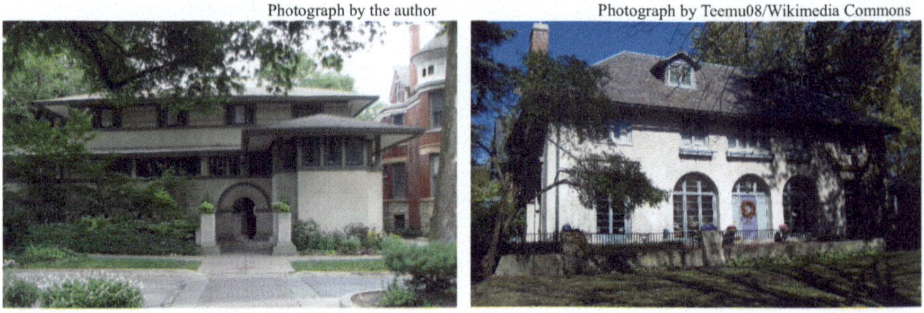

Fig. 5.2 Frank W. Thomas House (1901), Oak Park Illinois designed by Frank Lloyd Wright (left) and Carl E. Williams House, Evanston, Illinois (1909) designed by Howard Shaw (right)

Modernist high-rise building, Edificio Kavanagh (1934), in Buenos Aires. Dr Edith Farnsworth (1903–1977), a polymath, gave Ludwig Mies van der Rohe (né Ludwig Mies, 1886–1969), a leading Bauhaus master, a free hand in designing a weekend retreat for her. Political leaders in several formerly colonized counties sought Modernist architecture to break out of the "shackles of the past". More recently, a director of an avant-garde furniture company, Rolf Fehlbaum (b. 1941) was the client for the Vitra Fire station, the first design by Zaha Hadid to be implemented. The Luma Arles Tower (2021) designed by Frank Gehry is an artwork sponsored and imagined by the wealthy Swiss art collector and curator Maja Hoffman (b. 1956). It houses spaces that can serve a variety of instrumental functions, studios and a café. In contrast, the Neoclassical Atlantis design (1986–1992) prepared by Leon Krier (b. 1946), was sponsored by Helga and Hans-Jurgan Müller, art gallery owners who were interested in preserving European culture in a globalizing world (Verbruggen 2017).

Several clients for major architectural works have been dictators. Many architects have worked for regimes noted for their human rights abuses (Wainwright 2020). Dictators possess the economic and political power to get what they want done and they get significant schemes built. The world is full of buildings and urban designs sponsored by autocrats. President Ilham Aliyev of Azerbaijan commissioned Zaha Hadid to design the Heydar Aliyev Centre in Baku.

A close relationship between clients and architects who share the same values usually, but not always, results in fine designs for both. Gaps in understandings between the two do, however, sometimes develop. Despite approving the plans, Dr Farnsworth found the design of her retreat (1948–51) to be a "glib, false sophistication" (Craven 2019). The Villa Savoye (1929) was hardly ever used by the Savoyes but Farnsworth visited her glass and steel retreat until 1969. She heavily personalized the house to make it habitable by removing all the Mies-designed furniture. Both buildings are now uninhabited, much revered, works of Platonic art and museum pieces. They also reflect different perceptions of the fundamental purpose of a building: Is it to be a work of art or a habitation?

5.1.1 Buildings and Urban Designs as Public Art—Displays—And as Habitable Environments

Almost all architects state that the fundamental function of architecture is to contribute to society's well-being by providing pleasing settings for everyday lives. At the same time, every self-consciously designed building is a display of an architect's intentions as buffeted by competing demands in getting it built. Inevitably the twin ends of buildings being works of arts and serving other functions well tug on architects' aspirations (Tschumi 1998).

Sometimes the architect is purposely trying to create a work of art based on an intellectual idea. For instance, Bernard Tschumi (b. 1944), an internationally renowned architect, and Colin Fournier (1944–2024), an Anglo-French architect, based their competition winning scheme for the Parc de La Villette in Paris on the deconstructivist ideas in the study of texts as promulgated by Jacques Derrida (1930–2004), the French philosopher. The design is of considerable interest to the cognoscenti. It also functions as a place of entertainment for people who have no knowledge of, or interest in, the reasoning behind the design. They are concerned with what it affords them as a place to be (see Chap. 8).

Several architects who established new prescriptions for architecture have been inspired by contemporary movements in the fine arts from Jugendstil to Dadaism to Pop Art and Surrealism (Stouhi 2024). This link can be seen in the work of Piet Blom (1934–1999) in the Netherlands and Frank Gehry (b. 1929) and Daniel Libeskind (b. 1946) in the United States. Le Corbusier and Zaha Hadid spent time drawing and painting abstract images early in their careers. Le Corbusier's artworks were influenced by Cubism and De Stijl (Blau and Troy 1997); Hadid drew on the dynamic forms of the avant-garde Russian painter Kazimir Severinovich Malevich (1879–1935) in her effort to "reinvent modernism" (Jodido 2020).

Today, architects such as David Adjaye (b. 1966), the Ghanian-British architect, see their artwork as precursors to architectural designs. It is a form-first approach to design. Adjaye says that his sculpture Asaase III:

> … is in a way, a representation of how I imagine an ideal city – a city that is in symbiosis with the Earth, acknowledging it and honouring it in a very deep way, but also absolutely transforming it and creating new features. (Adjaye cited in Niland 2023)

This line of thinking is carried over into his design of buildings and places.

Some architects claim the function of buildings as works of art is of great, if not primary, importance in providing society with a social benefit. Other leading architects see the social benefit of their designs in a different light. To them a building as a display is a byproduct of other intentions. James Stuart Polshek (1930–2022), an American architect with a commitment to social justice, in his book *Build, Memory* (2014), observed:

> The true importance of architecture lies in its ability to solve human problems not stylistic ones. A building is too permanent and too influential on public life and private comfort to be created primarily as public art.

Polshek recognized that buildings are art works and that pleasing the cognoscenti is an important function of architecture but he believed that other functions should take precedence. He was not alone. Of architects practising today, Chilean Alejandro Aravena Mori (b. 1967), a 2016 Pritzker Prize winner is among those taking a similar position (Kimmelman 2016). The need is to integrate the meeting of social ends with the desire to create a work of art.

From the eighteenth century to the present, far-sighted owners of factories have been clients for company towns (Crawford 1995; Lang 2021, 54–64). They strove to not only provide healthy environments for their employees but to raise the pride workers had in themselves and the companies for whom they laboured. The social end of having well-behaved, conscientious, productive, and contented workers was met by the facilities provided and those, such as public houses and court houses (brothels), prohibited because they were deemed to be unhealthy or immoral.

Several celebrated architects have designed company towns. Arc-et-Senans (1774–1779) in France was designed by Claude-Nicholas Ledoux (1726–1806). Ledoux's goal was to create a work of art as much as a sound working and living environment for the employees of a salt works. In the twentieth century Echota Village in upstate New York (1901) was designed by McKim Mead and White, a firm whose work defined the American Renaissance in *fin de siècle* New York. 2018 Pritzker Prize winner B(alkrishna) V. Doshi (1927–2023), with his Vāstu-Shilpā Consultants, designed company towns in India that have striven to be culturally appropriate rather than in fashion aesthetically. In the unimplemented design for Vidhyadhar Nagar (1983) in Jaipur, Doshi's goal was to achieve a socially enlightened town that was also a work of art. Its geometry was based on the prescriptions presented in the traditional Indian *shastras*.

Architects at the Bauhaus and CIAM had strong social agendas. They sought to create healthy, spirit-enhancing, living environments for people. Unfortunately, their prescriptions often differed from what was required by those people whose lives they wished to improve (Michelson 1968; Brolin 1976; Blake 1977; Mikellides 1980; Ornstein 2022). They based their designs on too limited a model of people and of the functions of the built environment (Scott Brown 2004). Mies van der Rohe's 'Less is more' in architecture is a catchy aphorism but not a worthwhile statement of ends and means when applied to human environments.

Today, architects are circumspect in the claims they make of the social implications of their designs except when it comes to securing commissions. They are also careful in claiming architecture to be a pure fine art. Pierluigi Nicolin (b. 1941) in his *Architettura in quarantena* (2020) notes that while he sees architecture as art, architects must 'recognize that socially distanced architecture can lead to dangerous results'. Pritzker Prize winner in 2024 Riken Yamamoto (b. 1941) is a social advocate known for his consideration of buildings as the background for life and his commitment to community engagement in the design process. More broadly, what architects see to be their responsibilities depends on whether they are pragmatists, empiricists, or rationalists.

5.2 Pragmatists, Empiricists, and Rationalists

The twentieth century architectural prescriptions that continue to dominate architects' thinking in the first quarter of the twenty-first were a response to the Neoclassical, Neogothic, and Beaux Arts architecture of the nineteenth century. They were also a response to the polluted industrial city of the British Isles, continental Europe, and North America and the appalling working conditions and the dehumanizing environment in which workers lived. The devastations of World War One and the huge death toll of the Great Influenza Epidemic (Spanish Flu, 1918–1920) had an enormous impact on the thinking of both architects and their clients.

The end and means displayed in the various prescriptions of modern architects have been categorized in several ways. Charles Jencks (1939–2019), an American cultural theorist, and Karl Kropf (b. *ca* 1963), an urbanist, divided them into five categories: Postmodern, Postmodern Ecology, Traditional, Late Modern, and New Modern (Jencks and Kropf 1997). Several observers distinguish between Hypermodernity and Neotraditionalism (for example, Fraker 2010). Those ways are fine for describing architecture and urban design paradigms. Here, the responses are more broadly divided into three intellectual approaches—pragmatist, empiricist, and rationalist—that explain more in understanding the ways architects examine the world and design for it. The work of some major architects hovered between all three lines of thought; others have stuck to singular lines of thought. Much has depended on who the clients of their work have been.

5.2.1 The Pragmatists

The pragmatists among architects follow the thinking promoted by Jeremy Bentham (1748–1832), an English philosopher, and later by American philosophers William James (1842–1910), John Dewey (1899–1952), and Charles Saunders Peirce. The use of the word 'pragmatic' goes back to the way Greek historian Polybius (*ca* 200–118 BCE), described his writings, but it is Peirce who is credited with introducing the term 'pragmatist' to contemporary philosophical thought (Audi 2015).

Pragmatism, as well as empiricism, relies on experience as the basis for learning about the world. To Pragmatists, however, it is the usefulness, workability, and practicality of ideas and products that establishes their merit. Pragmatists stress the priority of actions over doctrines and of experience over principles in making decisions. Pragmatic approaches to life require people to consider the full future consequences of the actions they take. In business, pragmatism sometimes implies a ruthlessness in getting things done, but no such connotation is intended here.

5.2.1.1 The Architects and the Architecture

Pragmatic architects have a hands-on attitude to the design of buildings and urban settings. Buildings are designed to fulfil basic needs and not the intellectual aesthetic ideas of interest to the cognoscenti. Their designs can be highly innovative in the way fundamental ends are met within available, often tight, budgets. The focus of attention in their design is on way they fulfil their instrumental functions rather than the impressions they make. They are designed to operate efficiently and to provide a sound return on the capital invested in them. They are aesthetically satisfactory.

Pragmatic architects are not visionaries. They tend to accept existing zoning laws and building codes. They know them well and the loopholes that exist in them to achieve the design and financial ends of their clients. They accept current modes of artistic expression rather than strive for new ones. They embrace standards such as those of Leadership in Energy and Environmental Design (LEED) because in doing so the design process is shortened. They understand building typologies and often specialize in designing one type in practice.

Architectural firms that are fundamentally pragmatic in character range from international organizations designing standard office buildings clothed in an acceptable aesthetic garb to sole practitioners doing piecemeal alterations to existing dwellings. Some of the firms are highly pragmatic in all they do. Others may employ rationalist or empiricist thinking most of the time but are pragmatic in dealing with certain commissions. For instance, their work may be highly pragmatic when dealing with low-income housing where the end is to provide shelter for resource-poor people at a minimal cost and be highly rationalistic when working with avant-garde patrons striving to possess buildings with novel geometries.

5.2.1.2 The Clients

Clients seeking pragmatic designs see themselves as practical, no-nonsense people. They seek buildings that retain their value over time. They want buildings that are efficient in layouts and easy and inexpensive to maintain. They expect the aesthetic qualities to be socially acceptable but are uninterested in 'architalk' about ends and means. Clients expect their architects to listen to and understand their requirements. They want architects to carry out the design process openly and to clearly answer their questions as a design evolves.

The clients for substantial buildings tend to be large commercial firms, smaller organizations that seek no-frills buildings, and speculative property developers. The builders of standard suburban housing tracts, for instance, seek architects who provide them with a handful of basic house types that are straightforward to erect and meet the requirements of the contemporary marketplace. The types must be easy to sell. In a neoliberal economic world, many governmental agencies seek pragmatic designs with little concern for aesthetic issues. Taxpayers are increasing frugal in their demands of public sector buildings. They may, in contrasts, overcapitalize in buildings for themselves.

5.2.2 The Empiricists

Empiricism asserts that all true knowledge comes from observations of the world via the sensory systems. Knowledge is built up through everyday experiences, formal study, and research. The scientific experiment is the model—the ideal type—of research method. Prescriptive theories for action are based on evidence of what is perceived to work well. Some empiricists are copiers of past models; others imitate the principles underlying them.

Empiricism in European thought can be dated back to the writings of Thomas Aquinas (*ca* 1225–1274), Leonardo (di ser Piero) da Vinci (1452–1519), and Galileo Galilei (1564–1642) but its intellectual foundation was established later in *An Essay Concerning Human Knowledge* by John Locke (1632–1704), an English philosopher and physician. Empiricism is the line of thinking that is expressed in English Common Law (Lai 1988). For this reason, it is sometimes referred to as the Anglo-Saxon school of thought. Doing so is misleading because empiricist thinking has a long international history (Sen 2000). Today it is also often referred to as the Aristotelian line of thought because its roots can be traced back to Aristotle (394–322 BCE) and before him to Socrates (469–399 BCE) in ancient Greece. Islamic scholars of the tenth century and the Moorish Andalusian scholars of the twelfth also played major roles in the development of what we understand as empiricist thought today.

Systematic research on the functioning of the built environment and its place as part of the ecological world began in the nineteenth century. It was given a boost in the second half of the twentieth century with the formation of organizations such the Environmental Design Research Association (EDRA) and the International Association of People-Environment Studies (IAPS). The easily applicable finding of Kevin Lynch as presented in *The Image of the City* (1960), Oscar Newman in *Defensible Space* (1972), and William Whyte (1917–1944) in *The Social Life of Small Urban Spaces* (1980) had an immediate, if not lasting impact, on professional practice. The observations of Jane Jacobs in *The Death and Life of Great American Cities* (1961) still inspire creative empirical thought.

5.2.2.1 The Architects and the Architecture

Empiricist architects rely heavily on imitating precedents and existing building types. They are *regressive utopians*. When seeking ideals, they embrace what works and dismiss what does not (Boguslaw 1965; Tafuri 1973). Architectural design remains largely a type-based approach to the creation of buildings and urban designs (Heyda 2019). Many of the types that form the basis for design today are based on analyses of the functioning of buildings generated by rationalists, others by empiricists, and/or from the pragmatic requirements of clients.

The architects tend to be regionalists/contextualists (van Schaik 1985). They draw primarily on local precedents. The conservatives among them tend be formally trained in leading architectural schools. Many important early twentieth-century empiricists were products of the École des Beaux in Paris. The progressives, such as Frank Loyd Wright were less formally educated autodidacts but learnt much through their experiences as apprentices.

Several architectural movements of the past two centuries are clearly empiricist in nature. Others fall uneasily into the category. Arts and crafts architecture of the nineteenth century and today's Neotraditional architecture are examples of the former; Art Deco is an example of the latter. The empiricists look to the future with a messianic, revolutionary, and utopian optimism. The advocates for the Neoclassic, the Neogothic, and the architects associated with the École des Beaux all felt strongly about the ends they were promoting and the means for achieving them. The Garden City ideals as originally promulgated by Ebenezer Howard in *Garden Cities of Tomorrow* (1902) had ardent advocates. The Neotraditionalists of today such as DPZ CoDesign and Leon Krier are equally zealous advocates for their ideas.

5.2.2.2 The Clients

The clients for empiricist work have been and are real estate developers, other organizations, and individuals who want the buildings they work in, or inhabit, to reflect traditional ways of life and tastes and to be economically sound. They are unwilling to stick their necks out and experiment with new ideas. Property developers such as Robert Davies in Seaside, Florida and the Prince of Wales (b. 1948; now King Charles III) in Poundbury in England look back at traditional settlement and building patterns that they deem to function well as precedents for their own advocacies (Stueteville 2022).

Clients either want traditional building layouts that work well in affording their usual activity patterns, or, if they are progressives, they want to get their architects to produce more avant-garde designs that afford their ways of life. They tend to be practical, well-educated, but unscholarly people (Mann 1979). They, like pragmatists, expect their architects to work closely with them to get the designs they want rather than what their architects want (Eaton 1969).

The more conservative clients for bespoke houses tend to be people with inherited wealth or those aspiring to identify themselves with people of that status. They do not seek new aesthetic expressions to establish themselves as members of the avant-garde art world. Some are urbanists who seek the richness of life afforded by high density residential environments in which they have a degree of anonymity. Others seek suburban environments of single-family homes that express traditional aesthetic values. In common, they are unwilling for their notions of the good life to be challenged. Much depends on the experiences they have had of different types of environments during their childhoods and their adolescent years. Letting go of childhood experiences is difficult for both architects and their clients, both empiricists and rationalists.

5.2.3 The Rationalists

If empiricism is associated with the philosophies of Aristotle, then rationalism is with those of Plato (*ca* 427–348 BCE). As rationalists seek to establish ideal, artistic new worlds without references to the past, their view of design is often referred to as the Platonic approach to architecture. Rationalists believe that there are certain innate ideas and principles that are self-evident. Their views are derived from the philosophies of René Descartes in France, Baruch Spinoza (1632–1677) in what is now The Netherlands, and Gottfried Leibniz (1646–1716) in Germany. They believed that reason alone can be used to derive truths about the world, how it functions and how it should function if it were a good world. Empirical evidence is unnecessary to ascertain the veracity of their ideas.

Rationalism today is associated with the *Code civil des Français* (1804) established under Napoléon Bonaparte (né Napoleone di Buonaparte; 1769–1821) in France. Known as the Napoleonic Code, it proposed a legal framework based on *ideals* not on *precedents* (Lai 1988). For this reason, the application of rationalist thought to design is often considered to be the continental approach. The rationalists among architects established a way of thinking about the world and about how to improve it. They have had a profound impact on architecture and urban design promoting their views throughout much of Europe and Latin America and through expatriate architects to North America.

5.2.3.1 The Architects and the Architecture

If the empiricists are the *regressive utopians*, then the rationalists are the *progressive utopians* (Boguslaw 1965). Historically, they considered the empiricist architecture and the pragmatic work that they saw around them to be outdated (Le Corbusier 1923; Behne 1926; Giedion 1963, originally 1941; Sharp 1978; Peckham and Schmiedeknecht 2013; Frampton 2020). The rationalists opposed all historicist architectural thought and considered Art Deco architecture to be a disaster (see Le Corbusier 1925b).

Alberti's *De re aedificatoria: Libre decem* was the first to rationally prescribe a new architecture. For us today it has been the promotions of CIAM (1928–1959) that were instrumental in promoting an architectural paradigm reputedly uninfluenced by past buildings. CIAM members' focus of attention was on creating salubrious, efficient environments. The functions of the built environment to display individual aesthetic tastes and thus the status of its residents was downplayed in the effort to create a universalism of spirit through design.

CIAM formalized the principles of modern architecture in a series of consensual prescriptions, or manifestoes, on what buildings and urban design should be like. The organization's original members were all continental Europeans whose intellectual upbringing was steeped in the rationalist spirit. As timed passed not everybody in CIAM was happy with the declarations. Groups such as Team 10 broke away (Smithson 1968; Risselada and van den Heuvel 2005).

In 1977, an international group of architects, historians, and social scientists produced the *Charter of Machu Pichu* (1979). They argued for a better understanding of how places function before designing new ones. A year earlier another group of architects that included George Candilis (1913–1995), a Team 10 member, Moshi Safdie (b. 1938), Nadar Ardalan (b. 1939), José Luis Sert (1902–1983), and B. V. Doshi produced *The Habitat Bill of Rights* (Ardalan et al. 1976). They argued for a more empirical basis for architectural design than CIAM members sought. They wanted a softer, weaker, architecture. It was a line of thought advocated by the Spanish architect Ignasi de Sola Morales (1942–2002; de Sola Morales 1989).

A second phase of Modernist architecture appeared on the scene in the post-World War Two era. Several architects, including Le Corbusier, led the way to a more sculptural, artistically expressive architecture. The Postmodernists took these reactionary explorations a step forward by openly looking at precedents again.

5.2.3.2 The Clients

The clients of rationalist architects wanted to get away from what they perceived to be the strictures of the past. They were progressive people seeking to present an avant-garde social and artistic self-image through the buildings they inhabited and/or owned. Many were self-made successes in their fields. They did not identify with those who had inherited their success and high social status. Some were dictators who selected avant-garde architects to design buildings that would place them and their countries in the forefront of the modern world. They wished to be seen as progressives.

The clients were disproportionately people interested in the fine and performing arts. They, whether individuals or public agencies, were willing to accept what was regarded as good for them as seen by the architects who they commissioned. In the early twentieth century, many had read about the trends in the art world in journals such as *L'Esprit Nouveau*. That journal was founded in 1920 by Le Corbusier, Paul Dermée (né Camille Jansen; 1886–1951), a poet and literary critic, and Amédée Ozenfant (1886–1966), a painter. Some clients were innovative industrialists. Gabriel Voisin was typical of Le Corbusier's friends and clients. Voisin, who had once worked for an architectural practice, was a strong supporter of the new architecture, at least for others. The house near Tournus in eastern France that he chose as a retreat for his family was a traditional building.

5.3 Rationalism and Empiricism and Postmodern Architecture

A generation of architects of the late twentieth century rebelled against the puritan ethic of Modernist designs and the apparent lack of understanding, or rejection of, the potential richness of the symbolic function of designs. They sought a more empiricist architecture within a rationalist approach. The sally against the order and simplicity of the Rationalists' ideas was led by Bruno Zevi (1918–2000) in his book *Towards an Organic Architecture* (1950) but, more powerfully by Robert

Venturi in *Complexity and Contradiction in Modern Architecture* (1966) and in the book, *Learning from Las Vegas: The Forgotten Symbolism of Architectural Form* (1977), that he co-authored with Denise Scott Brown (b. 1931) and Steven Izenour.

The end that Postmodernists seek is an architecture rich in associations that are meaningful to both the cognoscenti and the lay public. The means is through the application of an abstraction of the patterns of past forms to current designs, particularly in the façades that buildings ("decorated sheds" in Scott Brown's terms) present to the world (Jencks 1977; Hartman 2022). In some cases, these abstractions are easily understood by both the public and the cognoscenti but at other times both groups, but especially the former, must have the abstractions explained to them. They are not immediately interpretable (Groat and Canter 1977).

5.4 Pragmatism, Rationalism, and Empiricism in Architecture Today

Many places have diverse populations. Clients for architectural services are equally diverse even if drawn from the wealthier segments of all societies. They range from conservative heads of corporations to astute businesspeople to members of the nouveau riche, and to communist party bosses of municipal councils (Olds 2001; McNeill 2008). The ends they expect their architects to meet differ considerably.

Many buildings are pragmatic responses to clients' demands. Simple, rectangular, Modernist office buildings coexist with bold hypermodern, geometrically varied structures clad in assorted skins. Computer-based algorithms enable the calculation of the complex structural requirements of innovative forms in a way that engineers were unable to do in the past. Artificial mechanisms are used to meet the metabolic comfort demands of their inhabitants. Such buildings satisfy the ends sought by many wealthy neoliberal clients. Urban districts that are full of diverse buildings end up having a unity through the very diversity of their individual building forms (for example the Doha skyline; see Fig. 3.7).

The design processes employed by Empiricist architects today range from the problem solving to the typological. The latter still predominates. Architects draw on a range of precedents but they have, in common, a desire to display a continuity in the architectural spirit of a locale in their designs. The approach goes under the label of Neotraditional architecture. As an aesthetic expression, such architecture meets the aspirations of a wide range of people. Clients are comfortable with it because it is what they are accustomed to seeing. They are neither challenged nor shocked by it.

The proponents of the Neoclassical, such as the Institute of Classical Architecture and Art (founded in 2002), argue that the Neoclassical is timeless and easily understood (Dowling 2004). The Institute promotes its ends through its *Classical American Series on Art and Architecture*. Many buildings being erected today could well have been constructed two hundred years ago. The Neoclassical also exists in a postmodern form. Elements such as columns appear in abstract forms. Sometime these forms are overblown to make mundane building types, such as social housing, appear heroic. They serve as the artistic expressions of their creators while the instrumental functions of buildings are pragmatically met.

The rationalists continue to generate bold ideas about the issues that they think architects should be addressing and the way they should be doing so. Pragmatists, rationalists, and empiricists are, in common, having to deal with the advocacies of diverse special interest groups seeking specific ends in new buildings and urban designs.

5.5 Advocacies for Specific Ends

All buildings and urban designs are the physical manifestations of advocacies prescribing the promotion of certain ends over others. Each design affords some activities better than others and each offers its beholders and occupiers some aesthetic experiences rather than others. Architects, critics, and observers of the built world note that certain functions of buildings and places are not being addressed and argue that they should be (Filippi and Hooper 2019). Architects, they believe, need to broaden their concerns and deal with meeting ends that they have avoided considering in the past (Fisher 2008). They need to have a broad view of the functions of buildings.

5.5.1 Climatically and Ecologically Sound Environments

The information on how to design for people's comfort without heavily resorting to artificial mechanisms has long been available to architects. The concern predates *Design with Climate* that was first published in 1963 (Olgyay 2015). Modern technologies have made many traditional ways of designing to deal with the vagaries of local climates obsolete. Air-conditioning and heating systems do have a place in modern architecture as observed in 1969 by architectural theorist and historian Reyner Banham (1922–1988) but they can be overused (Banham 1969). Hypermodernists today, with their heavy use of glazing on the façades of buildings, allow the aesthetics of their building to override local climatic considerations. Energy costs soar. Building can be designed so they are not extravagant energy users (Borràs 2024).

Manuals on how to design to avoid energy and carbon intensive environments abound (for example Giovani 1998). They deal with designing in regions as varied as those with hot-arid climates (Golany 1983) to the design of cities located in far northern latitudes (Pressman 1995). Shortage of water, particularly in hot-arid places, makes designing with the climate in mind even more urgent (Margolis and Chaouni 2015).

Advocates for climate sensitive architecture seek it not only for its sustainability but also its utility as a mechanism for creating regional aesthetics (Relph 1976; Chandavakar 2022; see also Lefaivre and Tzonis 2003). Designing for local climatic conditions is part of a broader set of prescriptions for creating ecologically sound architectures (Tjallingi 1996). The concern is with the health of the world's fauna and flora and the fragility of our highly populated planet. This concern, particularly in landscape design, goes back to the works of Rachel Carson (1907–1964) and Ian McHarg (1920–2001). Carson's *Silent Spring* (1960) and McHarg's *Design with Nature* (1969) made many designers think about their tasks anew (Jodido 2018). Their position was clearly stated by landscape architect Anne Whiston Spirn (b. *ca* 1948).

> My goal is to change the way people see and how human settlements are designed and built. That is why I write. I want to tell an engaging story and to inspire readers to shape the world into a humane habitat. (Spirn 1985)

The recognition of the world's changing climates has made architects more sensitive to the siting of buildings and the potential side effects they have on their surroundings (Cuthbert 2011).

5.5.1.1 Sustainable Architecture

While it has not been a self-conscious focus of attention in the architectures of the past, today a concern for the health of our planet has raised four issues in the design of buildings. The first is about the energy consumed in fabricating building materials. The second is for the costs in operating and maintaining a building. The third concerns the impact that buildings have on their surroundings and the fourth is for their utility as behaviour settings in a changing world.

A building has an embedded energy in the materials of which it is constructed and the transportation costs in getting to them to a building site. All buildings incur energy costs in maintaining them in a good condition. New buildings change the patterns of human, vehicular, and air movement in their vicinity. They can shed heat onto their surroundings as well as contributing to the formation of heat islands. They act as a catalyst for adjacent developments. On brownfield sites that may involve the demolition of well-functioning but less income-producing buildings. On greenfield sites the environment changes from being rural to being settled. How best should architects consider these matters?

The instrumental purposes of buildings change. If a building cannot accommodate a new set of activities easily enough it is abandoned or demolished. The Pantheon, a Roman temple built during the second century, survives because it has been in continuous use. Buildings erected to house World's Fairs and for Olympic Games can soon become white elephants if they were not designed to serve a new instrumental function once the reason for their existence passes. Cultures evolve. This observation raises the procedural question of how one evaluates such matters when considering potential designs for a building.

5.5.2 Culturally Appropriate Environments

Many architectural practices work globally and obtain commissions because they successfully convince potential clients—business corporations, government agencies, and powerful individuals—of the efficacy of their brand of work (Olds 2001; McNeill 2008). When designing in countries other than their own and in multicultural and multi-class societies, architects are dealing with groups of people who lead very different lives to their own. Internationally fashionable building and urban designs paradigms may not afford sub-groups of the population to develop the behaviour settings that respond to their distinctive ways of life and aesthetic values. Architects need, advocates say, to be able to design for a diversity of people and consent authorities need to accept different brands of architecture (Greenbie 1976). It is not an easy instruction to fulfil.

Differences occur among how people behave, how they say they would like to behave if it were possible, and how they behave when the environment affords them the opportunity to do so. Patterns of built form that afford ways of life well may not have desired symbolic aesthetic characteristics. Architects have found this out when designing for low-income populations in many countries. The Hodne/Stageberg Partnership had to redo their East River, New York scheme that had been carefully worked out to meet potential residents' ways of life. The potential residents rejected the design on aesthetic grounds. They wanted something that looked high class. Hassan Fathy (1900–1989), the Egyptian architect and a pioneer in applying appropriate technologies to design, confronted this situation in the design of New Gourna, a village in Luxor near the Nile (Fathy 1969). His design fitted villagers' ways of life very well but they believed that the traditional aesthetic qualities stigmatized them. They moved in reluctantly but never felt that they owned the design. They allowed the village to deteriorate (Aburawa 2012).

Cultures evolve as circumstances change. Advocates for designing for what they perceive to be emerging ways of life want to create different building programs than those whose position is to give people what they say they want without serious analyzing or discussing the implications of doing so.

5.5.2.1 Gendered Environments

Architects and city planners have been predominantly male and their images of ideal worlds have guided their views of what constitutes good environments. As a result, buildings and, more specifically, urban design have been devised largely with men's values in mind. As gendered roles and the very nature of gender is challenged so too questions arise about what the built environment ought to afford.

Detailed discussions on the gender-space relationship in the home and outside arose in the 1970s accompanying the feminist movement (see Rothschild 1999). Major changes in the interior design of buildings have taken places in cultures and sub-cultures where gender roles have changed radically. This observation is particularly true in the households of the middle- and upper-income groups where age-old heavily gendered spaces have given way to more neutral ones (Desai et al. 2012; Taylor and Lang 2016). Critics argue more should be done.

Urban historian and architect Dolores Hayden (b. 1945) advocates radical changes in the ways of life of men and women that would have major urban design and architectural consequences (Hayden 1980, 1981). Changes have been slower in coming than many proponents for new, less gendered, ways of life desire. The design of the built environment follows societal changes; its power to lead them is limited. It can, however, be designed to make change easier.

5.5.3 Barrier-Free Environments

Bicyclists, children, and frail people have many barriers, social and physical, in navigating buildings and the city but *barrier-free* environmental design has come to have a specific meaning. Operationally defined, barrier-free places are those that can be easily accessed by paraplegics in wheelchairs. Several advocates for such environments were active in the 1970s and 1980s (for an example, Goldsmith 1976). They succeeded in convincing public authorities and architects alike to accept some of their prescriptions. In the United States, their recommendations led to the American Disabilities Act of 1990. Advocates continue to fight for the rights of people with physiological and/or mental limitations to make their way independently through buildings and around cities with dignity.

The position widely advocated is that all public environments, at least, should be accessible to people with special needs (Imrie and Hall 2001; Skinner 2008; Swain et al. 2014). That end is achieved for people in wheelchairs by curb cuts and ramps, the use of lifts, push-button opening doors, and specifically designed toilets. Groups, such as the blind, also have strong advocates for their interests. The blind use tactile clues such as surface textures, the nature of the sonic environment, and the way temperatures are layered to recognize where they are (Nicolau et al. 2009). Buildings can be designed with differing surface textures but many difficulties remain for the blind. People suffering from dementia have spokespeople for their needs to be considered in the design of not only special housing accommodation but also in urban design (Mitchell et al. 2004). They have had some minor successes (Weaver 2023).

Barrier-free buildings and urban areas are a benefit to many people. Parents wheeling baby carriages and labourers moving goods find ramps helpful (Florian 2023). Paying attention to such variables shapes the details of both interiors and exterior spaces. They hardly infringe on a designers' rights to express themselves. Designing educative environments is a more divisive issue.

5.5.4 Educative Environments

Much learning takes place while engaging in everyday life. Educative environments are those that are rich in behaviour settings that people can choose to engage in actively or vicariously by observing the actions of others. Advocates for such environments worry not only about the lack of diversity of many recently designed urban and suburban precincts but also about the increased segregation of settings by age groups in modern life (Rogoff et al. 2010). There are fewer places today where children can learn from observing their elders and vice versa than in the past. In many societies parental fears lead to heavy restrictions on the independent mobility of children.

Advocates for educative environments prescribe building and urban precinct patterns that afford the crossing of different people's paths. Housing options, they argue, should be provided to attract diverse people to live in a locale. Multi-functional open spaces should be designed to accommodate both communal and formal events. Districts should accommodate diverse behaviour settings.

Many people regard such precincts, particularly residential neighbourhoods, as too busy, too messy, and too rowdy. In addition, individuals give up a degree of privacy in such places because they are frequently under the casual view of others. Few architects and landscape architects want to design environments rich in behaviours settings and few municipal authorities seek to have them. Design professionals, historically, have had a strong anti-urban bias and fight against designing lively places (see White and White 1964; Conn 2016). Many recent urban designs are, indeed, dull places although they make good compositions for photographs.

Educative environments are also ones where access to the natural world is easy to reach on foot. Nature can be artificially arranged as in parks and along streets or it can live as wild locales. In such places seasonal changes and a variety of bird, animal, and insect life can be observed. They also provide opportunities for expressive and self-testing activities such as those artificially provided in adventure playgrounds (Lambert and Pearson 1974).

5.6 Colour Advocacies

All the materials used in a building have a natural hue. Paint is used to create artificially coloured surfaces. Colours have been used to achieve different ends by different architects and architectural movements. The white marble columns and pediments of classical Greek temples that set a precedent for many subsequent Neoclassical designs were originally highly coloured. In some societies and places (for example, Jodhpur in Rajasthan) houses were colour-coded according to the social status of their residents. Over the course of history, architects too have prescribed colours for buildings based on their own predilections. They have little heeded what research has shown on the psychological impacts of colours as presented in Chap. 3. Anne Tyng's hypothesis may well hold for the use of colours in architecture as for the historical shifts in paradigms prescribing complex or simple forms.

Many architects have their own colour preferences. Frank Lloyd Wright, for example, preferred browns, yellows, greens, oranges, golds, and shades of warm reds (Barnett 2023). The Guggenheim Museum in New York was yellow brown—beige—before being painted white. The buildings of Frank Gehry are in shades of blues, greys, and reds. Many Modernist architects were predisposed to have their buildings white both externally and internally. White was seen as straightforward, clear, and honest. It presents a neutral background for the display of colourful objects.

The Weissenhofsiedlung—Settlement of White Houses—in Stuttgart (1918–1931), was a demonstration of what the Modernists considered to be good housing for workers. It was designed by seventeen European Modernists under the direction of Ludwig Mies van der Rohe. The buildings were predominantly white except for the two designed by Bruno Taut (1880–1938), a German architect. Taut noted "We must recognize colour as having the same rights as form" (cited in Kirsch 1990). His prescriptions for the use of colours were based on his view of the changing way the sun illuminates a building as it moves from east to west. He explained his position as follows:

> … to the clear light of day red becomes clearer, almost pink, and blue accompanies it with a darker and deeper shade; then the twilight comes and blue with the increasing darkness becomes more and more clear, almost white, while red becomes darker and darker, almost black. (cited in Ardizzola 2017)

Perhaps in reaction to the whiteness of much early Modernist architecture, Art Deco sought something different. Architects used bright, bold, daring colours—yellows, emeralds, peacock blues, reds, and purples against a white background. Chromes and blacks were further used to highlight details (Hillier 1968; Wolf 2013). Le Corbusier did not like that.

Le Corbusier presented his prescription for colour usage in *Polychromie Architecturale* (1931). He specified the appropriate colours for constructive, dynamic, and transitional spaces. For constructive contexts he advocated the use of ochres and browns. He said that they create a sense of harmony. For dynamic situations, he argued for the use of bold, vibrant, reds, blues, and yellows (the colours promoted by De Stijl artists). To Le Corbusier, they are the colours to use in highlighting the focal points in one's view. For transitional spaces, colours should be used in glazes to illuminate the features of surfaces. In the High Court in Chandīgarh (1954) he used bright colours to create a "vibrant and functional atmosphere" (Souza 2023). Colours thus not only shape the ambience of place but also give architects an identity although more subtly than the geometries exhibited in the buildings they design.

5.7 Designing for the Public Interest: Design Controls and Architecture

The characteristics of buildings and urban areas everywhere are partially controlled by zoning ordinances and building codes (Ben-Joseph 2005; Marshall 2011; Talen 2014). Their rules are enforced to serve the public interest by ensuring buildings are structurally sound, have access to sunlight, and do not infringe on the rights of other people. Some controls are contentious but others are widely accepted. In economically advanced societies, it is generally agreed that designing barrier-free environments is the morally correct thing to do. Attitudes towards aesthetic controls differ. Such controls can specify the materials that should be used and certain fenestration and surface patterns be followed to get a unity (or disunity) in the architecture of a locale. The willingness to accept such rules varies from society to society depending on the moral codes and attitudes of a culture. Pragmatic architects abide by them but many architects believe that they inhibit creative thought. They are seen to interfere with the rights of clients and architects to achieve their own design ends.

Thomas Mayne (b. 1944) and Daniel Libeskind (b. 1946), leading American architects, argue that a freedom of action is required in order to attain a radical heterogeneity and pluralism in the built environment (Lubell 2004; Green 2016). This view has been echoed by Patrik Schumacher (b. 1961), director of Zaha Hadid Architects. He has argued against all controls (Schumacher 2013; Frearson 2016). One reason that architects like to work in countries where the design controls are lax is that it gives them freedom to create unique, geometrically radical, eye-catching compositions.

The counter argument is that without legally binding controls, the risk of buildings catching on fire or being destroyed in earthquakes or creating a haphazard, chaotic world is high as architects and their clients strive to establish unique identities without caring for the rights of others. More generally, if each property developer works independently, the consideration of what are deemed to be public interest issues such as achieving a sustainable world is avoided.

The limits of public interest concerns are difficult to define. The problem with codes is that they are often poorly conceived and the logic behind them poorly explained. Public authorities are also slow in keeping pace with technological developments and societies' needs. Dealing with non-human species is one such area.

5.7.1 The Interests of Other Species

The history of architecture is a history of building, unselfconsciously or self-consciously, against the interests of species other than humans. Settlements have invaded the territory of and displaced many other genera, animate and inanimate. It is estimated that a billion birds die each year in the United States by flying into buildings (Singh 2023). Several species of animals around the world have, in return, invaded and settled in the human habitat. In search of food, monkeys invade many Indian cities, foxes live across London, bears climb fences in Alaskan cities, deer trample gardens in parts of North America and Europe, sloths can be seen in some Venezuelan urban parks, and baboons enter houses on the edge of Cape Town.

Songbirds, butterflies, and squirrels enrich the lives of many people in cities. Rats generally do not although their presence is revered in some temples and parts of some Indian cities. Mosquitoes are a nuisance. How should the human habitat function in relation to the lives of other species? Advocates for enhancing the quality of the environment for wildlife in cities argue for increasing the biodiversity of urban areas (Gallo and Fidino 2018). Increasing the biodiversity of cities not only serves the interests of wildlife but also contributes to making an educative environment. Many architects believe that dealing with such concerns is yet another imposition on their creativity but others see it as a challenge that can be overcome by creative thought. A few architects are showing the way in designing to minimize the death of migrating birds flying into buildings by the glazing patterns they use (Singh 2023).

5.8 Means

The same ends can be achieved using very different patterns of built form. The patterns can be created using different structural and construction methods and diverse materials. The range of types was hinted at in Chap. 4: *Types and Typologies.* Figure 5.3 shows a sample of different means for attaining specific ends. The development of prescriptive architectural theories has been much affected by the evolution of the means available to meet the *firmness* end of architectural design. That end is closely tied into the desire to *delight*. The structural dexterity shown in achieving ends can become an intellectual aesthetic goal. For some architects it is.

5.8.1 Designing for *Firmness*

The fundamental goal of designing structurally sound buildings and infrastructure systems is to provide habitats that protect their inhabitants from the vagaries of climates and the forces of nature. Many ways exist for designing structurally sound well-constructed built environments and the objects within them. Some structures are built pragmatically. Empiricists follow local customs and others rationally explore emerging technologies.

The means to achieve specific design ends is almost always restricted by the human and financial resources and tools available. These means vary considerably between places where designing and building are inextricably linked in an unselfconscious process and places where it is a highly self-conscious process carried out by professional architects and engineers but built by others.

5.8.1.1 Unselfconscious and Self-conscious Design

In societies and cultures where designing is an unselfconscious process, both the ends and means of design are constrained by the local materials and tools available to erect and shape structures. The diversity of the climates and the materials available at a locale has resulted in the diversity of vernacular architectures across the world and the similarities in the appearance of specific building types in any given

Photograph by the author

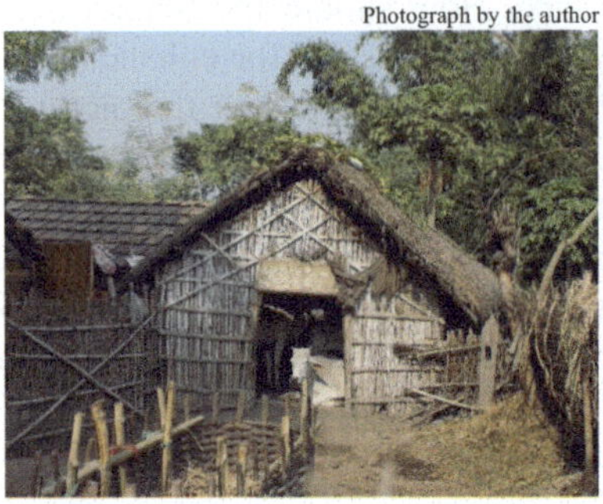

Photograph by the author

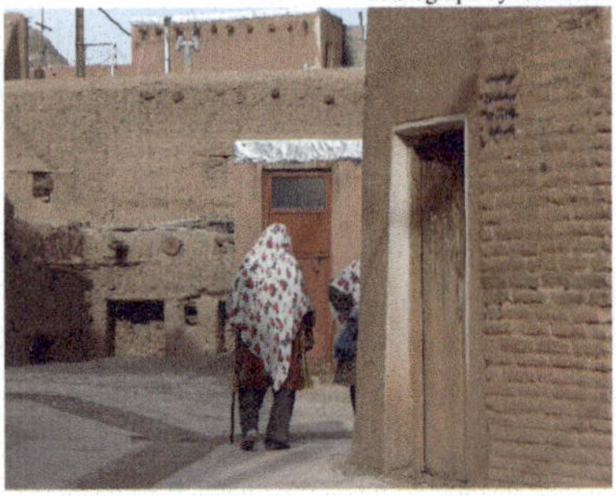

Vernacular architecture: Building in a rural, hot-humid, deltaic climate (left) and in a hot-arid setting using the materials at hand (right)

Watercolour by J. McNeven/Wikimedia Commons

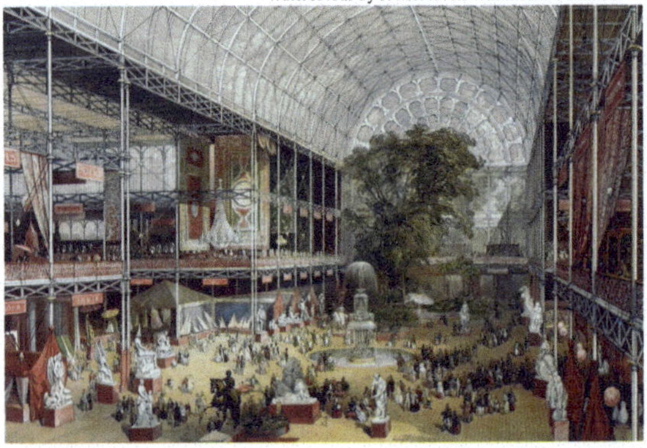

Source: Lang and Moleski (210, 126)
Courtesy of Taylor and Francis

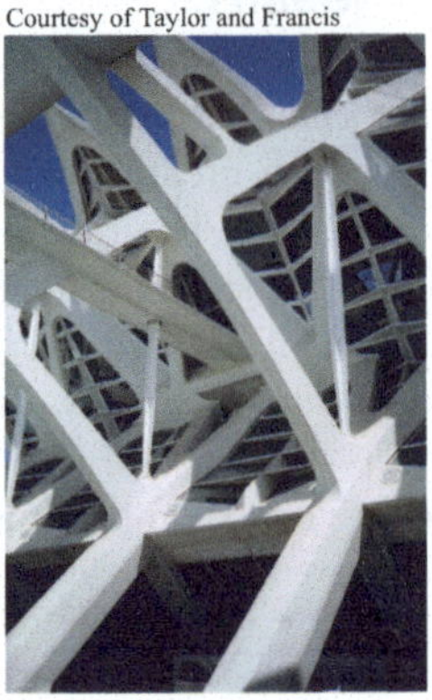

The architecture of structural dexterity: Crystal Palace, London (1851) (left) and Museu de les Ciéncies Principe Felipe, Valencia (2000) (right)

Photograph by the author

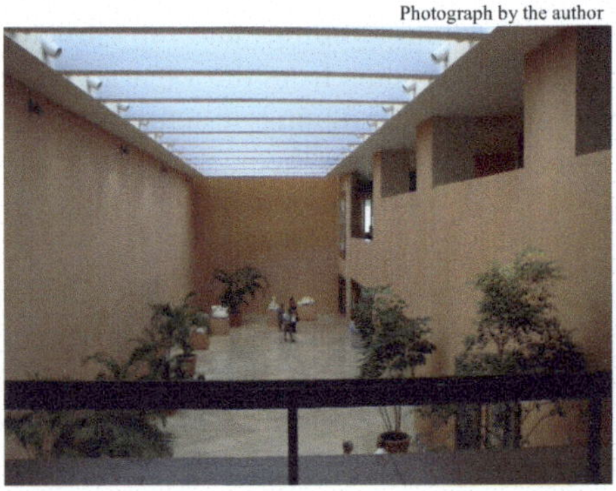

Photograph by Alix Verge: Courtesy of Alix Verge

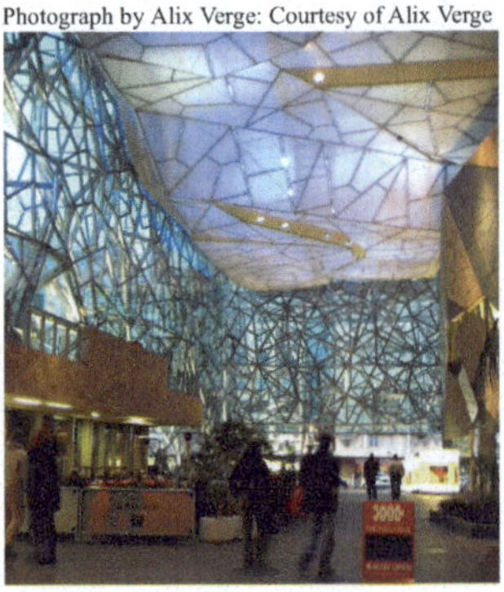

Architectural paradigms: Neo-Modernist – El Museo del Arte Thyssen-Bornemisza, Madrid (1992) and Deconstructionist – North Atrium, Federation Square, Melbourne (2000-2002)

Fig. 5.3 Means

location (Oliver 1997, 2003). It was a highly pragmatic architecture. In modern, technologically advanced societies where design is a self-conscious process, the array of ways of building is vast.

The history of architecture is a history of the development of the means of building. The development of the arch made more than a trabeated architecture possible. Arcuate structural forms are the basis of Romanesque and Gothic architecture. The development of concrete made large domes easier to build than when brick and stone were the available building materials. The technological inventions of the nineteenth and twentieth the century saw radically new building types emerge. The safety lift made the skyscraper possible. Wrought iron and the steel made it easy to span large spaces. The choice of structural system to use in the design of buildings is shaped not only by the pragmatic end being met but can become an end, a display, it itself.

5.8.2 The Architecture of Structural Dexterity

The desire to self-consciously *delight* observers through the exhibition of innovative structural mechanisms can be an aesthetic end as much as a means for achieving a safe and secure environment. The dexterity whereby ends are met becomes a means of meeting an experiential goal and, for the cognoscenti, an intellectual aesthetic end. The whole of architectural history has had individuals striving to get the structural system they have imagined implemented. Today, there are many different advocacies.

Light weight tent-like structures have been proposed to cover whole cities to enable their inhabitants to control the climates. Buckminster Fuller, with others, proposed a dome to cover Harlem in New York. The Archigram Group in England and the Metabolists in Japan proposed clip-in, plug in structural systems for cities and buildings to deal with the need for easily changeable built forms that cater to changing instrumental ends (Dahinden 1972; Sadler 2005; Pernice 2022). New geometrical forms made possible by advances in the use of computer-based algorithms have made hypermodernist architecture possible.

The structural pragmatists seek easy-to-implement architectures, the empiricists look at what structures work well, the rationalists seek to invent awe-inspiring new forms that they predict will work well in meeting the ends sought by architects and their clients. They are willing to take chances. In buildings designed by architects such as Pier Luigi Nervi (1891–1979) and Santiago Calatrava (b. 1951) the structural systems employed were developed as much for their sculptural effect as for their structural integrity. The means become ends in themselves.

5.9 Commentary

The ends of architecture and the means to achieve them are open to debate. Bernard Tschumi notes "Architects have been unable to reconcile their need to address everyday life with a wish to engage in abstract ideas" (Tschumi 1998). Histories show the efforts of architects to reconcile their social obligations and their role as creative artists. They have sought to meet the ends prescribed by their clients, their own aspirations, and perceptions of what is in the public interest.

Ways of life change and what is required of any building changes over time. Critics say that designing for the ends that specific populations seek is not a wise choice. Designing for ambiguity and flexibility is. By ambiguity is meant that the patterns of the built environment allow a variety of behaviour settings to form within them; they are not tailormade for a specific use. Flexible environments are ones that are able to be changed easily as the functions they were designed to serve change.

Major advances in information technology and artificial intelligence have changed and are changing patterns of life and ways of working. Architects' prescriptions for the qualities that the built environment should possess are being adapted to those changes. The history of architecture demonstrates their efforts to harmonize conflicting objectives and the means to fulfil them.

References

Filippi C, Hooper H (2019) The architect/the advocate: the role of advocacy and influence in architecture and design. Architecture Now. Accessed 10 Sept 2022

Fisher T (2008) Public-interest architecture: a needed and inevitable change. In: Wakeford K, Bell B (eds) Expanding architecture: design as activism. Metropolis Books, New York. Berkeley prize essay competition. Accessed 22 June 2023

Jencks C, Kropf K (eds) (2006) Theories and manifestoes of contemporary architecture, 2nd edn. Academic Editions, London

Lang J (2021) The Routledge companion to twentieth and early twenty-first century urban design: a history of shifting manifestoes, paradigms, generic solutions, and specific designs. Routledge, Abingdon and New York

Taschen (ed) (2022) Architectural theory: pioneering texts on architecture from the renaissance to today. Taschen GmbH, Cologne

Architectural Prescriptions: A History of the Development of Ends and Means

6

'The past is a foreign country. They do things differently there' (Hartley 1953). Understanding the shifting intentions of clients and their architects must be seen in their temporal context to comprehend our efforts today. The history of prescriptions for what constitutes good architecture has been a story of evolving social norms, design types, and technologies (Younés and Westfall 2022). It can be told in several ways. Manfredo Tafuri (1935–1994), an Italian architect and historian, has influenced the telling here. His view was that architecture serves the interests of capitalist clients (Tafuri 1968, 1973). The story also builds on Anne Tyng's belief that tastes for building forms become richer and richer until they reach a saturation point. A simple form is then favoured (Tyng 1975).

All histories have their biases. The story presented here follows conventions in the English-speaking world. It describes the evolution of building forms as products of contemporary attitudes. The story begins with the classical worlds of Greece and Rome. Their ideals underpin many current architectural prescriptions. The history then moves from the Romanesque to Gothic to the Renaissance's rediscovery of the ancient world and to the birth of Modernist architectural and urban design ideas. A postscript on architecture today concludes this historical overview of designers and their prescriptions. Throughout history, architects and their professional ancestors, have been expected to create better designs than those produced unselfconsciously by designer-builders.

6.1 Unselfconscious Design: Vernacular Architectures

The term *vernacular* was originally applied to the language of the masses in societies where elites spoke the accepted high-status language. In the mid-nineteenth century Sir George Gilbert Scott (1811–1878), the English Neogothic architect,

J. Lang, *The Scope of Architectural Theory*, Design Science and Innovation,
https://doi.org/10.1007/978-981-96-3690-7_6

used it to describe local architectures (Scott 1857). Today, we use the term to describe the buildings erected according to traditions by individuals with no formal architectural training. Historically the buildings they designed and built had to respond to local climates, ecosystems, and cultures using limited resources.

The importance for architects to understand vernacular building traditions was propounded by Bernard Rudofsky (1905–1988), an American architect and social historian, in his book *Architecture without Architects* (1964). It was furthered by the exhibitions he organized at New York's Museum of Modern Art. Architects, such as Frank Lloyd Wright, had long argued that an understanding of vernacular buildings should provide the theoretical introduction to architecture. He observed:

> Folk building growing in response to actual needs, fitted into the environment by people who knew no better than to fit them with native feeling. For us it is better to study it than all the self-conscious academic attempts at the beautiful in Europe. (cited in Oliver 2003, 19)

Later, Paul Oliver (1927–2017), an architectural historian, brought attention to the variety of the world's vernacular building types (Oliver 1997, 2003).

Some architects were already learning from vernacular designs and methods of construction. Hassan Fathy learnt from local modes of construction when designing rural villages in Egypt (Fathy 1969). Laurie Baker studied the indigenous use of brickwork for his designs in Kerala (Bhatia 2003). Today, architects are exploring vernacular architectures to understand how best to create eco-friendly buildings that cater to local climates without using artificial heating and/or cooling equipment (Borràs 2024). Rammed earth structures, for instance, have become the focus of much attention. It has been as much for their plasticity of form as for their ecological credentials (Krahn 2019) (Fig. 6.1).

6.1.1 Clients and Designers

The notion that vernacular buildings are hand-built by their owners is misleading. While some buildings undoubtedly are, many are erected by craftspeople qualified in local ways of building and the use of the materials close at hand. Even in pre-literate societies there were well-known individual designers. Certainly, the builders of American log cabins used an extensive array of tools but they followed an unselfconscious design process (Glassey 1968). The clients and craftspeople of vernacular buildings share cultural values and expectations of what a building should be like whether it is an individual home, a place of worship, or a public spaces. They also understand the correlation between a building's characteristics and social status.

The craftspeople learn about how to build from their predecessors. They serve an apprenticeship under the guidance of experienced workers. In doing so, they internalize prescriptions for good designs and ways of designing that guide the decisions they make. The rules may not be written down but they are often, nevertheless, rigid.

Photograph by the author

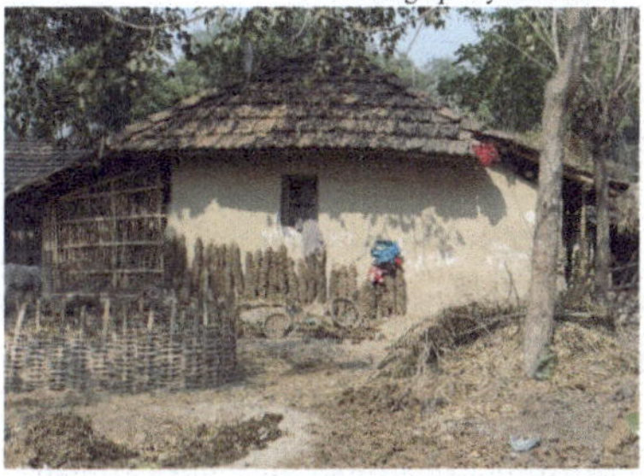

Photograph by the author

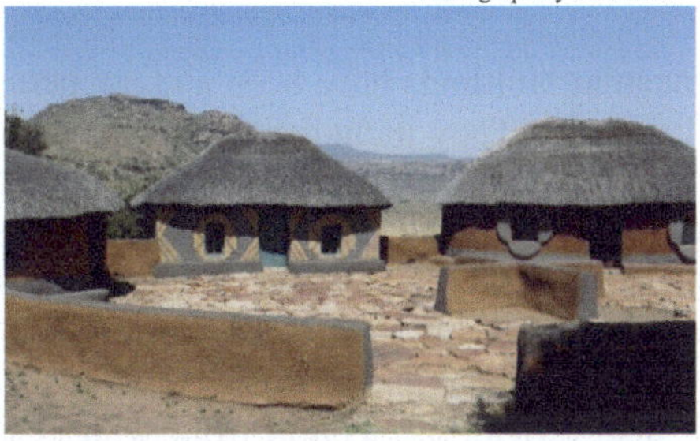

A Bengali bungalow, Malda, West Bengal, India (left) and a Lesotho village, Orange Free State, South Africa (right)

Photograph by the author

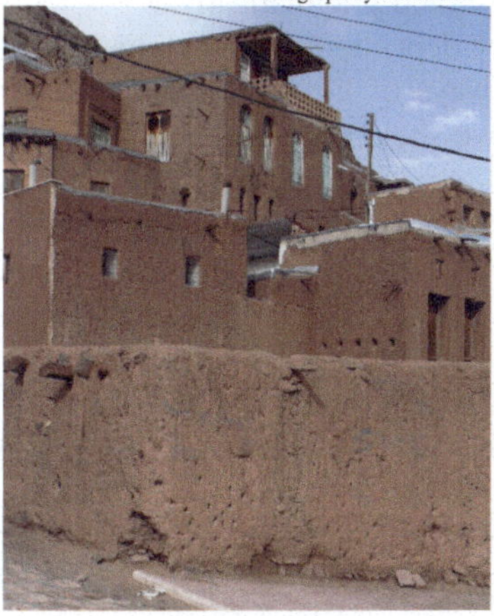

Photograph by the author

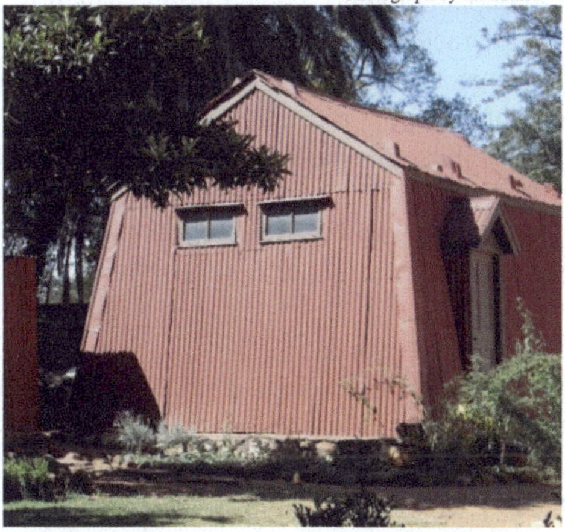

Mud and mud-brick construction, rural, Iran (left) and Corrugated iron construction, South Africa (right)

Photograph by the author

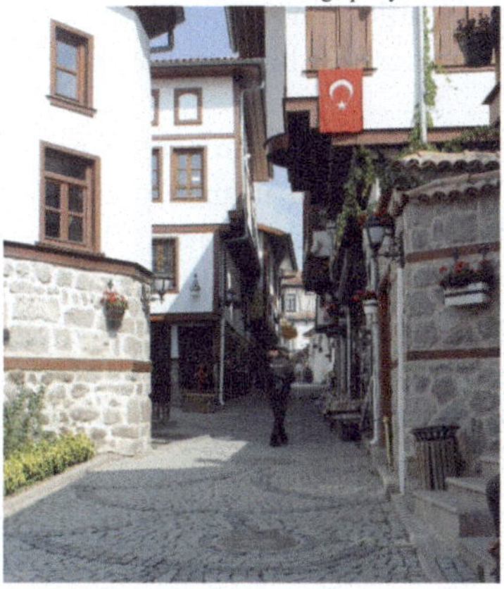

Photograph by the author

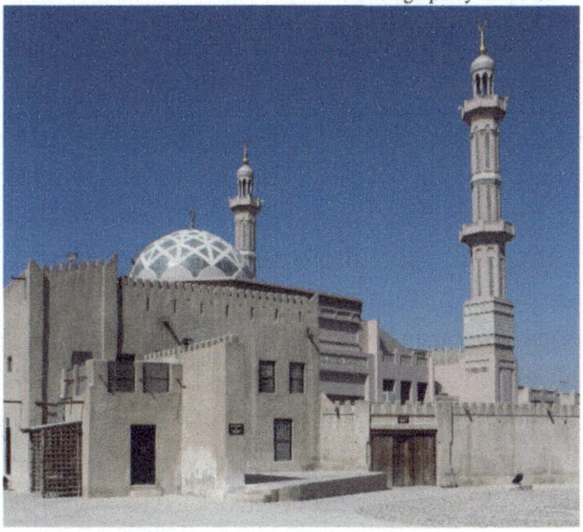

Traditional houses, Ankara, Turkey

A traditional Persian village form, Dubai

Fig. 6.1 Unselfconscious design: vernacular architecture

6.1.2 The Work: Types and Functions

Vernacular buildings have been and are being built to serve a variety of instrumental functions. They include religious, commercial, and residential purposes. Some are crudely designed but others are highly sophisticated and precisely crafted. They vary in size. The mudbrick mosques in Mali are substantial structures. In Shibam Hadramawt, Yemen, residential building types are often multi-storeyed. Their fundamental function, as for any building, is to provide shelter from the elements for their inhabitants and, in places, farm animals and pets. Sometimes they also provide protection from wild beasts, human enemies, and supernatural spirits. Resources are usually limited, so buildings function well enough to allow their occupants to survive and thrive. Much is designed pragmatically.

The characteristics of vernacular buildings vary depending on the climatic conditions, materials available at a locale, and the ways of life and wealth of their inhabitants (Rapoport 1969). Nomadic people have shelters of different structural types than those people whose lives are restricted to a single locale. Their shelters need to be easy to move and erect. Often, they are lightweight, tent-like structures that have been the precedent for the self-conscious explorations of what light weight structures might afford in technologically advanced societies today (Hamm 2015).

Buildings in warm climates are constructed of light materials and allow for cooling air to flow through them as in the deltaic villages of Bengal. In hot-arid areas of the world houses of the wealthy have courtyards with water features. Some buildings in such zones have wind catchers that capture breezes to move air over water to cool their interiors. In cold climates buildings have heavily insulated walls and small windows to retain heat. Igloos are windowless. Houses are built on stilts in flood-prone locations. Roofs profiles depend on the nature and quantity of precipitation. Flat roofs only occur where there is little rain. In hot-arid locales such roofs often provide living space for families after the heat of the day has dissipated.

Buildings are constructed of locally available materials. Mud, in some form or other, remains the most widely used material. It is even used in monsoon zones where houses must be repaired annually when the rainy season is over. The mud walls dividing building compounds in those regions are roofed. Mud, along with cow dung, is often used as a plaster veneer for walls. When corrugated iron sheets became available in the nineteenth century, they often replaced local construction materials. They are waterproof and easy to erect but they have the negative effect of absorbing heat and cold. They are employed, nevertheless, as a symbol of modernity and high status. Local natural materials become stigmatized. Designing to maximize shelter and comfort gives way to building patterns that meet culturally defined expectations (Rapoport 1969).

The similarity of house forms in a locale gives a clear identity to a community. The identity is reinforced by the way interiors cater to family structures and the roles of men and women in a society. Privacy expectations for specific activities are reflected in the plans of houses although in the single room houses of the poor, the inhabitants need to use behavioural mechanisms to attain a modicum of privacy. The zones of penetration of outsiders into a house vary from culture to culture but depend on the kinship status and social standing of visitors. The size of a house, its material quality, and its external decoration and, in some places, as already noted, its colour are hallmarks of status.

Vernacular settlement morphologies are very much shaped by the nature of the topographies where they are located. Those along hill slopes and shorelines tend to be longitudinal in form; those on plains are more shaped by social necessities unless ridges and strong interfluves make linear forms easier to build. In hot-arid areas, buildings are clustered; in humid zones they are spread out and the form of settlements is less well-defined.

Vernacular design types evolve slowly. Buildings are designed by habit following customary structural and constructional techniques. The self-expression of their creators is generally restricted to the outside decoration of buildings as in the painted villages in Burkina Faso and the Ndebele houses in southern Africa. Architect-designed buildings are a self-conscious effort to improve on what designers of the vernacular produce.

6.2 Self-conscious Design

In the western world, the self-conscious discourse on what constitutes a good building can be traced back to at least the second century BCE with a poem written by Antipater of Sidon, a critical architectural tourist (Borngãsser and Bednorz 2008). He identified seven marvels of the world that, to him, represented superior design. The grand pyramid of Giza is the only one that still stands. Since then, if not before, in India and other Asian countries, the discussion of and prescriptions for good design also thrived. It must also have been so in meso-America, although no records exist.

The history of indigenous design in the former colonized countries of Africa, Asia, and the Americas is built on empires and civilizations of pre-colonial times. These empires produced rich architectures whose functions and the means of accommodating them are only beginning to be studied. The colonial powers, in contrast, self-consciously brought ideas of 'home' with them and strove to supplant local design traditions. Only now, in the post-colonial era, are architects in former colonies who were trained in European and American scholarly traditions are beginning to study with pride their pre-colonial histories and their implications for designing for the future.

6.2.1 Classical Architecture

Five of the architectural marvels identified by Antipater of Sidon were Greek. The most influential self-consciously designed structures of the era between 850 BCE and 300 CE were temples and the theatres built according to clear prescriptions.

Rome introduced some new building types and forms. Equally important were public urban spaces such as the agora in Greek cities and the forum in Roman. While the Temple of Artemis at Ephesus was on Antipater of Sidon's list, it is the Parthenon and Erechtheion on the Acropolis in Athens that have set precedents for later architects in western countries.

6.2.1.1 The Architects

The architects of ancient Greece and Rome received both formal and informal educations (see Vitruvius 1914, Chap. 1). They developed a competency in geometry, surveying, and drawing. In ancient Greece texts on design and construction were required reading although their nature is unknown. None survive. We know that architects were educated to deal with clients and workers. They learnt much, including masonry skills, under the tutelage of experienced architects.

The names of some ancient architects are known but not necessarily their backgrounds or what each designed. Theodore of Samos (d. 540 BCE) was one of the earliest of Greek architects. He designed the Heraion, a Doric order temple on Samos. We acknowledge Hippodamus (active in the fifth century BCE) as the ancestor of rationalist urban design for his grid-iron plan for Miletus. In the fourth century BCE, Polykleitos the younger, designed the Theatre and Tholos of Epidaurus. Paeonius of Ephesus (*ca* 480–*ca* 400 BCE) designed the Temple of Artemus there. A Greek architect, Hermodorus of Salamis, designed the temples of Jupiter and of Mars in Rome where he was active between 146 and 102 BCE.

Vitruvius is the best-known Roman architect because of his treatise, *De architectura, Libri decem.* Dedicated to his patron Caeser Augustus (63 BCE–14 CE), it prescribed how buildings should be designed. He drew on the work and writings of Arcesius, a critic of Doric architecture and the designer of the Ionic sanctuary of Aesculapius, as well as the work of Hermodorus of Salamis. Vitruvius designed and constructed a basilica in Fano and completed works for Julius Caesar (100–44 BCE) and Caesar Augustus but we know little else about him. The architect of the Pantheon (126 CE) was, probably, Apollodorus of Damascus, the favourite architect of Emperor Trajan (53–117). Apollodorus was a prolific engineer and author of *Siege Engines* as well as texts on building techniques. He designed and oversaw the construction of the Forum in Rome, the Temple of Trajan, Trajan's Column, and the Stadium of Domitian. That structure was commissioned around 80 CE by Emperor Titus Flavius Domitianus (51–96). Elsewhere in the Empire Apollodorus designed bridges across the Danube and the Tagus in Spain. He also designed the triumphal arches of Trajan at Benevento and Ancona (Chisholm 1911).

6.2.1.2 The Clients

The clients in ancient Greece were the rich and powerful families and politicians of the Athenian state. Pericles (485–429 BCE), for example, was a politician and a patron of the arts. He used the surplus of the war treasury of the Delian League and other public funds to enhance the prestige of Athens, and his own, by creating public works. He, as a member of the educated elite, was familiar with the classical buildings of the Hellenic era (900–323 BCE). He initiated the buildings on the

Acropolis and it was he who hired Ictinus, a fifth century BCE architect, along with Callicrates (*ca* 470–*ca* 428 BCE) to design the Parthenon. Callicrates also designed the Temple of Nike on the Acropolis. We know little of the specific clients for other projects.

The major clients in Rome were the emperors and the Roman Senate. They sponsored most of the memorable buildings. The end was to demonstrate the power of the state. They sought an architecture of grandeur that also served utilitarian functions. Marcus Vipsanius Agrippa (b. *ca* 63 BCE), a military commander, and a friend of Emperor Augustus (63 BCE–14 CE) was the supervising client for the Pantheon (*ca* 25 BCE). In 126, Hadrian (76–138) was responsible for rebuilding it. The emperors of the Flavian Dynasty were the clients for the Colosseum (*ca* 75 CE) with Emperor Vespasian (9–79) initiating its construction. The designers are unknown but Jewish slaves built it directed by Roman engineers and craftsmen (Evans 2023). Leptis Magna in North Africa, a significant urban development, was the birthplace of Emperor Septimus Severus (145–211). He was a patron and a major investor in the development of the city. The mass housing that dominated cities across the empire was sponsored by the wealthy elite.

6.2.1.3 The Work

We know little about Greek domestic architecture and how its forms related to gender roles, family structures and social standing. The major building types that are still influential were temples, amphitheatres, and agoras. Although they evolved in character, each was an imitation of one of a set of standard types (Younés and Westfall 2022). The temple was a shrine dedicated to a deity. It stood on a base reached by steps. It consisted of a *cella* that provided shelter and privacy for the statue of a deity and an altar for sacrifices or the burning of incense. The cella was not a space for communal worship. The public's focus of attention was on the temple's external appearance. Hence the importance paid to it.

Columns and entablature surrounded the cella. By the fourth century BCE, the orders—Doric, Ionic and Corinthian—for the columns and capitals had been developed (Fletcher 1954; Conway and Roenisch 2005; Borngässer and Bednorz 2008; Emerson 2018). Pediments at the front and back of the temple formed part of the type; some contained sculptures. The number of columns and the nature of sculptures adorning the pediments functioned as a symbol of the temple and its deity's importance.

The temples were windowless. Interior illumination came from clerestories between roofs and the upper portion of walls, from skylights of thin translucent marble, and from the east facing doors. To correct optical illusions architects introduced a series of refinements to the geometries of temples when seen from a distance six times their columns' height. Horizontal lines were built slightly convex to correct for an otherwise sagging appearance; vertical features were inclined inwards to avoid the appearance of falling outwards (Blankenbehler 2015). The buildings were polychromed with details highlighted in reds, blues, and greens. Only traces of the colouring remain so the temples and sculptures as seen today are monochromatic.

The lesson learnt is that an almost infinite variety of compositions can be based on a building type consisting of a few elements. Another lesson arose from the grouping of buildings to make a composition as in the agoras of towns and on the Acropolis in Athens. They are not only places for activities and gatherings but provide a sequence of views for a person moving through them. The vista changes as an observer moves from behind one set of occluding building surfaces to another (Martienssen 1956; reissued 2013). The Greeks took these ideas to their colonies spreading the design lessons they had learnt. Many aspects of Grecian architecture were thus carried forward in Rome.

Although it was fundamentally Greek in origin, Etruscan antecedents influenced Roman architecture during the republican era (509–27 BCE). Building forms also differed from the Greek because the materials used, particularly concrete, allowed structural innovations. Concrete made possible the Pantheon and the Colosseum in Rome and the construction of the Pont du Gard (first century CE) in Nimes (Bromwich 2006). Domes, rounded arches, and barrel vaults were major Roman structural innovations. Their use afforded the creation of internal spaces that could shelter large gatherings. The spaces made communal worship possible. The Romans also developed the Tuscan and the Composite orders for columns. The former was based on the Doric but had unfluted columns and a simple entablature; the latter mixed the volutes of the Ionic with the acanthus leaves of the Corinthian (Fletcher 1954; Borngässer and Bednorz 2008; Rodgers 2013).

Rome was highly populated and wealthy. Much of the housing consisted, however, of crowded tenements—*insulae*—pragmatically designed at a high-density to function as shelter with a minimal concern for their residents' health. Wealthy families invested in them. They themselves lived in large houses—*domuses*—with atriums that were the centre of family life. They often also owned a suburban *villa* outside the polluted city.

The Romans not only developed new building types accommodating new instrumental functions, but also rationally laid out urban designs. They established settlements, military and civic, throughout the empire. Their towns were based on a generic plan and were replete with public spaces, palaces, memorial arches, temples, amphitheatres, baths, and, when located on the coast, ports. Among the best preserved of Roman cities is Leptis Magna, a second century development within the Carthaginian Empire. It lies in modern day Libya. The more easily accessible Italian examples have, however, been the most influential in shaping the thinking of later Italian architects. They were essential destinations for foreign architects and their patrons on their grand tours of European cities (Cheney 1998). They sketched the ruins and wrote about them in laudatory terms (Fig. 6.2).

6.2.2 Medieval Architecture

In Europe, the medieval period lasted for a millennium from the fall of the Roman Empire until the emergence of the Renaissance. The influential architectures during the period were the Byzantine (Rodley 1994), the Romanesque (Toman 2002; Prina

Source: B. Jankuloski/ Wikimedia Commons

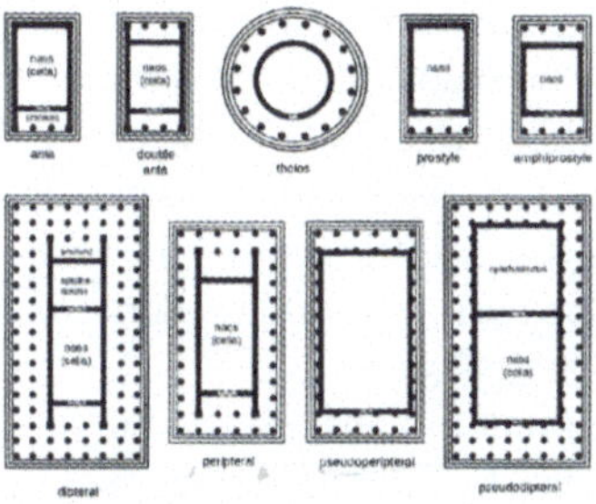

Temple plan types

Attributed to various artists/Wikimedia Commons

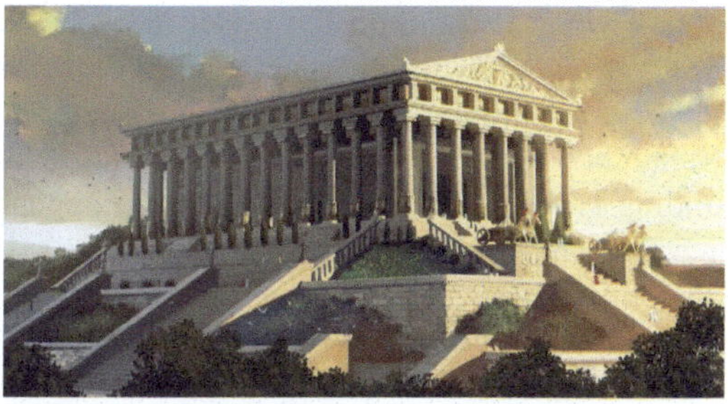

The Temple of Artemis, Ephesus, a reconstructed view

Adapted from an 1846 painting by Leo von Klense/ Wikimedia Commons

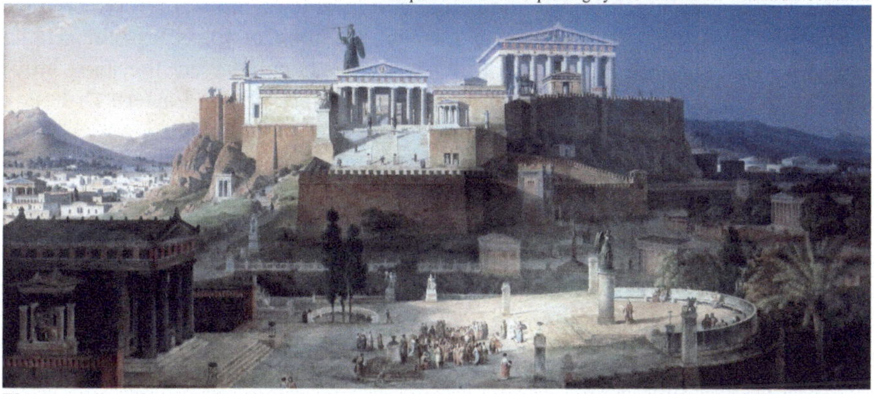

The Acropolis, Athens: An idealized reconstruction by Leo von Klense (1784-1864), a German architect

Adapted from an 1882 etching by I. Buhlmann/Wikimedia Commons

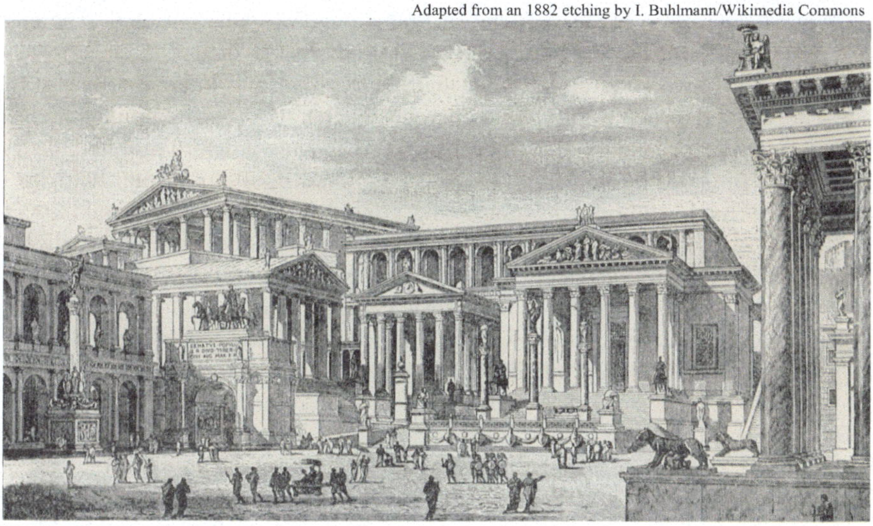

The Forum, Rome: A reconstructed view looking towards the Capitol in the time of Domitian (51-96)

Fig. 6.2 Classical architecture

2012), and the Gothic (Frankel 2000; Prina 2011). The Byzantine era began around 330 CE when Byzantium fell under the control of Emperor Constantine (*ca* 272–337) and ended in the fifteenth century when the Ottomans captured the city, by then Constantinople. The Romanesque is a term applied principally to the character of churches of the tenth to twelfth centuries but also to many other instrumental building types. Gothic designs began to appear in the twelfth century in France. They overlapped the Renaissance and continued to be built well into modern times. Neogothic buildings are still being erected.

Byzantine buildings of consequence, unlike the predominant use marble in the Greek and Roman, were largely built of stone and brick. Like them they had thin alabaster sheets for windows. They were embellished with detailing visually richer than in Classical architecture. Buildings featured marble columns, coffered ceilings and interiors decorated with colourful mosaics of stone or glass tesserae. Hagia Irene (532) and Hagia Sophia (537) in what is now Istanbul were the most influential examples of Byzantine religious architecture especially in Eastern Orthodox countries.

The Romanesque evolved out of the barrel-vaulted Roman basilicas. Its building interiors have heavy columns with carved capitals. Buildings have rounded arches, weighty stone masonry walls, and little decoration. These characteristics give interiors a stark appearance. Perhaps, as Anne Tyng might have suggested, it was a reaction to the visual richness of Byzantine architecture (Toman 2002; Prina 2012). The Gothic was very different being lighter and visually richer. It was more influential in the long run although the simplicity of the Romanesque has parallels in modern architecture.

The Gothic resulted from a series of structural experiments under the patronage of Abbot Suger (1081–1151), a friend of both Louis VI and Louis VII. The end was to increase both the seating capacity and quality of light in churches. It was met by an innovative structural system that made broad, tall internal spaces possible. The skeletal structure of Gothic churches is characterized by pointed arches that had Islamic precedents and ribbed vaults. The outward thrust of the skeleton structure imposed by the vaulted roof is countered by flying or pier buttresses that carry the load safely downwards to the ground (Fletcher 1954; Frankl 2000; Prina 2011). Although constructional experiments had started a century earlier it was Abbot Suger who systematically applied the new patterns when rebuilding the church of Saint-Denis in the early twelfth century. This structural system frees up the enclosing perimeter to allow large, filigreed windows to be inserted in the walls. Stained glass windows filled the interiors with coloured light.

6.2.2.1 The Architects

The architects of the late Roman and Byzantine period were military engineers. Some of them had strong academic backgrounds. Those of the Romanesque and the Gothic differed. They were drawn from stonemasons, craftsmen, and builders. They had in common high aspirations but a pragmatic approach to achieving them.

Hagia Sophia (opened in 537), the prime example of Byzantine architecture, was (Fig. 6.3):

… the creation of two Greek architects from Anatolia, Isidore of Miletus and Anthemius of Tralles. Hagia Sophia successfully marries the old Greek science of theoretical geometry to Roman skills of practical engineering. (Beaton 2021)

Photograph by the author

Photograph by Iman Fakhri/Wikimedia Commons

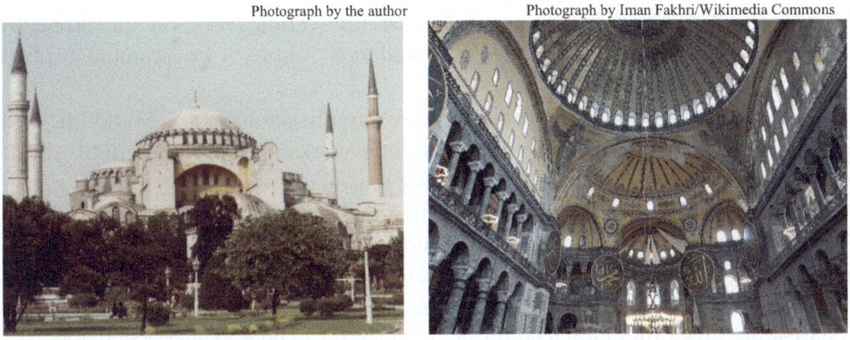

Byzantine architecture: Hagia Sophia, Istanbul (532-537)

Photograph by Nino Barbieri/ Wikimedia Commons

Photograph by the author

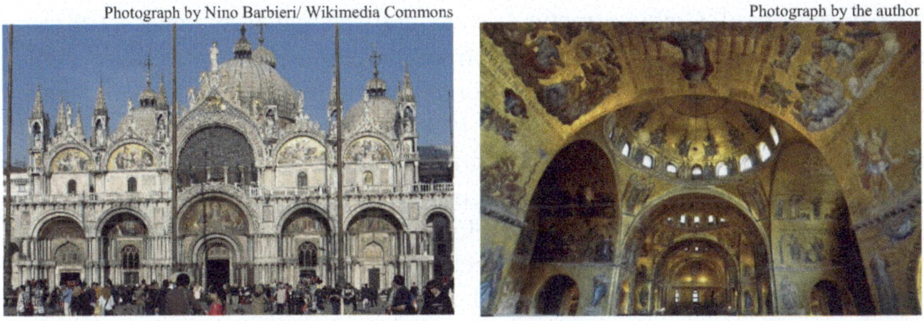

Byzantine architecture: Basilica San Marco, Venice (878 and 1069-1094)

Photograph by the author

Photograph by Ninarus/Wikimedia Commons

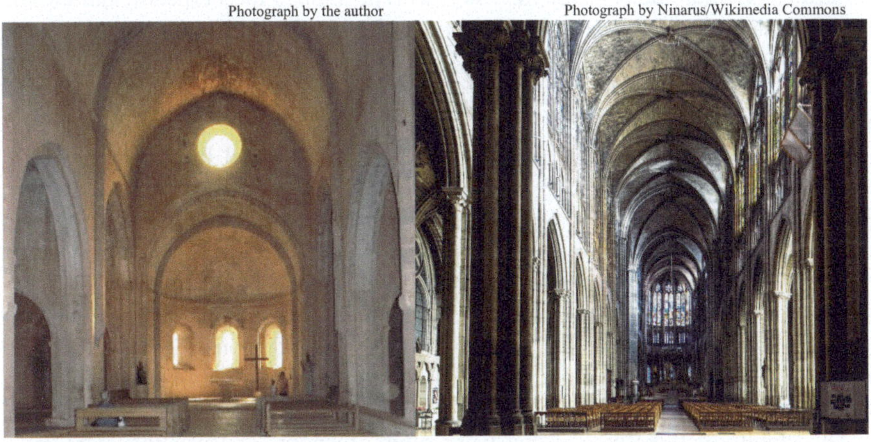

Interiors: Romanesque L'abbeye du Thoronet, Var (1190s) and Gothic St Denis, Paris (completed in 1144)

Fig. 6.3 Medieval architecture

Isadore of Miletus (b. *ca* 475) was a mathematician who taught at universities in Alexandria and Constantinople. He, apparently, wrote a treatise, *On Vaulting*, which explains the geometry of vaults and parabolas. Anthemius of Tralles (*ca* 474–*ca* 550s) was a professor of mathematics and an expert on ellipses. Either they, or people like them, worked out how to build domes on square rather than round bases as in the Roman types. Trdat (*ca* 940–1020), the chief architect of the Bagratid kings of Armenia, did some renovation work on the dome of Hagia Sophia as, in 1355, did George Synadenos Astras, a government official and provincial governor.

The architects of the Romanesque era are largely anonymous. Sometimes all that is known is their names as inscribed on their works. The 'churchwright' for St Peter's Abbey in London was a master stone mason, Leofsi Duddeson. His name, along with those of masons Godwin Gretsith and Tienfrith Duddeson, is associated with the later designing/building of Westminster Abbey, a Norman Gothic edifice. "Built solely by Buscheto's genius" is inscribed on Pisa cathedral (1063). It was he who developed Pisa Romanesque architecture in which Romanesque forms are mingled with Byzantine and Islamic elements. The cathedral was funded by the spoils of wars against Moors in Sicily.

The designers of Gothic structures were craftspeople (see James (1985) on Chartres). They received their training from master masons who were skilled geometers (Harvey 1971). They often worked in guilds or workshops associated with the erection of a specific building. Pietro Baseggio, the fourteenth-century architect, who along with Filippo Calendario (executed for treason in 1355) supervised the building of the Doge's Palace in Venice (1350+). Jacob van Thienen, a mason, was the designer of Brussels' Hôtel de Ville (1402) although the belfry (1415) was designed by Jan van Ruysbroeck, another mason. Colin Biart (b. 1460), the French architect was a stonemason. Jean Métezeau (d. 1600), an architect of the late Gothic infused with early Renaissance ideas was another from a family of masons. William de Ramsey (d. 1349) who developed the Perpendicular Gothic that dominated English ecclesiastic architecture for three centuries was the scion of a family of master masons.

6.2.2.2 The Clients

The clients for medieval architecture were, as in Greece and Rome, the powerful members of society. For the Byzantine it included the emperors of the Eastern Empire. Haga Sophia was commissioned by Justinian I (Byzantine emperor from 527 to 565). His ambition was to restore the Roman Empire to its former glory. He hired the architects and paid for the building out of public funds. Domenico Contarini (d. 1071), the thirtieth Doge of the Venitian Republic, the descendent of a leading family, was a prolific builder of churches and monasteries. The design of St Mark's is often attributed to him but the building is more likely the work of several architects and craftsmen.

The clients of the Romanesque were rulers, bishops, and wealthy merchants. The rulers, especially in European regions, such as what is now Germany, that had many monarchies, sought palaces. Many bishops did too. Merchants wanted bold townhouses that displayed their high social status. Edward the Confessor (*ca* 1003–1066), the last Anglo-Saxon King of England, was the client for St Peter's Abbey, on the site of what now is Westminster Abbey. The sponsors for the church were merchants Godwin and Wendleburh Gretsyd (meaning 'fat purse') (Jenkyns 2004). William I, the Conqueror (*ca* 1028–1078), the first Norman king of England, had the White Tower in the Tower of London (1078) built. Bishop Anselme III da Rho (d. 1093) commissioned the Basilica of Sant'Ambrogio in Milan.

The clients of the Gothic were political leaders, the clergy, and the aristocracy. They were well educated, knowledgeable about the arts, and had elite tastes. They strove to demonstrate their status and power through the designs they commissioned. Louis IX (1214–1270) had Sante-Chapelle (1240–1248) built in Paris to outdo the clergy who were building magnificent Gothic edifices such as Notre Dame (completed 1250). Some of the aristocracy, such as Philippo le Bon, Duke of Burgundy (1396–1467) retained their own architects.

Members of the clergy continued to solicit the work of architects. As noted, Abbot Sugar, was the patron of the Gothic works at Saint-Denis (completed 1144). The building inspired the archbishop of Ruen, Hugh of Amiens (d. 1164), to initiate the construction of the cathedral there. It was not only for religious buildings that the clergy hired architects. High ranked, they wanted handsome dwellings. Georges II d'Amboise (1488–1550), archbishop of Rouen, for instance, was the client for the Château de Verger (demolished in 1776).

6.2.2.3 The Works

The types of self-consciously designed buildings erected during medieval times were palaces of the wealthy, castles, administrative buildings, community gathering places, and, most notably, religious buildings. The emergence of Christianity as the dominant religion in Europe resulted in a proliferation of churches and monasteries. City squares began to replace agoras and forums.

Unlike the Greek and Roman temples, but like Roman basilicas, the Byzantine church interiors functioned as settings for communal gatherings. Hagia Sophia, built in six years, is a large domed structure, exploiting the Roman use of concrete and brickwork and fusing domical and classical columnar construction. The interior is decorated with mosaics and the exterior with striated brickwork. The building served as an Orthodox and then Catholic Church until 1453 when it was converted into a mosque. Its Christian symbols were removed or plastered over. From the 1930s it was a secular museum before being returned to function as a mosque in 2020. The changes reflect contemporary Turkish political attitudes. The building has served as an ideal type for many later ecclesiastical buildings. Hagia Sophia (1240s) in Trebizond (now Trabzon) is a direct descendent in name and form. Many mosques being built today are geometric imitations.

St Marks (Basilica Cattedrale Patriarcale di San Marco) in Venice, built in 828 and remodelled between 1063 and 1094, has Romanesque and Gothic as well as Byzantine characteristics. It reflects Venice's political ties to Byzantium. The church has a central plan rather that of a Roman basilica. Its transept was adapted to the site's configuration and to its need to function as a setting for the ceremonies of the Venetian Republic (Vio 2000). It, like Hagia Sophia, displays the technical advance of placing a dome on a square base by corbelling the supporting brickwork. Its interior surfaces are replete with gold ground mosaics of saints and prophets. As such the church is a display of the wealth and power of the Venetian state and its doges.

Gothic buildings evolved in boldness and in detail from an early, restrained, form to a more decorated version, and finally a flamboyant form. In France, Saint-Denis, Sainte-Chapelle and Notre-Dame-de-Paris are Early Gothic edifices. Chartres (1210–1250) is a High Gothic church. Rouen Cathedral (1488–1506) is an example of the livelier type. Town and guild halls, and some private dwellings were Gothic structures. Ypres Cloth Hall (1304; reconstructed after destruction in World War One) is a prominent Gothic secular building that became a model for civic buildings.

Two functions were emphasized in the Byzantine and Gothic churches. The first was the instrumental need to house a congregation that was preached to by a priest and the second was to strike a sense of awe in worshippers. The pews were arranged in a socio-fugal pattern so worshippers would focus their attention on the priest. While inspiring, the mosaic decoration, the sculptures and the representations in the stained-glass windows had another purpose. It was to educate a largely illiterate population about their obligations as Christians. The end was also to control their behaviour with the promise of eternal life if they behaved properly and believed in God. The quality of light filtered through the windows was a means to sense the presence of God in the church (Frankl 2000; Prina 2011).

All three—the Romanesque, the Byzantine, and the Gothic—served later architects as precedents for their work. The late nineteenth-century Westminster Cathedral in London designed by John Francis Bentley (1839–1902), is Neo-Byzantine. Le Corbusier's work reflects his admiration for the simplicity, the bold forms, and the qualities of light and shade that appear in the Romanesque L'abbey du Thoronet that he visited in 1953 (Sternberg 2012). Countless Neogothic buildings have been built but the Classical remains the more significant referent for architects. It certainly was for the architects of the Renaissance.

6.2.3 The Renaissance

The Renaissance was a period of looking forward to a new world and looking back to establish a 'Golden Age' akin to classical antiquity. Architecture was a topic for explicit discussions about prescriptive theories. Beginning in Florence, architects proposed a new architecture. Its evolution can be divided into four phases. The High Renaissance began early in the sixteenth century when Classical compositions were applied with clarity to create harmonious works. Seeking harmony was deemphasized during the Mannerist period (1520–1600). Using freer forms was the means for attaining a more relaxed architecture. An even greater sense of sculptural freedom was attained by Baroque architects and, from the 1730s, by practitioners of the Rococo style where the idea of symmetry was abandoned for decorative effects.

Renaissance architects rediscovered the dome, a structural type missing from Gothic architecture although it had been a major feature of the Byzantine. The Gothic concern with geometry was taken a step forward with plans based on circles and squares. High Renaissance buildings were symmetrical in plan with the façades possessing a flat classicism. The decoration of walls, whether with pediments, arches, columns, pilasters, or string courses, had little physical depth to it. The visual organization of compositions based on the proportional relationship of parts to the whole, imitated the configuration of Classical architecture. Architect referred to their work as *all'antica*—in the ancient manner.

The patterns of the interiors of buildings also followed proportional systems but their visual character was very different to the exterior. Decorative motifs drew on ancient themes with great elaboration. Clients acting as patrons commissioned frescoes of imaginary classical scenes. Although appearing in Asian frescoes much earlier, linear perspective as developed by Filipo Brunelleschi (1377–1446) was used by artists and architects to give a sense of depth to drawings and to lead an observer's eye to specific points of interest in buildings.

For Renaissance architects and polymaths, such as Leon Battista Alberti and Leonardo da Vinci, the ideal urban environment consisted of straight streets lined with buildings built to the property line. The buildings would be of the same height and possess uniform door and window details. The goal was to create a sense of nobility. The ideal urban space, as depicted in a painting, probably by Luciano Laurana (1420–1479), possessed a foreground building on display. Other structures formed a backdrop to it. The work illuminated Alberti's prescriptions. A host of fortified, ideal city plans further illustrated Renaissance architects' ideals while pragmatically having to deal with struggles for regional power among competing aristocratic contenders.

From the 1520s, High Renaissance architecture started to give way to Mannerism, a non-conformist approach to design. Buildings began to possess varied spatial relationships often with elongated proportional systems used to organize the façades. Interiors began to have a more tunnel-like character; the lighting was harsher. The decoration consisted of mixed antecedent stylistic elements to cater for the more individualist tastes of clients in each building. The Baroque took Mannerism a step further (Fig. 6.4).

Baroque architecture flourished between the early 1600s and the mid-1700s before being superseded by the Rococo. It was an architecture of exuberant detail, deep colours, and an awe-inspiring grandeur. Highly ornate, it was a political response of Roman Catholic Church officials to the austere churches of the protestant breakaway. Columns became twisted reflecting light in several directions. Cartouches (carved plaques, usually oval, with decorative features or text highlighted in gold) were a common decorative device. The texture gradients of walls, floors, and ceilings were distorted to give illusions of great depth to spaces when only a little existed (Martin 1977; Neuman 2013).

Photograph by Commonists /Wikimedia Commons

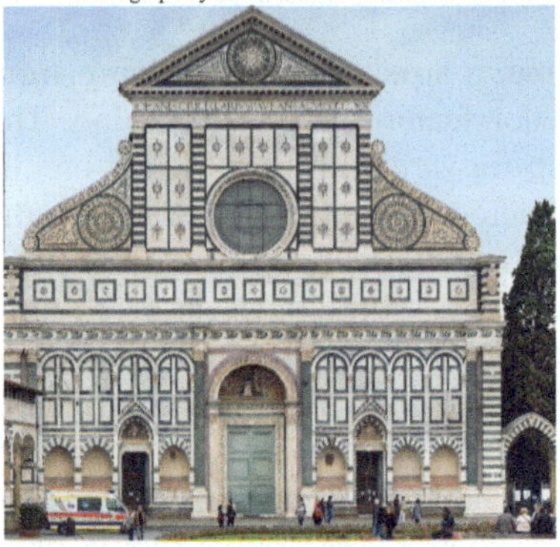

Illustration by Straffarello Gustavo/Wikimedia Commons

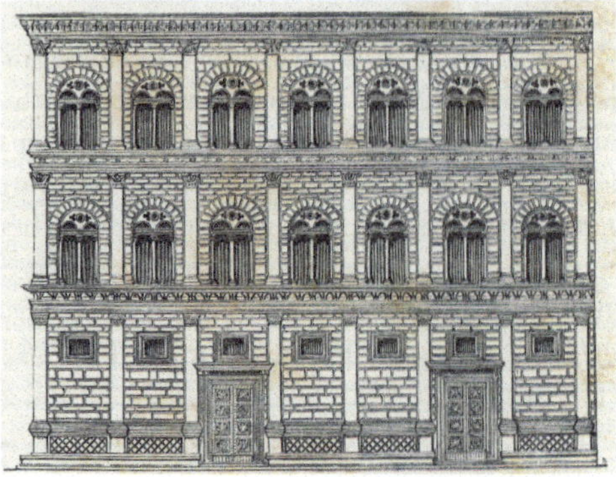

Early Renaissance architecture: Santa Maria, Novella, Florence (left) and the Palazzo Ruchellai, Florence (right)

Photograph by the author

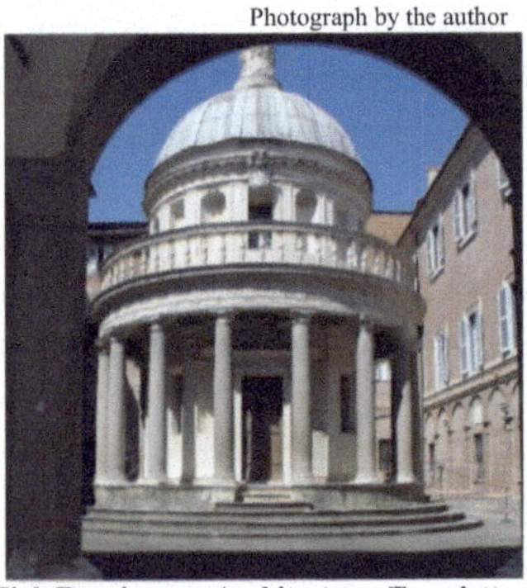

Etching by Guiseppe Vasi/Wikimedia Commons

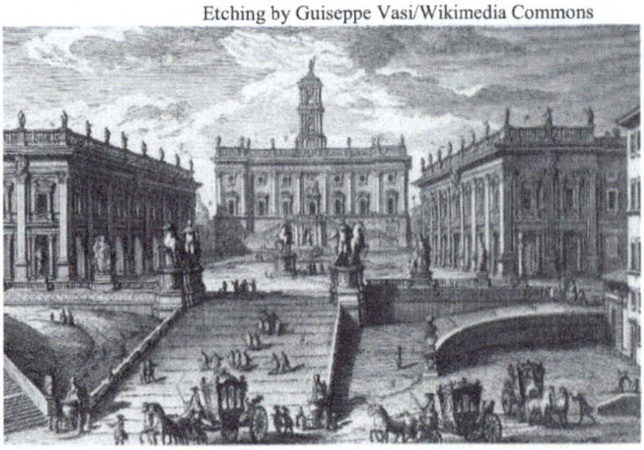

High Renaissance Architecture: Tempietto, Rome (left). Mannerist Architecture: Piazza de Campidoglio, Rome (plate 80 in Guiseppe Vasi's book of etchings (*ca* 1750) (right)

Collection of the Urbino National Gallery/Wikimedia Commons

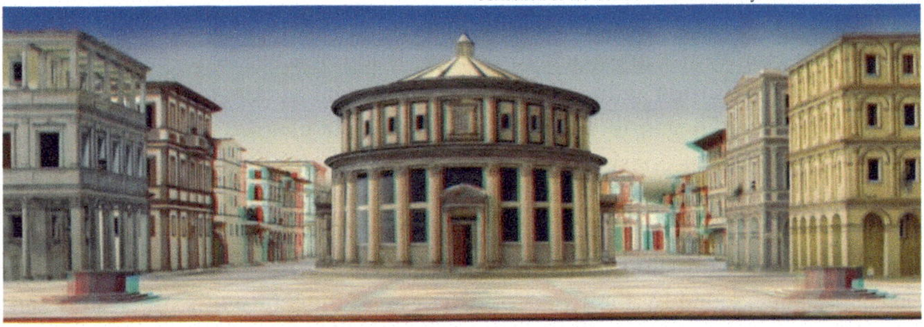

The ideal Renaissance city: *La Città Ideale* usually attributed to Luciano Laurana

Fig. 6.4 High Renaissance and Mannerist architecture

Beginning in France in the 1730s, Rococo, was a highly ornate style of interior design. The exterior of buildings remained relatively simple. Architects rejected the symmetry of the High Renaissance and combined asymmetrical, complex floor plans with interior decorations full of curves, sculpted mouldings, festoons of flowers, and trompe-l'œil ceilings replicating the sky. Angels and cherubs floated on the edges of ceilings. Japanese and Chinese motifs, such as dragons, began to appear in the decorations. They created a sense of drama. Palaces had grand staircases that when ascended presented a sequence of different views of the interior details. What began in secular buildings influenced later churches in Europe and Latin America (Charles and Carl 2010).

6.2.3.1 The Architects

During the Renaissance architects began to be clearly identified and celebrated individuals. Some of them were formally educated in mathematics and philosophy at the emerging tertiary educational institutions. Primarily, however, architects continued to be trained through working with practitioners (Cheney 1998). They all, in common, studied the classical buildings of ancient Rome and Greece. Drawing relics made them pay attention to the proportional systems and the details of ancient designs.

Filippo Brunelleschi and Donatello (Donato d'Niccolò Betto Bardi; *ca* 1386–1466) self-consciously instigated Renaissance architecture through their advocacies and work. Alberti's treatise established the key elements of architecture and the ideal plans for different building types based on mathematical proportional schema. Imitating Roman prototypes, Alberti advocated clean lines in building and argued for the parts of buildings being related mathematically to the whole (Rowe 1976). He demonstrated his ideals in his designs as exemplified by the church of Santa Maria Novella (1470) in Florence. His prescriptions were applied to all building types, many serving instrumental functions totally different to those served by their Classical antecedents (Moore 1905). Antonio Sangallo (1485–1540) and Raphael (Raffaelo Sanzio da Urbino, 1485–1520) were among the architects who followed.

The most important architect of the High Renaissance because of his impact on subsequent architecture was Andrea Palladio (1508–1580). His designs and his sumptuously illustrated *I quattro libri dell'architettura* (1570) along with Alberti's earlier *De re aedificatoria* (1452), included wide-ranging discourse on urban design and architecture. They solidified High Renaissance architectural attitudes. Palladio's works and reputation led to the spread of Renaissance architectural ideals throughout Europe (Reid and Farber 1980). They also spurred architectural tourism with Rome as the principal destination. It is Michelangelo Buonarroti (1475–1564), however, who is *the* exemplar of the Renaissance polymath. His early work was Mannerist, but it is his Baroque architecture, his sculptures, artwork, and mechanical inventions that established his renown.

Francesco Borromini (1599–1667) was the chief formulator of the Baroque and architect of San Carlo alle Quattro Fontane (1634) in Rome. The other major architect of the Baroque was Gian Lorenzo Bernini (1598–1689) although he is better known for his sculptures. Many architects produced Rococo works but no individual spokesperson for the approach stands out (Neuman 2013). Germain Boffrand (1667–1754) designer of the Hotel de Soubise in Paris (with some interiors by Juste-Aurèle Meissonier (1695–1750), an Italian born French artist) was the strongest promoter of the style. His book, *Livres d'Ornemens Inventés and Dessinés* (1745), was responsible for the spread of the Rococo throughout Europe. It along with *The Genius of Architecture, or, The Analogy of that Art with our Sensations* (1780) by Nicolas Le Camus Mézières (1695–1768) served as a Rococo design copybook.

6.2.3.2 The Clients

The Renaissance was characterized by a close tie between the intellectual elite and the architects they commissioned. The clients consisted of the royalty, nobles, powerful guilds, members of the clergy, including politically astute popes such as Julius II (1443–1513), and wealthy merchants. Among the royalty were King Ferdinand (1452–1516) and Queen Isabella (1451–1504) of Castile and León (Loth 2012). Philip II of Spain (1527–1598) commissioned El Escorial outside Madrid to demonstrate his imperial power. In Florence, the merchant families were great patrons of architecture.

The introduction to Europe, via Toledo, of Indo-Arabic decimal arithmetic had made accounting, banking, and the accumulation of wealth by merchants, such as the Medici family, easier. The families strove to outdo each other and spurred artists and architects to do the same (Cooper 1996; Asch and Birke 1991; Hebron 2018). The Medicis, who really ruled Florence although it was republic, commissioned works by architects such as Brunelleschi. The Strozzi family in Florence and Milan and the Gonzagas in Mantua were among other powerful patrons. Although they were not in a social position to do so, some clients for grand architectural works were, nevertheless, powerful women. Queen Isabella in Spain and Isabella d'Este (1474–1539), the highly educated Marchioness of Mantua led the way in Italy. Isabella d'Este sponsored works by Michelangelo and other architects.

Commissioning art works and substantial buildings was fun. As Giovanni di Paolo Ruchellai (1403–1481), a wool merchant, who hired Alberti to design the Palazzo Ruchellai (1494–1451) and the façade of Santa Maria Novella, noted:

> All the above-mentioned things have given me the greatest satisfaction and pleasure, because in part they serve the honour of God as well as the honour of the city and the commemoration of myself. (cited in Kilroy-Ewbank and Graham 2023)

Artists often depicted their patrons in the paintings and sculptures they produced but it was the architects who set the pace. It was they who formulated the principles of Renaissance designs. It was they who turned to Mannerism. It was they who invented the Baroque and it was they who sought a richer style in the Rococo and it was they who got tired of it.

6.2.3.3 The Works

The Tempietto in the Cloister of San Pietro in Montorio (*ca* 1502) is the exemplar of High Renaissances architecture. Built on the spot where Saint Peter was said to have been crucified, it was designed by Donato Bramante (1444–1514) for Ferdinand and Isabella (Loth 2012). Its instrumental function is as a reliquary—a commemorative martyrium—but it became more important as the display of an architectural idea. Its dome being supported on a two-tier drum, the lower part of which consists of a Doric peristyle, topped by a balustrade, was an innovative design. Based on the domed Hellenic tholos form it became *the* model for many later domed buildings of importance.

The Villa Almerico Capra Val Marana—La Rotunda—(1556–1590), designed by Palladio was designed for a retired Vatican priest, Paolo Almerico (1514–1589) and not for family life. As Johann Wolfgang von Goethe (1749–1832), the German polymath, noted in his *Italian Journey 1786–1788* (Fig. 6.5):

> The interior can certainly be described as habitable, but not made to be lived in. The main hall's proportions are beautiful, just like the floors and the bedrooms, nonetheless it could barely be accepted as a summer residence for a distinguished family. (Goethe 1982, originally 1816)

While La Rotunda is a four-sided imitation of the Pantheon, the triumphal arch motif for windows distinguished Palladio's work from its precedent and from the designs of his contemporaries (see Fig. 3.10). La Rotunda remains the exemplar of Renaissance design. The Laurentian Library in Florence (1523) begun by Michelangelo and his design of Piazza de Campidoglio and its buildings with their oversized columns are examples of Mannerist architecture. Touring foreign architects took the style from Italy to other parts of Europe. The Grandmaster's Palace by Girolama Cassar (*ca* 1520–*ca* 1592) in Valetta, Malta is one of many Mannerist projects around Europe.

The interiors of St Peter's in Rome, as designed by Michelangelo, and the Basilica of San Juan de Dios in Granada (1730s), designed by José de Bada y Navajos (1691–1750), are hallmarks of Baroque architecture. In London, St Paul's Cathedral (1710) designed by Sir Christopher Wren (1632–1723) is the prime British example of the Baroque. The style was taken by Spanish and Portuguese administrators and clergy to their colonies in the Americas and Asia. It symbolized their presence.

The Rococo work was even richer in detailing. Examples abound (Neuman 2013). They include the Hotel de Soubise in Paris, designed by Juste-Aurèle Meissonier (1695–1750), Amalienborg Pavilion (1734) in Munich designed by François de Cuvilliés (1695–1760), the French trained court architect to Duke Maximilian II Emanuel of Bavaria, Zwinger Palace in Dresden designed by Matthäus Daniel Pöppelmann (1682–1736), best known for his Baroque designs, the Helbing Haus, Innsbruck (1732) designed by Anton Gigi (1700–1769), and the Pineau room, Hôtel de Lagrangeville (1736). A reaction to such rich detailing set in during the

Photograph by the author

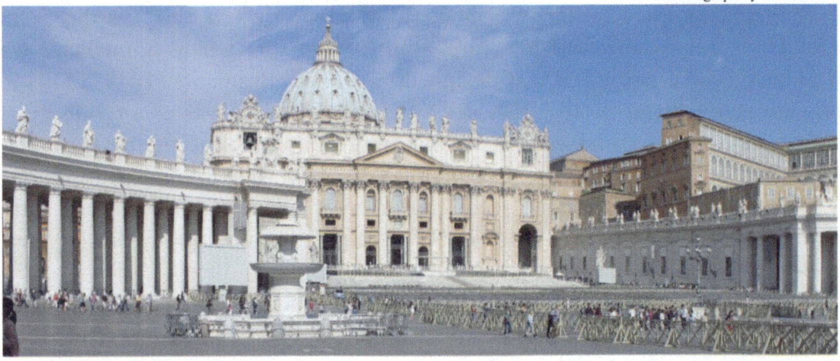

The Baroque: St Peter's Basilica, Rome

Photograph by Karel /Wikimedia Commons

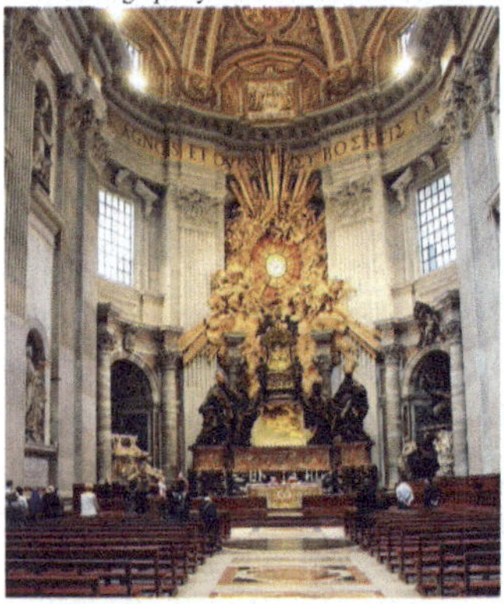

Photograph by Dan /Wikimedia Commons

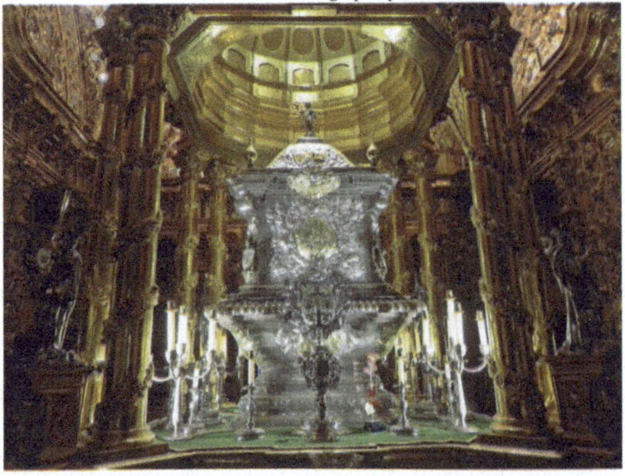

Baroque interiors: St Peter's Basilica, Rome (left) and the Basilica of San Juan de Dios, Granada (right)

Photograph by Kawneer/Wikimedia Commons

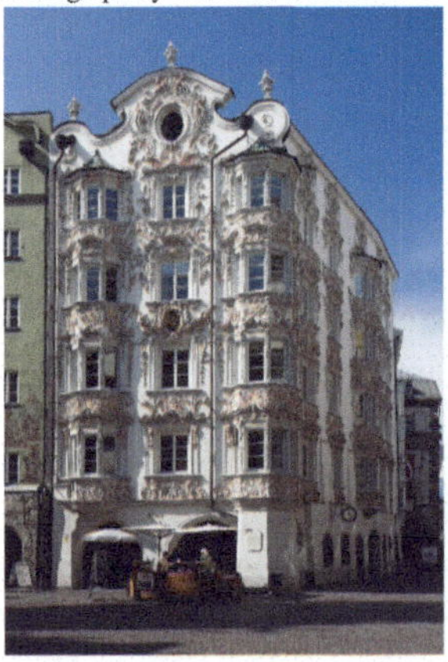

Mr and Mrs Charles Wrightsman Gift, MMA/Wikimedia Commons

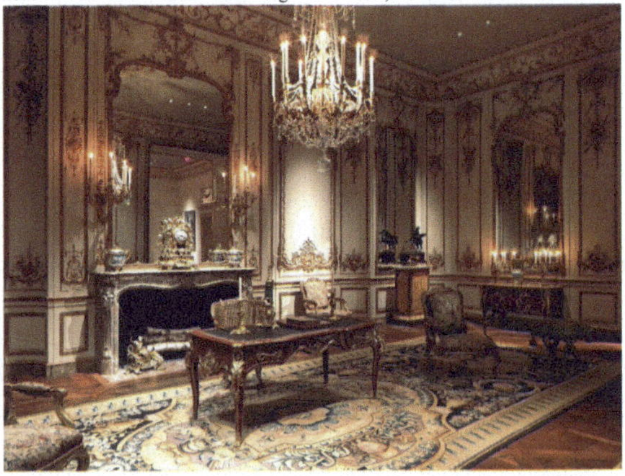

The Rococo: Helbing Haus, Innsbruck (left) and Pineau room, Hôtel de Lagrangeville, originally in Paris (right)

Fig. 6.5 The Baroque and Rococo

eighteenth century with a taste for the simpler forms of pure classical architecture. The Rococo, like the Gothic earlier, began to be regarded as degenerate.

6.2.4 Revivalism: Neoclassic, Neogothic, and Beaux Arts Architecture

In the mid-eighteenth century, architects, tired of the visual richness of the Baroque and Rococo, looked back to the classicism of Rome and Greece for inspiration in developing a new direction for architecture. Palladio's designs, his *I quattro libri dell'architettura*, and the discoveries of ancient ruins in Italy stimulated a new generation of architects. The interiors of the villas unearthed at Pompeii and Herculaneum became models to be imitated (Middleton and Watkin 1980). Published drawings of the buildings of ancient Greece were also readily available for architects to study. The Neoclassical ensued. Other architects sought inspiration in the structural integrity of Gothic architecture. Architecture as taught at the École des Beaux Arts in Paris overlapped the Neoclassic and Neogothic chronologically from the mid-nineteenth century. The school's prescriptions were for a mixture of Classical and Baroque elements in both buildings and urban designs (Drexler 1977).

Alberti's and Palladio's manifestos inspired a host of new books that reflected the values of contemporary societies. The most influential in the English-speaking world was *Vitruvius Britannicus* (1715) by Colen Campbell (1676–1725), a Scottish architect. Neoclassical architecture, to many, symbolized Enlightenment empiricism. At its outset it drew on Roman antecedents but from the 1800s the inspiration was largely the columned, unpainted, primarily Doric, Greek temple. Buildings began to be symmetrical and simple in geometrical form. Surfaces were planar and projections flatter than their classical antecedents. Neogothic architecture, in contrast, retained a visual richness.

When exactly the Gothic became the Neogothic is open to conjecture but the movement saw its heyday in the nineteenth century. Gothic architecture had persisted throughout the Renaissance and the eighteenth century everywhere in Europe. Many clients considered it to be the appropriate style not only for churches but also for university buildings, city halls, and other public structures. In England from the 1740s the Neogothic was associated with Anglo-Catholicism's response, like that of the Roman Catholic earlier, to the puritan architecture of nonconformism. It was also romantic in nature and paralleled contemporary literature's presentation of modern themes in medieval settings.

The Architectural Antiquities of Great Britain (1807) written by John Britton (1771–1867), an English antiquarian, popularized the Neogothic. The use of step roofs, pointed arches, vaults, filigree work, and decoration were types of elements taken directly from the Gothic buildings erected between the twelfth and sixteenth centuries. Neogothic buildings were not, however, copies but richer imitations of medieval precedents. Debates existed as to whether iron could be used in conjunction with stone in buildings. In France, Eugène-Emmanuel Viollet-le-Duc

(1814–1879) had no qualms about using iron work in his restoration of Gothic churches to what he believed they should be rather than to what they once were (Bressani 2014). He presented his views in *Dictionnaire Raisonné de l'Architecture Française du XIe au XVIe Siecle* (1854).

Among the Neogothic's strongest advocates was John Ruskin (1819–1900), the English art critic and polymath. His books, *The Seven Lamps of Architecture* (1849) and *The Stones of Venice* (1853) remain influential in academic circles. Ruskin considered the Venetian-Gothic Doge's Palace to be *the* exemplar of good architecture. Not everybody agreed with him. To its detractors the Neogothic was a tragedy. Scholars at the École des Beaux Arts certainly thought so.

In the 1830s professors at the École des Beaux Arts broke away from the strict formality of the Neoclassical in seeking a clearly French architecture drawing on French precedents. They formulated a typology and composition-based method of designing (Carthian 1979; Wozniak 2016). They argued for a restrained Baroque architecture that was influenced by the character of the Paris Opera House (1860–1875) designed by Charles Garnier (1825–1898). The Opera House is essentially a Baroque building symmetrical in plan, façade, and exterior decoration (Drexler 1977).

6.2.4.1 The Architects

The architects of the era were increasingly educated at formal institutions. The Académie d'architecture in Paris was founded by Louis XIV in 1671. Francois Blondel (1618–1686), an architect with a civil engineering background, was its director. In the nineteenth century L'École Nationale Supérieure des Beaux Arts, to give it its full name, in Paris (1817) became the most influential architectural school. It attracted elite, politically well-connected, wealthy students from around the world. They disseminated the school's ideas across Europe and the Americas. Some leaders in the field were self-taught through visiting and sketching important buildings and their details. Most architects, however, were still educated by being apprenticed to a practitioner.

Ange-Jacques Gabriel (1698–1782), a graduate of the Académie d'architecture, designed the Petit Trianon (1762) in Versailles. He was also responsible for the Neoclassical Place de Louis XV (1748; now Place de la Concord). In Britain, the Neoclassical movement was led by architects as prominent as Robert Adam (1728–92) and John Soane (1753–1837). In *fin de siècle* New York, the firm of McKim, Mead and White promoted Neoclassicism as an expression of progressive architecture. In Germany, Carl Gottfried Langhans (1732–1808) and Kark Friedrich Schinkel (1781–1841) produced fine foreground buildings. Public Works Departments in many countries saw the Neoclassic as the epitome of modern architecture.

The leading architectural proponents of the Neogothic were Viollet-le-Duc in France and his contemporary Augustus Welby Pugin (1812–1852) in England. Pugin argued that "buildings should have no features which are not necessary for convenience, construction or propriety" and "all ornament should consist of the enrichment of the essential construction of the building" (Pugin 1841; Yorke 2017).

Pugin learnt drawing and design from his father, an architect, and by visiting places of architectural interest on the continent. Viollet-le-Duc, who travelled to Italy to learn about architecture, studied at the Collège royal de Bourbon d'Aix.

In the United States architects such as Richard Morris Hunt (1827–1895), the first American to study at the École des Beaux Arts, and Daniel Burnham (1846–1912), were producing Beaux Arts buildings. Louise Blanchard Bethune (1856–1937), one of the first female architects in the United States, designed the Hotel Lafayette (1904) in Buffalo, New York, in the contemporary Beaux Arts manner when she was a partner of Bethune, Bethune and Fuchs (Hartman 2022) (Fig. 6.6).

6.2.4.2 The Clients

The governments of the United Kingdom, Canada, and Hungary commissioned Neogothic parliament buildings. Many universities hired architects to design buildings in what is called 'collegiate gothic', but it is the clerical bodies who are the best-known patrons of the Neogothic. The Episcopal Church and the Church of England as well as the Roman Catholic Church in many countries were major clients. Some individuals were both important clients and promoters of the Neogothic. Among them was Horace Walpole (1717–1797), an English writer and politician. He was both a client and part-designer of his mansion, Strawberry Hill House (1749+), in London.

The Neoclassic appealed to clients wishing to express vastly different values through the buildings they commissioned. Democratic and republican governments wished to symbolize their values of reason, order, and liberty in built form. In the United Sates they hired architects to design state capitol complexes. Other clients were members of royalty and the aristocracy who wanted to exhibit their power, wealth, and good taste through the buildings they possessed. They included the kings of France and the Russian aristocracy. Louis XV (1710–1774) commissioned the Petit Trianon. Peter the Great (1672–1725) and, even more so, Catherine the Great of Russia (1729–1796; empress 1762–1796) were responsible for modernizing St Petersburg; Napoleon III (1808–1890) was actively involved in seeking a modern capital. He was responsible for the rejuvenation of Paris.

The people who commissioned Beaux Arts works were civic leaders, wealthy and influential individuals, and institutions who wished to display their status and good taste with lavish and ornate buildings. The French government commissioned the École des Beaux Arts building designed by Felix Duban (1790–1870) (Van Zanten 1978). George Washington Vanderbilt II (1862–1914), an art collector from a wealthy family, was the client for the châteauesque Biltmore Estate designed by Richard Morris Hunt in North Carolina. Other wealthy American families built competing estates. They saw themselves as the aristocratic elite.

6.2.4.3 The Works

Neoclassical buildings abound in Europe and the Americas. The Petit Trianon, commissioned by Louis XV for his mistress Madame de Pompadour, ended up being a retreat for Marie Antoinette, the queen of his son Louis XVI (1754–1793).

Photograph by the author

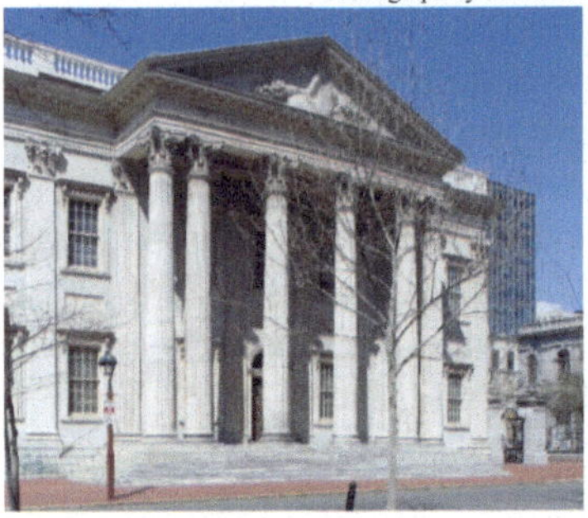

Photograph by Moneycantbuy/Wikimedia Commons

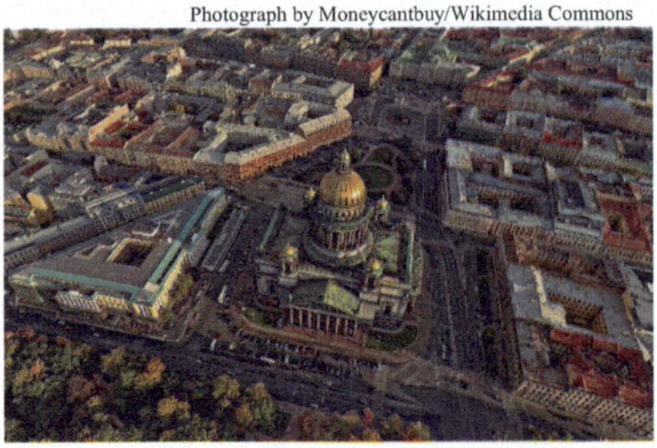

Neoclassical architecture: First Bank of the United States (1797), Philadelphia and St. Petersburg, Russia (1770s +)

Photograph by Marc Ryckaert/Wikimedia Commons

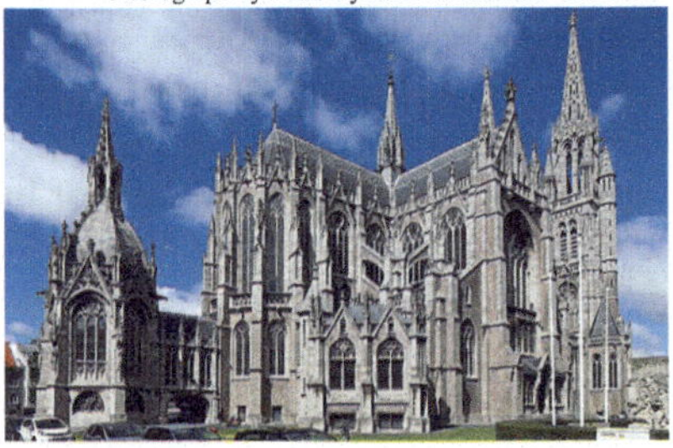

Photograph by Joe Ravi/Wikimedia Commons

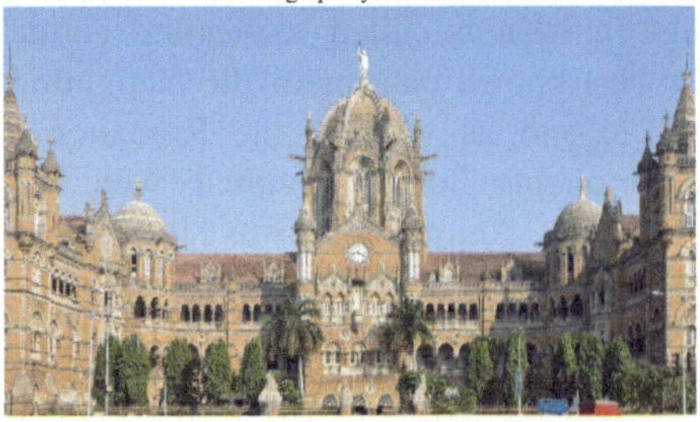

Neogothic architecture: Sint Petrus-en-Pauluskirk, Ostend (*ca* 1900) and Chhatrapati Shivaji (Victoria) Terminus, Mumbai (*ca* 1898)

Photograph by the author

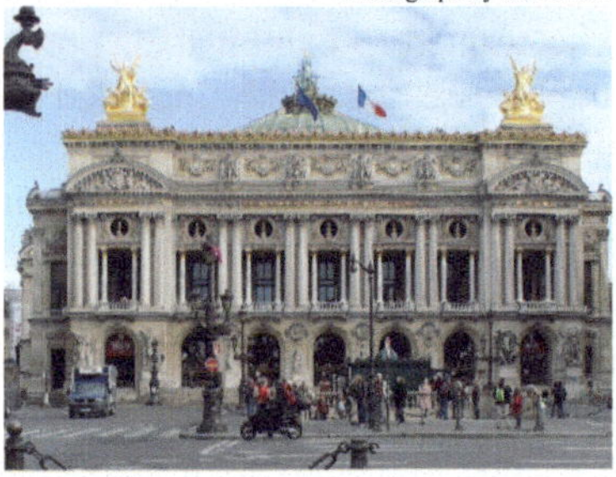

Photograph by Benh Lieu Song/Wikimedia Commons

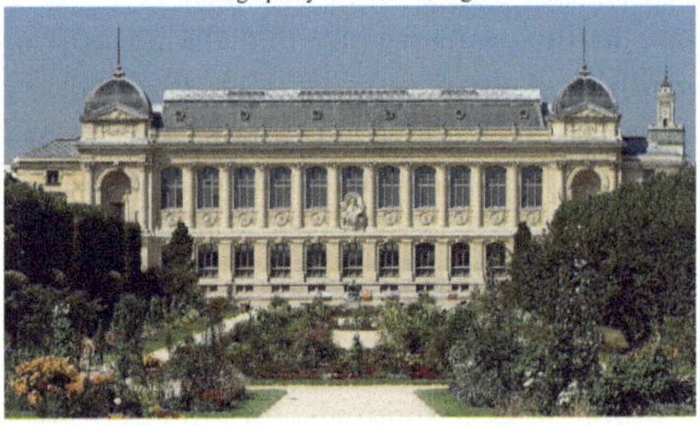

Beaux Arts: The Opera House (Palais Garnier), Paris (1861-1875), the precedent (left), and Museum national d'histoire naturalle (1887+), Paris (right)

Fig. 6.6 Revivalism: Neoclassical, Neogothic, and Beaux Arts buildings

The building is an exemplar of both the form and spirit of the Neoclassical. Modernized St Petersburg had an urban design featuring long axial vistas that focused on distinctive points. The streets were lined by neo-classical buildings. St Petersburg's design was one precedent for Napoleon III's renewal of Paris under the direction of Baron Georges-Eugène Haussmann (1809–1891).

In the United States the White House (1792), designed by James Hoban (1755–1831) was based on the earlier Neoclassical Leinster House (originally Kildare

House, 1745) in Dublin and the design of the Capitol building (1800+) in Washington by Dr William Thronton (1759–1828), had Neoclassical antecedents that can be traced back to the Tempietto. Many Neoclassical state capitol complexes in the country followed suit. In addition to housing their instrumental functions, the buildings symbolized America seeing itself as a descendant of democratic Rome. Banks looked towards the perceived solidarity of Rome in commissioning Neoclassical buildings. The Lincoln Memorial (1914–22) in Washington is a twentieth-century Neoclassic design. Inspired by the Parthenon, its architect was Henry Bacon (1866–1924), a Greek Revival and Beaux Arts architect.

While British architects debated whether the Neoclassic or Neogothic was the architecture appropriate for new buildings, many Neoclassical country mansions were built in England and Ireland. In the colonies of the European powers, the Neoclassical served a variety of functions. Colonial governments used it to assert their presence on foreign soil. For those colonized it had different meanings. Several mansions of wealthy Indian businesspeople and landowners in Kolkata, for instance, were built with Neoclassical façades to show that their owners were modern in spirit. Behind the façade was a traditional gender-segregated, multi-courtyarded building (Taylor and Lang 2016). British military engineers designed the façades; local *mistris* designed what was behind them.

A vast array of Neogothic buildings was built during the eighteenth and nineteenth centuries. Churches predominated, but they included building types as diverse as high-rise commercial structures and railway stations (Yorke 2017). One of the grandest Neogothic stations is the Chhatrapati Shivaji (formerly Victoria) Terminus (1880s) in Mumbai. Frederick William Stephens (1847–1900), a British engineer was its designer. He was influenced by the drawings of Gothic buildings by Axel Hermann Haig (1835–1921), a Swedish architect working in England. The taste for Neogothic buildings has dwindled. The taste for Beaux Arts architecture has also waned. Neither architecture has been extinguished.

The Beaux Arts urban and building designs represented an architecture of grandeur. Clients sought that end in the works they commissioned. Buildings were symmetrical and consisted of a highly rusticated base, with the façades above containing pilasters or columns. At the top was an attic level and a flat roof. The buildings had large arched or pedimented entrance doorways, arched windows, and sculptures (Drexler 1977; Middleton 1982; Borngässer and Bednorz 2008). Designed by Louis-Jules Andre (1818–1890), the Museum national d'histoire naturelle (1877) in Paris is an example. Internally, Beaux Arts buildings had a hierarchy of spaces from their grand entrances to mundane service areas, ideally laid out in a pattern as symmetrical as Palladio's Villa Rotunda. In buildings that required large internal volumes such as railways stations and museums, iron spans with glass infill panels were used extensively.

In urban design the Beaux Arts prescribed what were essentially Renaissance designs of wide streets lined with monumental buildings and vistas that focused on important buildings or specific points in squares. The urban renewal of Paris during the stable, authoritarian government of Napoleon III was directed by Haussmann between 1854 and 1873. Designed with gracious boulevards, unified architecture, plazas, and commanding vistas, it was impressive in its imagination and execution. It has been an inspiration for urban renewal works in other cities. Haussmann, in turn, had taken inspiration from the Piazza del Populo and the streets radiating from it, designed a century earlier by Giuseppe Valadier (1762–1839) Italy's leading Neoclassicist. The plan for Washington designed by Major Charles L'Enfant (1754–1825) was another precedent. It, in turn, was based on designs going back to the plan for Rome formulated for Pope Sixtus V (1521–1590).

Beaux Arts buildings and partial urban designs were built around the world in the latter half of the nineteenth and early twentieth centuries as part of the City Beautiful Movement. Many such developments remain much admired and loved for their grandeur. Beaux-Arts architecture gave way to the Arts and Crafts movement, Art Nouveau, and Art Deco architecture and more thoroughly to the rise of the Modernist. They tugged in different directions. The first mentioned strove to revive artisan work in the face of the industrial production of environments and objects (Davey 1980). The last, in contrast, looked forward to the creation of a brave new world.

6.3 Self-conscious Design: Twentieth-Century Modernism

The industrial revolution produced technological innovation after innovation. They led to changes in settlement patterns, people's occupations, and their ways of lives. It also produced the "dark satanic mills" as William Blake (1757–1827), the British poet and artist, described the factories. If the Renaissance was a response to Gothic architecture and the cramped, medieval city, Modernism was a response to Revivalist architecture and the cramped, polluted industrial city. The responses of the Empiricists and Rationalists among architects differed. The former were cautious; the latter were for a totally new world. Much architecture was simply a pragmatic response to capitalist clients' demands.

Architects' prescriptions are much influenced by their backgrounds. Le Corbusier and Ebenezer Howard (1850–1928), instigator of the Garden Cities movement, came from very different worlds. Le Corbusier was a Calvinist and his early architecture shows the strictness of that faith. He was also a music critic and an artist from the inception of his career (Brooks 1997). Howard was part of the London Congregationalist community but his ideas were honed by Theosophy. He worked as a farmer and journalist in the United States (Moss-Eccardt 1973). He did not see himself as an artist but as a reformer. Le Corbusier saw himself as both.

Some architects, consciously or subconsciously, changed the philosophical underpinnings of their ideas as their careers developed. Those who were Rationalists found that their designs simply did not suffice and chose to follow a more empiricist route. Empiricists realized that the world was changing and that imitating precedents no longer sufficed. In common, the two broke away from the strictures of the Victorian era. In common, their philosophies had links to the social and philanthropic movements of the nineteenth century. They strove to do good works.

6.3.1 The Pragmatists

Pragmatic thinking is necessary when dealing with a range of problems including those of meeting substantial ends while minimizing costs. Dealing with issues such as the sustainability of buildings and how to deal with climate change requires much pragmatic thought. The aesthetic quality of a pragmatically designed building is, however, largely a by-product of dealing with other functions. The observation 'Form follows instrumental and financial functions' applies to it.

Most architect-designed buildings are largely pragmatic works. Many are based on generic designs and a society's prevailing aesthetic attitudes. They meet the expectations of the marketplace while meeting the stipulations of building regulations and codes. They are not designed to meet the aesthetic interests of the architectural cognoscenti. Although we celebrate the buildings designed by leading architects in the world's major cities, it is the pragmatic work of unheralded architects and the way buildings relate to streets that gives cities their character.

The characteristics of some building types such as train stations and airports evolve as technologies evolve. While the design of transport aeroplanes has not changed significantly over the past sixty years, airports have changed dramatically in size and layout as domestic and international travel accompanied by fears of terrorist attacks and anti-social behaviour in general have increased. Most stations and airports fulfil the requirements for their instrumental functions and solve the need for large spans with a minimum of fuss. The geometrically deviant examples attract our attention.

6.3.1.1 The Architects

Pragmatist architects were educated at an extraordinary array of academic institutions from the most prestigious to the least well-known. They have either been pragmatists by nature, educated to be so, or have become so in responding to professional demands. Some architectural firms employ several hundred such skilled architects. The design departments of large property developers producing highly pragmatic work are headed by such architects.

Some architects are highly pragmatic in all the work they produce. As noted earlier, others may usually be Rationalists or Empiricists but are Pragmatists when dealing with certain problems. In the twentieth century, architects such as Howard Van Dorn Shaw (1869–1926) in Chicago and Richard Seiffert (1910–2001) and Partners in London were noted for their pragmatic work. Derided by the Rationalists, the buildings they designed are now being re-evaluated. They met the needs of clients usually within budget limitations and on schedule and contributed positively to the quality of the streets on which they were located. Among those architects only dealing with certain problems pragmatically are designers such as Alejandro Aravena Mori (b. 1967) when dealing with designs such as housing for low-income populations. His design goal then is to provide accommodation for families without stigmatizing them through the qualities of the architecture employed. The work of Shaw, Seifert, and Aravena is described in Chap. 8.

6.3.1.2 The Clients

The soliciting clients for pragmatic buildings everywhere are property developers, large corporations and institutions that wish to obtain buildings as speedily as possible with the minimum of fuss. Increasingly, they are government agencies seeking to build as inexpensively as possible to show taxpayers that their money is not being wasted on frivolities. The focus of concern of clients is on obtaining buildings that meet their instrumental needs expeditiously within a specified budget. They are concerned with buildings as financial investments and little concerned with the catalytic effect of their buildings unless regulations demand it. They want their buildings to look good but are little interested in the logic of their aesthetic character.

In the 1930s, Empire State, Inc. (now Empire Reality Trust) was the developer of the Empire State Building in New York. Real estate interests were the developers of much of the commercial building stock built in the twentieth century in the City of London. William Levitt and Sons built vast tracks of suburban housing in post-World War Two United States. The houses consisted of a handful of standard types that proved so easy to adapt that the original types are now, a half-century later, barely recognizable. Government agencies have been responsible for vast tracts of social housing around the world. In the twenty-first century the large development companies are international in character and invest in projects across the world.

This characterization of the client needs to be taken with caution. Some only want to maximize the interest they receive on capital invested; others, the less optimistic, want to avoid financial failure at all costs. What they expect of buildings differs. Some do care about the catalytic effect of their buildings and wish them to meet more than minimum quality standards. What property developers have in common is the desire for actions be done their way—that is, the way they are used to dealing with them. They are primarily pragmatic prima donnas.

6.3.1.3 The Works

Cities around the world are replete with pragmatically designed buildings. Their Central Business Districts are full of office buildings designed by what are regarded as commercial architectural firms. Pragmatic building types include public buildings, railway stations, shopping centres, housing, and houses. Many pragmatically designed buildings have the aesthetic quality of the currently acceptable architectural style. It is a safe option. The Empire State Building (1930) in New York is an example.

The building was designed by William Lamb (1883–1952), a Columbia University and École des Beaux Arts trained architect, of the firm Shreve, Lamb and Harmon. For long the tallest building in the world at 104 floors in height, it was designed to be and was built rapidly during the economic doldrums of the depression. It took a little over thirteen months to complete. It is a pragmatic symmetrical building with setbacks as stipulated in the New York City building code. Clad in limestone and granite sheets typical of its era, it has Art Deco touches. Its design stands in strong contrast to the nearby Chrysler building, designed by William Van Alen (1883–1954) that is *the* exemplar of an Art Deco commercial building.

Gateway House (1969) in Manchester, England is Modernist in appearance. Pruitt-Igoe, a mass housing scheme in East St Louis, mimicked aspects of Le Corbusier's Unité d'habitation in its design. Similar projects in the United Kingdom also imitated the Unité but in a much-diminished form (Marmot 1981). The affordances of the housing types chosen failed to match their residents' needs and resources. The histories of many twentieth-century mass housing projects reflect the degree to which they matched the ways of life of their residents (Fig. 6.7).

Urban Design

Settlements from Miletus until today have been laid out in pragmatic grid iron-like patterns to make easily saleable and buildable lots. Perhaps the exemplars of pragmatic architecture are the suburbs of tract housing built around the world. Levittown on Long Island, New York (1947+), consisted of 17,000 houses each sitting detached on its own site. A team of workers built them as if they were labouring on an assembly line. Speedy construction enabled a rapid recovery of front-end construction costs. The houses provided starter homes for military veterans and their families. To ensure rapid sales, covenants excluded the sale of houses to African Americans, a pragmatic, if illegal, decision at the time. African Americans did not move into any of the Levittown houses for twenty years.

The design of mass housing is often driven by pragmatic necessities. The houses built for Africans during the Apartheid era in South Africa provided the very basic need for shelter. They could be modified as wanted by their inhabitants. The post-apartheid South African government now follows essentially the same once-derided model to meet housing demands of a rapidly growing population. It is a practical solution to a persistent problem. Much of the mass housing being built on a vast scale in China today is a highly pragmatic version of Bauhaus ideals. It is easy to design and build.

6.3.2 The Empiricists

The Empiricists of the twentieth century made several inventive imitations, or reinterpretations, of urban and architectural precedents. The work of individual empiricist schools of thought differed because their members were inspired by different antecedents. It also differed because architects had different clients carrying differing images of a good world in their heads.

The prescriptions of the Arts and Crafts Movement of the nineteenth century lingered into the twentieth and influenced the work of architects as diverse as Frank Lloyd Wright and Howard van Doren Shaw in America and Alvar Aalto (1894–1949) in Finland. Arts and Crafts buildings, in getting away from the mass production processes of industrialization, are characterized by structural authenticity, the use of native materials, earth tones, and attention to types of wood and their grains (Davey 1980; Livingstone and Parry 2005). The founder of the movement,

Photograph by Reading Tom/Wikimedia Commons

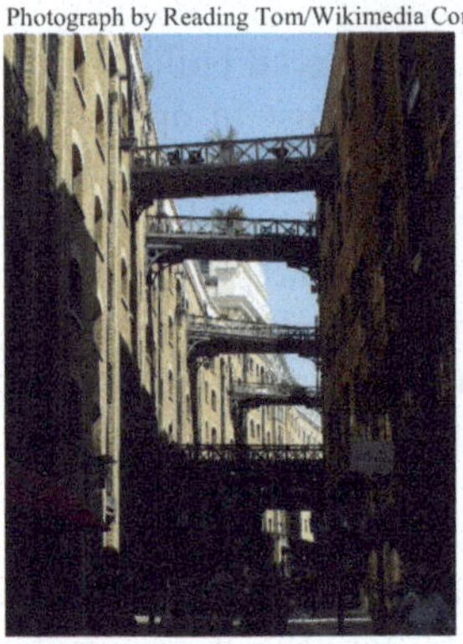

Photographer unknown/Wikimedia Commons

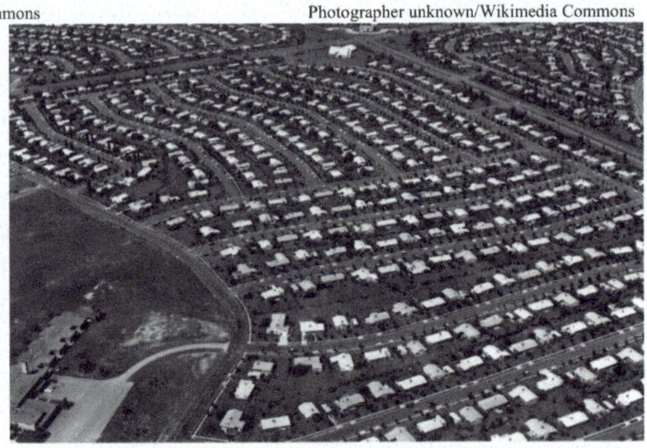

Butlers Wharf, London (1880s) (left) Levittown, Pennsylvania (1957) as originally laid out (right)

Photograph by Stephen Richards/Wikimedia Commons

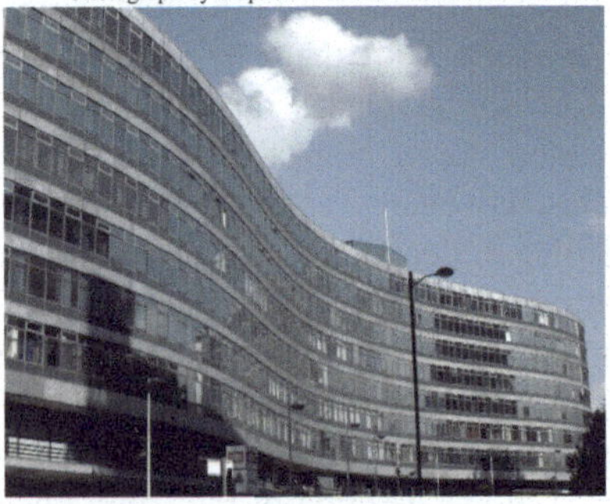

Gateway House (1969), Manchester

Photograph by Baycrest/Wikimedia Commons

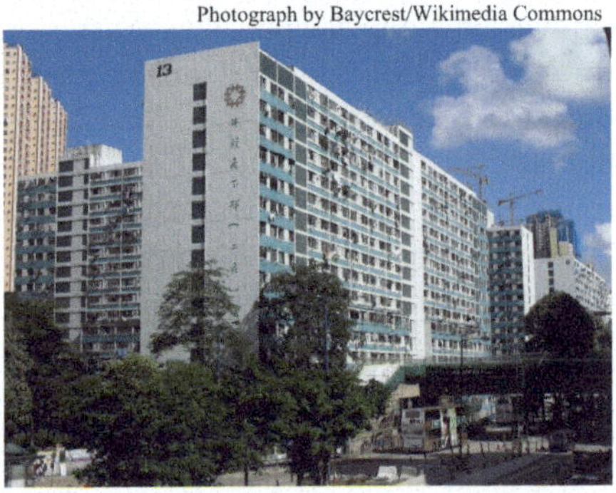

Lower Ngau Tau Kok (II) Estate (1960s), Hong Kong

Source: US Geological Survey, Department of the Interior

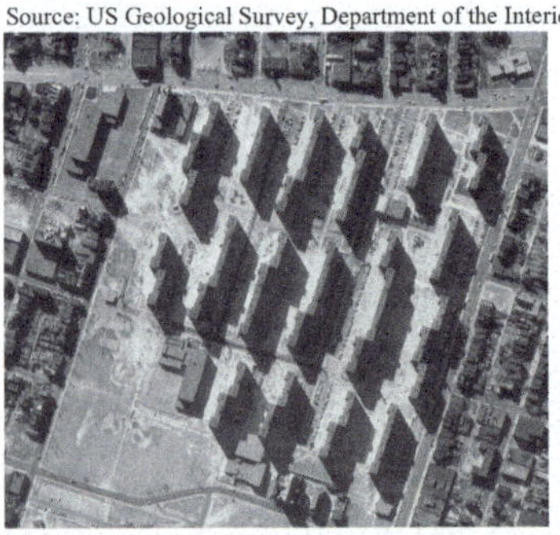

Photograph by the author

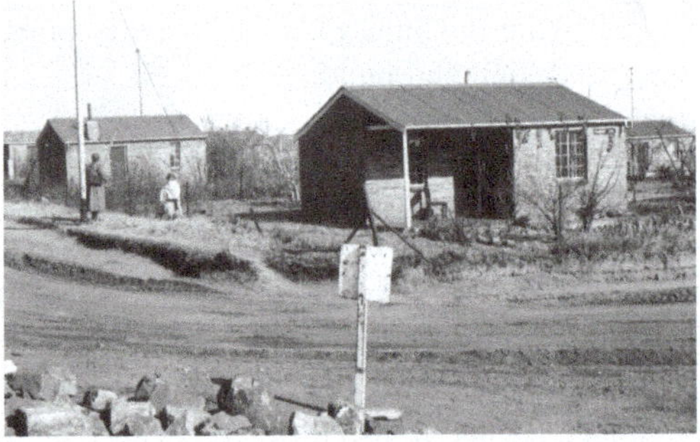

The modernist paradigm applied pragmatically: Pruitt Igoe (1950s), East St Louis, Missouri (left) and contemporary basic housing units in Soweto, Johannesburg (right)

Fig. 6.7 Pragmatic architecture

William Morris (1834–1896), was inspired by nature and the emerging socialist spirit in Europe. In contrast, Camillo Sitte was attracted by the medieval city, the liveliness of its settings, and its intricate built qualities. He presented his urban design ideas in *City Planning According to Artistic Principles* (1889).

Other architects looked at what was being pragmatically produced in cities and strove to shape the trends that they observed to obtain a better future. They sought an architecture that symbolically placed a design in its time and locale while fulfilling its instrumental function well. Hugh Ferriss (1888–1962), an American architect and illustrator, presented a speculative extrapolation of architectural trends and urban life in *The Metropolis of Tomorrow* (1929). It was more a prediction than a prescription for the future. Francisco Mujica (1899–?), a Mexican architect seeking an American architecture presented his prescriptions for the future city in *100-storey City in Neo-American Style* (1929). In the same vein Frank Lloyd Wright developed his ideas for a multi-layered skyscraper city in his 1926 article 'Skyscraper Regulation' (Wright 1969; see Lang 2021, 112–113).

Ebenezer Howard's prescription for future cities was different. He presented it in *Garden Cities for Tomorrow* (1902). He sought to synthesize what he regarded as the best aspects of country and of urban life into an ideal model. He drew his inspiration from the English country towns that he admired. Howard not only promoted a design idea but also a way of getting it implemented in the United Kingdom. He advocated the formation of limited dividend development companies financed by wealthy people of good will.

Howard's ideal town had a population of 32,000—the then population of Salisbury, his model. It would be self-sufficient in terms of employment opportunities and would have all the amenities of daily life easily at hand. It would, however, rely on access to the facilities that only a large city can supply. The generic model evolved into one that has a central core of commercial facilities surrounded by neighbourhoods each served by its own centre. The Neighbourhood Unit concept, as introduced in Chap. 4, was based on Clarence Perry's observations of life in Forest Hill Gardens (1909), a New York suburb on Long Island. Forest Hills had been designed by Frederick Law Olmsted Jr (1870–1957) to be the prototype for suburban development in the United States (Klaus 2004). It remains a much praised and much-loved neighbourhood.

Frank Loyd Wright, after his 1926 explorations on how to improve the pragmatically developing city, prescribed a radical agrarian urban type—Broadacre city—in *When Democracy Builds* (1945). His architecture drew on several antecedent ideas in striving for something American. His Prairie style houses drew on local architectural forms but were also inspired by Meso-American architecture. Art Nouveau and Art Deco architecture, which developed contemporaneously, differed (Huxtable 2004).

Art Nouveau was influenced by nature, Art Deco by the ocean liners and the cars of the modernizing world (Bayer 1999; Wolf 2013). Art Deco had its roots in France in the first decade of the twentieth century. It was a break from past building styles but drew on a variety of precedents. Its bold geometric forms, flowing lines, and colours reflected the art and architecture of the Vienna Secession. It flourished in the 1920s and 1930s and continued into the 1950s when it was overtaken by the International Style being promoted by CIAM and at the Bauhaus as *the* architecture for the future.

6.3.2.1 The Architects

Architects of the twentieth century brought very different, albeit masculine, middle-class backgrounds, to their work. Some studied architecture formally; others took the apprenticeship route. Both learnt much from mentors before starting their own practices. Edwin Lutyens studied architecture at the South Kensington School of Art and worked for Ernest George (1839–1922) and Harold Peto (1854–1933), an architect best known for his garden designs. Frank Lloyd Wright worked as a draughtsperson and studied civil engineering at the University of Wisconsin before working for Joseph Lyman Silsbee (1848–1913) and Louis Sullivan (1856–1924). The latter is often regarded as the father of American modernism but his high-rise buildings also included art nouveau motifs. His 1891explorations of skyscraper city designs were followed by Wright's studies of 1916.

Ebenezer Howard was not an architect but had a major influence on architects' thinking about urban design. He quit school at the age of fifteen. He later learnt much about the realities of the lives of people as a court stenographer. Along the way he was a farmer and journalist in America. He was much influenced in his thinking about cities and what they could be from the utopian time-travelling novel, *Looking Backward: 2000–1887* (1888) written by Edward Bellamy (1850–1898) an American political activist. Based on rationalist thinking, it described an ideal socialist America in 2000.

Architects increasingly became academically trained. Louis Kahn studied at the Public Industrial Arts School and the University of Pennsylvania where he was influenced by Paul Phillippe Cret (1846–1945), a French-born Philadelphia architect. Cret's own work was in a parsimonious Beaux Arts style with classical overtones. Christopher Alexander, the most empiricist of architects, studied architecture and then mathematics at the University of Cambridge. He completed his doctoral studies at Harvard University. He presented, as noted later in Chap. 7, his prescriptions for architecture and urban design in *The Timeless Way of Building* (1979) and, with colleagues, in *A Pattern Language: Towns, Buildings, Construction* (1977).

The major architects of Art Deco work included such luminaries as Finnish-born Eliel Saarinen (1873–1950) and French architect Hector Guimard (1867–1942). Guimard was the designer of the first Art Nouveau building—Castel Béranger (1899) in Paris and the Paris Metro station entrances. Raymond Hood (1881–1949), architect of the Neogothic Chicago Tribune Tower as well as Art Deco buildings such as those at Rockefeller Center in New York, and William Van Alen designer of the Chrysler Building are among other American architects of note.

A variety of architectural firms produced memorable Art Deco buildings during the 1930s. The Hoover Building in London was designed by Wallis, Gilbert, and Partners. Leading Italian architects produced many fine Art-Deco works in Asmara, Eritrea, now a UNESCO World Heritage Site. Both Briton- and Indian-headed architectural practices designed Art Deco cinemas in Mumbai. Several firms co-operated to build Art-Deco Napier in New Zealand after a 1931earthquake.

6.3.2.2 The Clients

The clients were generally conservative although progressive in spirit. Members of the British landed gentry employed Edwin Lutyens (Brown 1996). Other wealthy people invested in the limited dividend companies proposed by Ebenezer Howard. Mary Emery (1844–1927), a member of the American elite, believed it was her social responsibility to use her inherited wealth constructively. She sponsored the development of Mariemont (1923), a new town near Cincinnati designed by John Nolan (1869–1937). It catered to all income groups. Frank Lloyd Wright's clients varied but, in common, they had artistic tendencies or had been persuaded by others who had them. Some of his designs were self-funded. He had no client for his Broadacre City proposal or for his design of a mile-high skyscraper, The Illinois, that he proposed for Chicago. The exhibition of Broadacre City was sponsored by the National Alliance of Art and Industry, a Rockefeller family funded group. The model was paid for by Edgar Kaufman, the owner of Kaufmann's Department Store in Pittsburgh.

Wright's client for the Malcolm Willey House in Minneapolis was really his wife, Nancy. She wanted a dramatic avant-garde work of art but settled for a more realistic design. She worked closely with Wright and supervised the progression of the project. She was "a prototypical modern woman" (Sikora 2018). Liliane, an art collector, and Edgar Kaufmann were the clients for their weekend retreat, Falling Water. Their son, Jonas, who had read Frank Lloyd Wright's *An Autobiography* recommended that they commission Wright to design it. Solomen R. Guggenheim (1861–1949), a wealthy art collector who made his fortune in mining, was Wright's client for the eponymous museum in New York. Guggenheim wanted to have a temple to display the modern art collection that he had amassed under the guidance of Baroness Hilla von Rebay (1890–1967), an abstract artist. It was she who pressed Guggenheim to hire Wright.

The clients for Art Deco architecture were trendsetters. They wanted their buildings to be glamourous, innovative, and luxurious. Walter B. Chrysler (1875–1940), head of the Chrysler Corporation financed the Chrysler Building but it was William H. Reynolds, a real estate developer, who hired William Van Alen to design it. Griffith Jenkins Griffith (1850–1919), a notorious mining tycoon, funded the building of the Griffith Observatory (1935) for the City of Los Angeles. He was seeking to salvage his reputation. The Eden Atrium Sociedade de Cinemas funded the Eden Theatro in Lisbon, a noteworthy example of the popular world-wide approach to cinema design in the 1930s. When the Depression hit the Americas and Europe, clients' attitudes became more subdued. Architectural tastes followed.

Some clients acted as patrons seeking new solutions to recurring problems. The Russell Sage Foundation funded the development of Forest Hills Gardens in New York (Klaus 2004). While many company towns were pragmatically designed to house workers, industrialists such as the Krupp Company in Germany, and Cadburys in England built model towns drawing on richly functioning antecedents. The American new towns, after a brief period of Federal government support in the 1930s, were built by large scale property developers who saw a niche in the housing marketplace. Some of them, such as the Rouse Corporation, developer of Columbia, Maryland, were idealists (Mitchell and Stebenne 2007) (Fig. 6.8).

The people in Tel Aviv who hired Patrick (né Peter) Geddes (1854–1932), a Scottish biologist and sociologist and author of *City Development* (1904) and *Cities in Evolution* (1915), were idealists. They were familiar with the 'constructive surgery' rather than the demolish-and-build-anew approach that Geddes employed in his urban renewal work in Glasgow. Zionists Israel Zangwill (1864–1926) and David Elder (1865–1936), a Fabián socialist, pressed for Geddes' employment by the Zionist Commission in Palestine to design the extension to Tel Aviv because of his reputation for designing places that were modern but in context historically.

6.3.2.3 The Works

Empiricist designs varied. The early designs of Edwin Lutyens, for instance, were in the Arts and Crafts style with Tudor elements but his later works were more conventionally classical with contemporary Art Deco and even Modernist touches. His designs for buildings in New Delhi for the Imperial Government of India led to commissions for war memorials from the Imperial War Graves Commission and from commercial clients such as for the Midland Bank Building in Manchester (1928). Lutyens invented a new Indianized classical order for the Governor's Palace (now Rashtrapati Bhavan) in New Delhi. For its garden he drew on Mughal precedents so that it would be seen as Indian (Irving 1981; Ridley 2003). The bank building's simple proportions were calculated by Lutyens. The sculptures by local artist John Floyd are subdued.

Frank Lloyd Wright's contemporary residential designs have strong horizontal lines inspired by the flat landscape of the American west (Knight 2001; Huxtable 2004). The ground floors had open plans and the windows were large. As the Arts and Crafts movement advocated, much of the detailing of his designs was hand crafted. The Arthur Heurtley residence (1902) in Oak Park is an early example of his intentions. The design, highly adaptable, has undergone numerous changes starting from the moment that the Heurtleys moved in. His best-known house design is, however, Falling Water in Mill Run, Pennsylvania, designed for the Kaufmanns (see the frontispiece).

Many Art Deco buildings were built around the world. The Chrysler Building New York is among the best known. The Eden Theatro (1931) in Lisbon designed by Cassiano Branco (1897–1970) and Carlo Florencio Dias is one among many Art Deco cinemas in the world. The Griffith Observatory in Los Angeles shows the mixture of antecedents that many Art Deco buildings possess. It has elements of Greek, Roman, and Moorish forms while its plan is a Beaux Arts one. Its architects were John C. Austin (1870–1963) and Frederick M. Ashley (1870–1960).

The decorations on Art-Deco buildings varied depending on their location. It gave rise to such variants as Pueblo-Deco in Arizona (Breeze 2009) and Indo-Deco

Source: Sullivan (1891) Source: Mujica (1929, Plate CXXIV) Source: Ferriss (1929, 63)

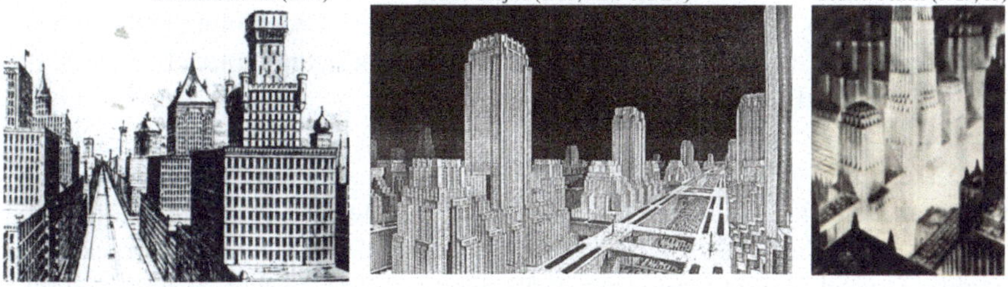

Louis Sullivan's design for a tiered city (left), a neo-American urban proposal by Francisco Mujica (left) and Hugh Ferriss' city of crowded tower (right)

Courtesy of the Chicago Architecture Center Photograph by the author

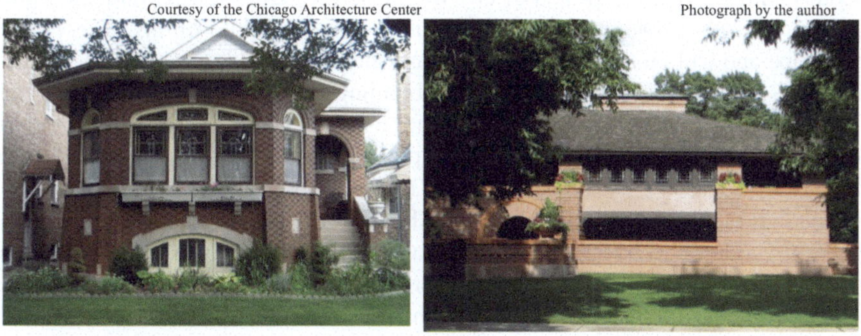

An Arts and Crafts house, Chicago **Arthur Heurtley House, Oak Park, Illinois**

Photograph by Limno/Wikimedia Commons Author's Photograph Photograph by Chris Simpson/Wikimedia Commons

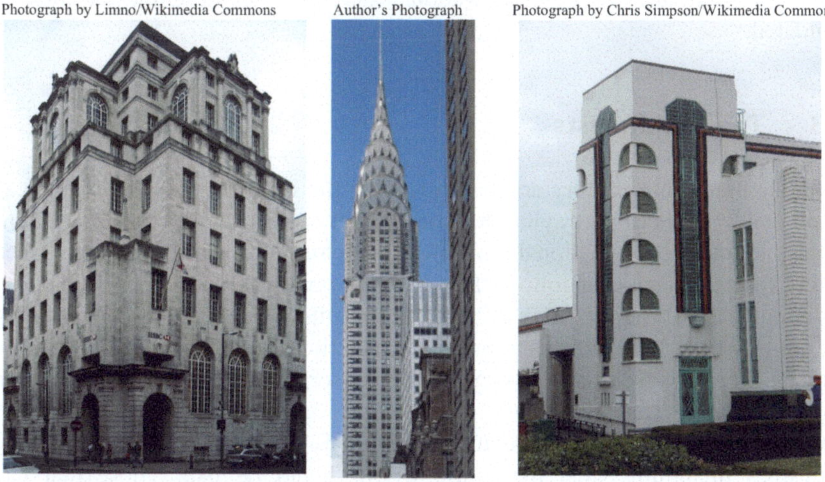

Midland Bank building, Manchester, Chrysler Building, New York, and Hoover House, London

Fig. 6.8 Early twentieth-century architectures

in Mumbai (Ramani 2007; Lang et al. 2022). The Hoover building in London borrowed its lines from the Art Deco work of Erich Mendelsohn (1887–1953) in Germany and its colour scheme from the Meso-American approach as presented at the 1925 Paris Exhibition (Betjeman 1974).

Urban Design

Individual Empiricists differed on what they selected as models of good urban environments to imitate. The urbanists produced designs such as Frank Lloyd Wright's Crystal Heights proposal (1939; Duran 2017). Earlier, Wright sought to improve on what was being built in Manhattan, New York (Wright 1969). Patrick Geddes who enjoyed the drama of city life prescribed a dense environment with a hierarchy of roads design for Tel Aviv. His design was based on his observations in Europe and Colonial India of what urban patterns worked well. Although much modified, his plan and the architecture of buildings by Bauhaus-trained immigrant architects fleeing Nazi Germany give the city its renowned International Style character (see Lang 2021, 126–127). Geddes' prescriptions differed considerably from those of the Garden City advocates. They had a different image of urbanism.

The first of the English new towns was Letchworth (1903+) and the second was Welwyn Garden City (1920+). Their impact as model types was worldwide. The two towns with their individual houses, tree-lined streets, and large open spaces became a model for the British new towns that were built as government policy after World War Two and many other new towns built in countries as diverse as Japan and Argentina. Contemporary Rationalists rejected all these lines of thinking as outdated.

6.3.3 The Rationalists

Late in the nineteenth century artists challenging the Beaux Arts orthodoxy began to explore new ways of portraying the world. Their searches led to the abstractions of cubist art and the international movement in architecture with its unembellished building forms. "Form ever follows function", an observation made by Louis Sullivan about the way objects look became "form follows function" as the working prescription of the Rationalists (Behne 1926; Giedion 1963, originally 1943). The Rationalists presented the world with exciting prescriptions for building and urban designs.

During the early twentieth century, individual Rationalists authored several manifestoes imagining what the future world should be like. Those that had the greatest influence on what was built were *Cité Industrielle* (1917) by Tony Garnier (1869–1948), the three books *Vers Une Architecture* (1923), *La Ville Contemporaine* (1923) and *La Ville Radieuse* (1933) by Le Corbusier, *Großstadt Architektur—Metropolis Architecture* (1927), by Ludwig Hilberseimer (1885–1948), and Walter Gropius' *Scope of Total Architecture* (1955), promoting his and

Bruno Taut's ideas from the 1920s. Their impact was strongest during the post-World War Two years from 1945 to 1970 but it has persisted into the twenty-first century.

Gropius and like-minded colleagues established the Bauhaus in 1919. It replaced the École des Beaux-Arts as Europe's leading architectural academy. Emigrant Bauhaus-trained architects fleeing Nazi Germany took functional architectural concepts across the world. It was an architecture of undecorated buildings that efficiently housed activities conducted efficiently. The structure and construction methods used were also efficient. Contemporaneously, the belief that architects had a social responsibility was promulgated by architects such as Bruno Taut and the *Gläserne Kette* (Glass Chain) group in Germany. The ability to solve social problems by substituting slums with salubrious and uplifting buildings and designs seemed intuitively obvious.

The prescriptions developed at the Bauhaus and, more particularly by Le Corbusier, found their home in CIAM. They were presented in a series of charters. The best known is the 1933 *Charter of Athens* (Le Corbusier 1933; Gold 2019). Sigfried Giedion, in *Space Time and Architecture* (1941), was another promoter of Rationalist ideas. Contemporaneously, bold prescriptive statements for buildings and urban designs came from the Futurists in Italy and the Constructivists in Russia (Lang 2021, 131–160, 171–203). The individuals and groups were unified in their embrace of modern technology, industrialization techniques, and by their rejection of applied decoration. In spirit, they looked back to the idealism and purity of Palladio's ideas (Rowe 1976).

What emerged was an architecture that rejected the ornamentation of the Neogothic, the historicism of the Neoclassical, and the Baroque qualities of Beaux Arts architecture. Embracing minimalism, it was an architecture of clean lines, consistent surfaces, floor to ceiling windows or strip fenestration. The roofs were flat, with broad overhangs and the compositions asymmetrical. It was all narrowly functional and formed a paradigm that could be, was and still is imitated internationally.

6.3.3.1 The Architects

The key figures in establishing rationalist modernist architectural thought were Walter Gropius and Le Corbusier. The supporting cast was vast. Ludwig Mies van der Rohe and Marcel Breuer (1902–1981) were associated with the Bauhaus before emigrating to the United States. Arne Jacobson (1902–1971) led the way in Denmark. Oscar Niemeyer (1907–2015) and Italian-born Lina Bo Bardi (1914–1992) produced fine designs in Brazil, and Minette de Silva (1918–1998) in Ceylon (now Sri Lanka). In the Soviet Union the architects included Konstantin Melnikov (1890–1974), Moisei Ginsburg (1892–1946) and Nikolai Ladovsky (1881–1941) (Reisch 2013). They advocated a constructivist design of angular, dynamic shapes employing bold technological possibilities. Ginsburg's manifesto, *Style and Epoch* (1924), presents a prescription for architecture similar to Le Corbusier's year earlier in *Vers Une Architecture*.

A second generation of Rationalist approaches to design differed from the first. Some architects were members of both generations as they learnt from experience and their ideas matured. Le Corbusier is a prime example. The design for the chapel at Ronchamp differs considerably from the rigidity of his earlier work. B. V. Doshi in India began his career as an acolyte of Le Corbusier. His early designs reflected it. Later, the realities of the country led to his more empiricist view of architecture. Alvar Alto, Eero Saarinen (1910–1961), a Finn, but one who made his career in the United States, and Danish architect Jørn Utzon (1918–2008) are among others who broke from the strict orthodoxy of the earlier Rationalists.

Contemporaneously, subgroups with less rigid agendas produced manifestoes on good architecture and urban designs. The *Team 10 Primer* (Smithson 1968; see also Risselada and van den Heuvel 2006), the *Habitat Bill of Rights* (1976), and *The Charter of Machu Pichu* (1979) were among them. They were responding to CIAM's dismissal of historic precedents and CIAM's strong focus on creating salubrious environments in which circulation was easy to the detriment of other concerns. Specific groups such as the Metabolists in Japan, and Archigram in England, strongly rationalistic, promoted their own agendas. None had the impact of CIAM on what was built.

6.3.3.2 The Clients

The clients for rationalist designs, in common, sought something new. They included individuals interested in architecture, government authorities who were not, ecclesiastical organisations, museum trustees, industrialists, and commercial firms. Some clients acted as patrons allowing architects to put the ideas they produced on paper into practice. Much housing work was produced for government agencies after heavy lobbying by architects. The Weissenhofsiedlung built for the 1927 Deutscher Werkbund Exhibition was paid for by the City of Stuttgart. The French Government paid for the Unité d'habitation in Marseilles after Le Corbusier finally convinced them of the rectitude of his ideas.

Many explorations of radical design prescriptions were sponsored by architects themselves. They defined the problem they wanted to solve and explored ways of doings so. The Wolkenbügel proposal of Lazar Markovich Lissitzky (1890–1941) was his personal project. The radical proposals of Archigram were financed by other architects, including David Rock (b. 1929), an influential practitioner. Rock later nominated the group for the Royal Institute of British Architects' Gold Medal, which they received in 2002. The explorations of the Metabolists were self-funded.

The early clients of Le Corbusier's, as already noted, gave him a free hand to develop both the program and the designs of their houses (Eaton 1969; Soth 1983). Pierre and Eugénie Savoye gave him and Pierre Jeanneret an opportunity to apply his "five points" of architecture. The resulting building is a landmark in the development of the International Style. Henry Fruges (1879–1974), a sugar refinery owner, after reading an article by Le Corbusier in *L'Esprit Nouveau* gave Le Corbusier and Jeanneret total control over the design of the housing for his workers in Pessac (Boudon 1972).

Clients were progressives. Carl Benscheidt (1858–1947) wanted a radical design that broke from the past for his Shoe Last Factory. Truus Schröder-Schräder (1899–1965), a socialite, had an unconventional view of family life. She wanted an avant-garde house without internal walls. She co-designed it with Gerrit Rietveld (1888–1964), a leading De Stijl furniture designer and architect. The close alignment of their values is reflected in the design. Edith Farnsworth, as noted earlier, was not as lucky.

Another of Mies' projects was paid for by a commercial corporation but it was really for an individual client, Phyllis Lambert (b. 1927). She, an architect, philanthropist, and daughter of the president of the Canadian liquor company Joseph Seagram and Sons, wanted a landmark building for the company in New York. She got it. The means was to set a Miesian building as a foreground element at the back of a plaza. It is surrounded by background buildings. Other clients inspired designs that have had a major impact on the subsequent direction that architecture has taken. Gabriel Voisin, Is an example. He, a friend of Le Corbusier, sponsored his Plan Voisin explorations. The two shared a common vision of the modern future.

National, state, and local governments in several countries developed major architectural and urban design schemes based on modernist principles. The Corporation of the City of London was the client for the Brutalist Barbican development (completed in 1982) in the City. Imitating many of Le Corbusier's ideas, but with a more empiricist touch, it, despite early criticism, has turned out to be a popular place in which to live. The United States Housing Authority, as subsidizer, and the Philadelphia Housing Authority as client, as described in Chap. 4 were less successful with their endeavours in the city. The gap between residents' requirements and what was provided was too large.

6.3.3.3 The Works

The work of the first generation of Rationalist Modernists appeared in Europe before World War One. Buildings such as the Fagus Shoe Last factory at Alfeld an der Leine (1911) designed by Walter Gropius and Adolf Meyer (1881–1929), and the urban design ideas of Tony Garnier (1869–1948) as shown in his *Cite Industrielle* drawings led the way (Garnier 1917). Rationalist architecture had, however, evolved piece-by-piece over the last quarter of the nineteenth century. The work of architects such as Charles Rennie Mackintosh (1868–1928) and his design for the Glasgow School of Art (1896+) and Antoni Gaudi (1852–1926) and his Catalonian modernism in Barcelona showed that new design attitudes would find clients who favoured them.

Le Corbusier's early designs, as noted earlier, were driven by the spartan, puritanical view of life in which he was raised (see also Chap. 8). In *Vers Une Architecture* he reduced building forms to rectilinear, plane surfaces, cubes, and, occasionally, cylinders. The building stood on pilotes, had strip fenestration patterns and flat roofs. They had no applied decoration. It was a powerful, easy to imitate, model.

Many fine functional buildings following Bauhaus and CIAM principles were built in the world. The Bauhaus in Dessau (1926) designed by Walter Gropius, the Villa Tugendhat (1930) by Mies van der Rohe and Lilly Reich (1885–1947), an interior and textile designer, the Unité d'habitation in Marseilles (1953) designed by Le Corbusier, and the Harry and Penelope Seidler House (1966) designed by its owners in Sydney are among them. In them, the aspirations of architects, property developers, and users coincided.

The least successful rationalist designs have been the mass housing projects built for low-income families beset by financial and social problems. While the Rationalists understood that salubrious, environments are the basis for a good life, they failed to consider the richness of settings required to provide for truly functional environments (Jacobs 1961; Brolin 1974; Blake 1977; Wolfe 1981; Scott Brown 2004). Sponsors were unwilling to pay for them to meet more than basic functional needs (Figs. 6.9 and 6.10).

Team Ten's prescriptions for achieving more liveable environments than those advocated by the mainstream of CIAM's members, although of great interest to the cognoscenti for the internal logic of their designs, did not achieve the ends that the group sought. Robin Hood Gardens, a housing estate in London completed in 1972, was demolished in 2017. It neither afforded the quality of life that most of its residents wanted nor functioned in the way predicted by its designers, Alison (1928–1993) and Peter Smithson (1923–2003). Many leading architects provided staunch support for its retention because of the central architectural idea on which its design was founded.

A second phase of rationalist modernism saw more diverse artistic expressions in building design but no significant reform in urban design. It also displayed a greater self-conscious concern for the symbolic function of built forms and the meanings they potentially communicated. In his design of Chapelle Notre Dame du Haut (1953) in Ronchamp, Le Corbusier strove, as a puritan, to show a sensitivity to the values of Catholicism. The chapel's design is rich in symbolism, both sacred and profane, available for the cognoscenti to contemplate (Samuel 1999). Lay people can respond to the qualities of its form, spaces and light.

Several other buildings of note showed more free flowing sculptural qualities than earlier rationalist prescriptions allowed. Among them are the TWA terminal at New York's Kennedy Airport designed by Eero Saarinen (1910–1961) and the Sydney Opera House designed by Jørn Utzon (1918–2008) and completed by Peter Hall (1932–1975) with considerable engineering assistance from Jack Zunz (1923–2018) of Ove Arup Associates. These two buildings sufficed in meeting their instrumental functions but it is their sculptural quality that attracts attention. The TWA building, with extensive modifications, is now a hotel lobby and the Sydney Opera House has undergone renovations to enhance its instrumental functional qualities. The most successful function of the Opera House has been its catalytic effect. It placed Sydney in the world's eyes.

Source: Wagner (1912)

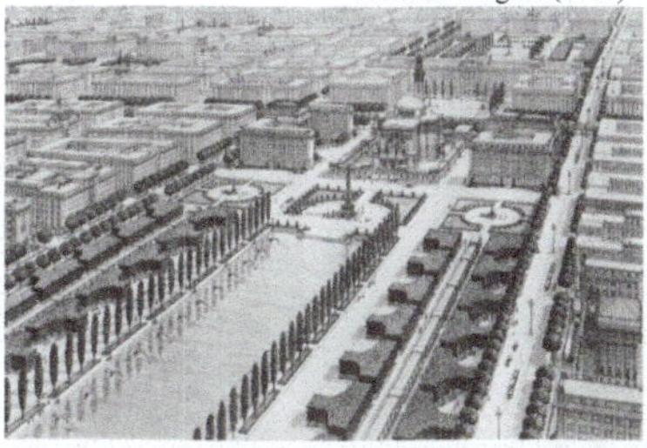

Source: Garnier (1917)/ Wikimedia Commons

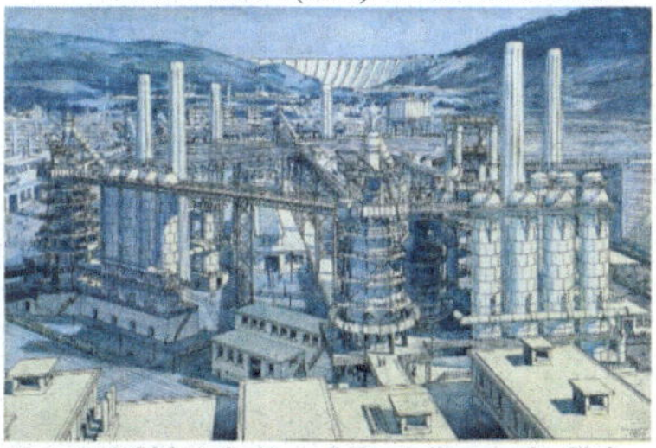

Early Rationalist urban designs: Otto Wagner's Großstadt (1900s) (left) and Toni Garnier's *Cité Industrielle* (1906+) (right)

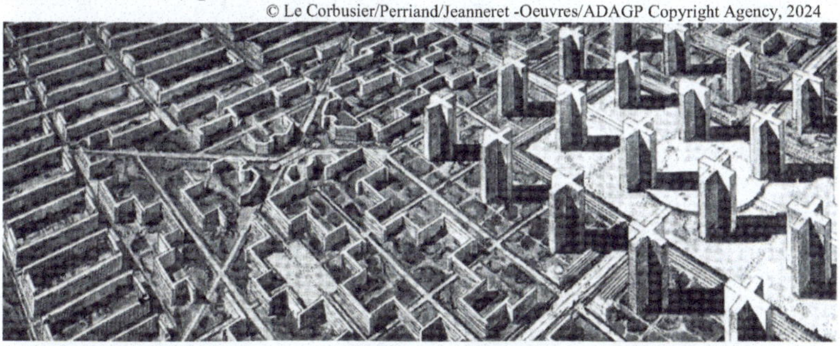

Le Corbusier's La Ville Contemporaine (1922)

Photograph by Carsten Jansen/Wikimedia Commons

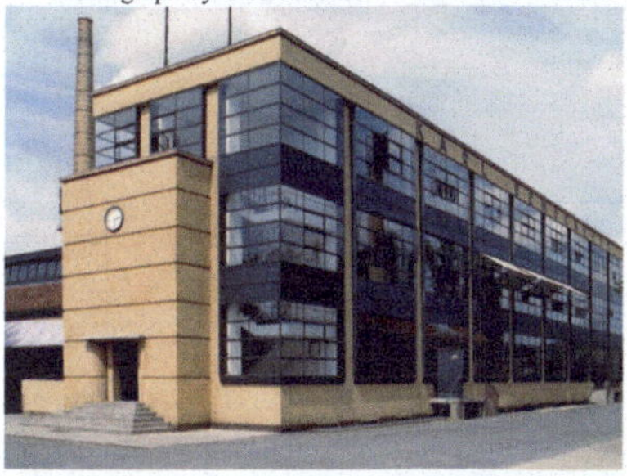

Photograph by Svetlov Artem/Wikimedia Commons

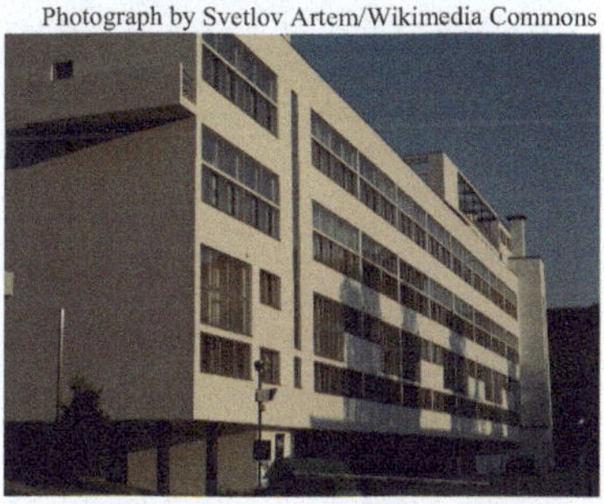

Early Rationalist architecture: The Fagus Shoe-Last factory, Alfeld an der Leine (1911) (left) and Narkomfin, Moscow (1926- 1928) as rehabilitated (right)

Photograph by the author

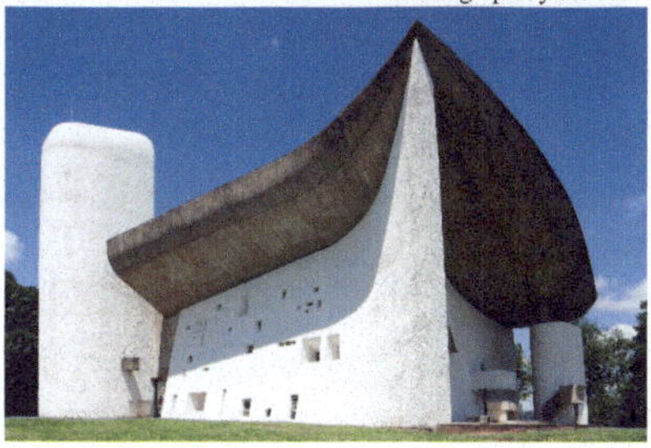

Photograph by Wladyslaw/Wikimedia Commons

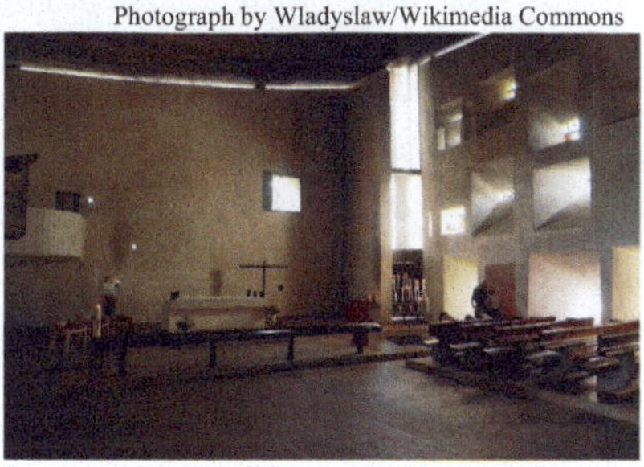

Second generation rationalism: Notre Dame du Haut, Ronchamp (1950): Exterior (left) and interior (right)

Fig. 6.9 Rationalist prescriptions for urban designs and architecture

Photograph by the author

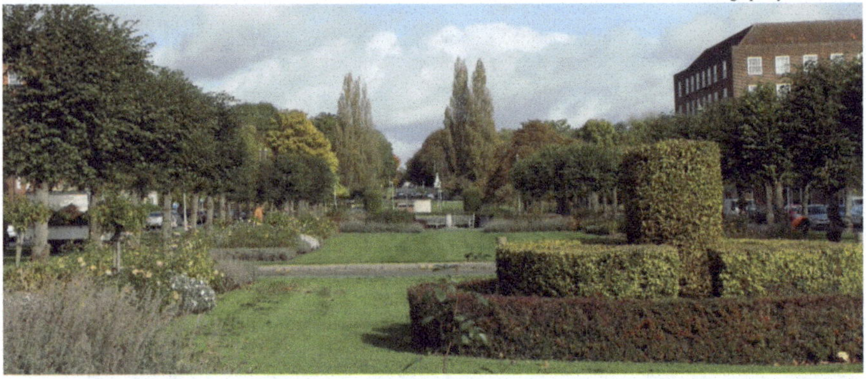

Parkway, the central axis, Welwyn Garden City

Photograph by the author

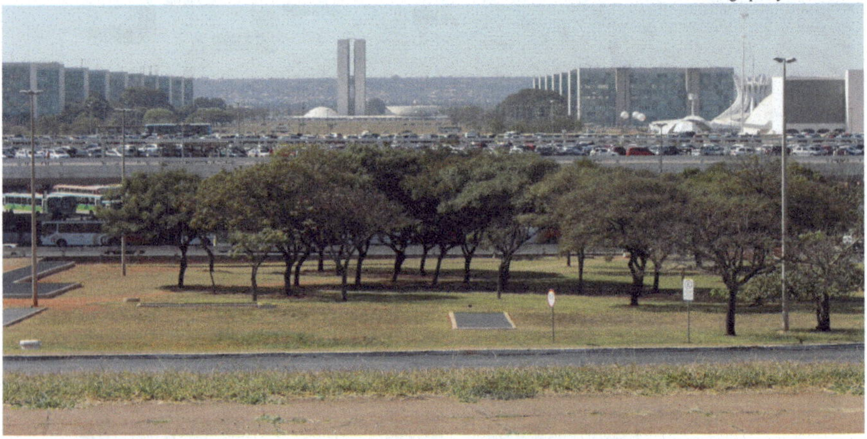

The Monumental Axis, Brasília

Photograph by the author

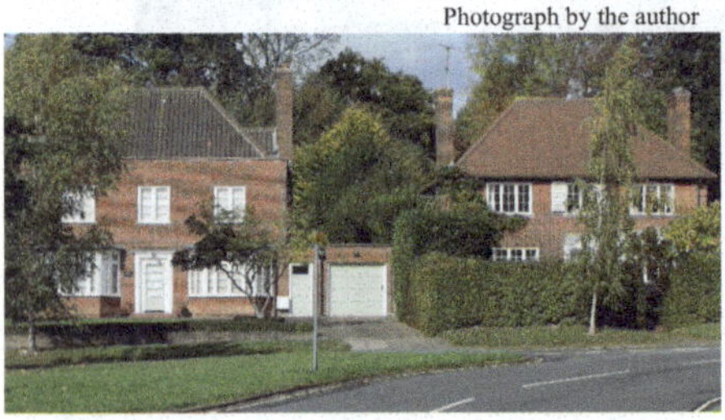

Houses, Welwyn Garden City

Photograph by the author

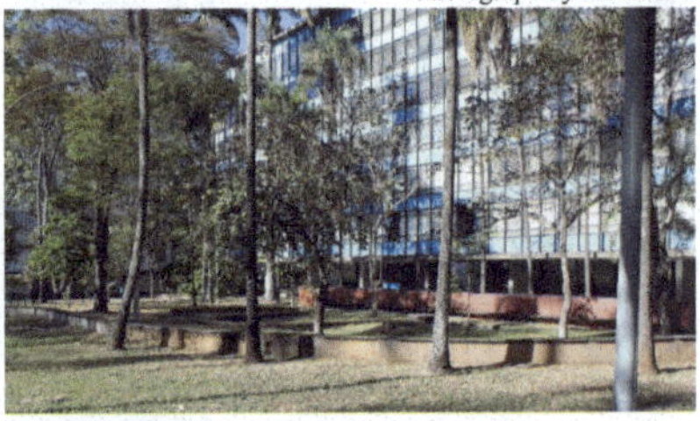

Housing, Pilot Project, Brasília

Fig. 6.10 Empiricist and rationalist views of urban design

Urban Design

Le Corbusier's urban design ideas, as presented in *Urbanisme* (1925) and *La Ville Radieuse* (1933) were bold and severe. Their orthogonal grid of streets was designed for the movement of traffic not as the location for daily life. He imagined he could drive his Voisin C7 speedily along them. Pedestrians had their own largely independent paths. Buildings were in slab or tower block form. Recreation facilities abounded. It was an ideal system for ideal people (Ellis and Cuff 1989). The exemplar of Rationalist urban design is the Pilot Plan for Brasília.

The plan was a competition winning entry by Lúcio Costa (1902–1998). It features two major axes. The monumental axis lined by government buildings has at its heart the capital complex located on the Praça de Três Poderes. The cross axis links the residential superblocks that contain housing, schools, shops and community facilities. The housing in the southern part of the plan was designed by Oscar Niemeyer. His design owes much to Le Corbusier's thinking. It is laid out in an orthogonal pattern with the apartment buildings sitting on *pilotes*. Each superblock is divided into four quadrants by a service road that was designed to be used for provisioning local shops. That vehicular service corridor instead of serving the shops is now, pragmatically, a pedestrian spine. The backs and fronts of the shops have been reversed to form a traditional ribbon shopping strip along the axis (Lang 2021, 186–189). The open spaces between the apartment blocks are planted with trees and lawns and have children's playgrounds. As the population has aged in place the children attending the schools are no longer primarily residents but those of the servants working there.

Substantial housing estates following Bauhaus principles were built in the Post-War years. They have met with varying degrees of success depending on the extent to which they match their residents' desired ways of life, concepts of privacy, aesthetic values, and financial abilities. The 1950s Pruitt-Igoe housing scheme in East St Louis (see Fig. 6.7) did not fare well. It has been demolished. The vast 23 de enero housing scheme (1950s) in Caracas survives although all the open space between apartment blocks that was characteristic of the Rationalists' ideal of cities set in parks is now crowded with squatters' houses. Their builders saw the opportunity to make a home for themselves on unused land. Such schemes still are being built pragmatically in many places including on a vast scale in East Asia and being exported to countries such as Angola in which the Chinese government is investing.

6.4 Self-conscious Design: Post-modernism

Challenges to the Empiricists' and Rationalists' prescriptions came from the pragmatic designs sought by property developers, the Neoclassicists, and from the Nordic Classicism work of Scandinavian architects. The last-mentioned group included Kay Fisker (1893–1965) in Denmark, Gunnar Asplund (1885–1940) in Sweden and Alvar Aalto in Finland. Very differently, the design of Port Merion (1925–1976) in Wales by Sir Clough William Ellis (1883–1978) set a precedent for a more eclectic architecture and urban design. The popularity of its concatenation of styles and colours made architects rethink their architectural and urban design ideas (Rowe and Koetter 1979). During the 1960s and 1970s other architects began to react to the pure geometries and the use of steel and glass as symbols of modernism and the general placelessness that such designs had promulgated (Relph 1976; Ghirardo 1996).

Two books clearly proposed a new prescription for architecture. In *Towards an Organic Architecture* (1950) Bruno Zevi argued for an architecture of complexity not simplicity. The key "gentle" manifesto rebuking Rationalist ideas was, however, Robert Venturi's *Complexity and Contradiction in Modern Architecture* (1963). Venturi prescribed a diverse, ambiguous, and expressive approach to the design of buildings. Architecture, he argued, should embrace complexity and contradiction rather than simplicity and purity. It should reflect the multiplicity of human experiences. Denise Scott Brown, his partner, added that popular cultures should be the springboard for creativity. Venturi considered Mies van der Rohe's "less is more" to be "less is a bore". He noted:

> In the 1960s the use of history as allusions was exciting and new. History had been called bunk by Hitler and Goring and irrelevant by Modernists whose slogan was "We're starting again". We were not starting again, we were evolving. But because modernism had become an old revolution, we were revolutionary in being evolutionary. (cited in Dickinson 2023)

Charles Moore (1925–1993) took this argument a step farther in the essays he wrote for *Perspecta.* He prescribed an architecture of "everything not nothing" and an urbanism of "juxtaposed fragments". Port Merion's and Disneyland's conglomeration of architectures were the precedents. Neither Venturi nor Moore were looking at past forms with nostalgia. They were developing ideas about the communication function of architectural forms and the decoration of buildings.

By imitating in an abstract manner past forms, postmodern architects sought to create buildings using a richer and wittier set of patterns. The referents they chose were, however, seldom obvious and had to be explained to even the cognoscenti. More easily understandable were the less-abstracted designs that drew on Classical elements as in the work of Taller de Arquitectura led by Ricardo Bofill (1939–2022). The Taller used classical forms at an exaggerated scale to make mundane buildings heroic.

Other approaches to architecture were also contradicting the tenets of modernist architecture. One was the architecture of structural exposure where the framework of a building is displayed on its façades. It was mainly, but not entirely associated with the New Brutalist movement in the United Kingdom. Another approach to enriching architecture was shaped by the deconstructivist philosophy of Jacques Derrida (1930–2004), a French philosopher. Derrida challenged the status quo of taken-for-granted interpretation of meaning in literary texts (Wigley 1993). Applying his views to design, led to a deconstructivist architecture of unease appearing around the world (Fig. 6.11).

Source: Scott Leonard for the Kellogg Dolittle Estate: Courtesy of Kellogdolittlehouse.com

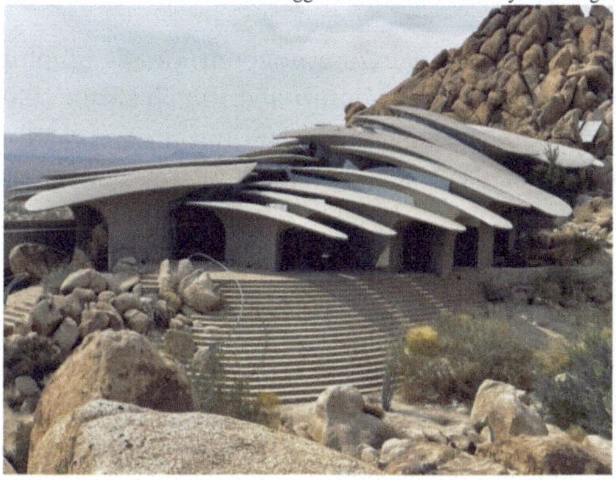

Doolittle House, Joshua Tree, California (1988+)

Photograph by the author

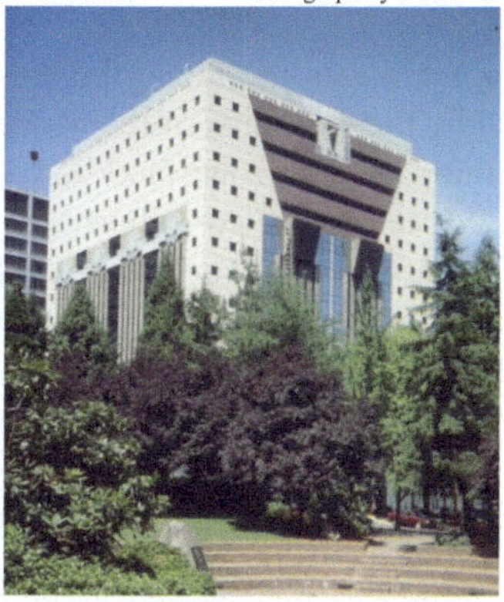

Municipal Services building, Portland (1980)

Photograph by Colros /Wikimedia Commons

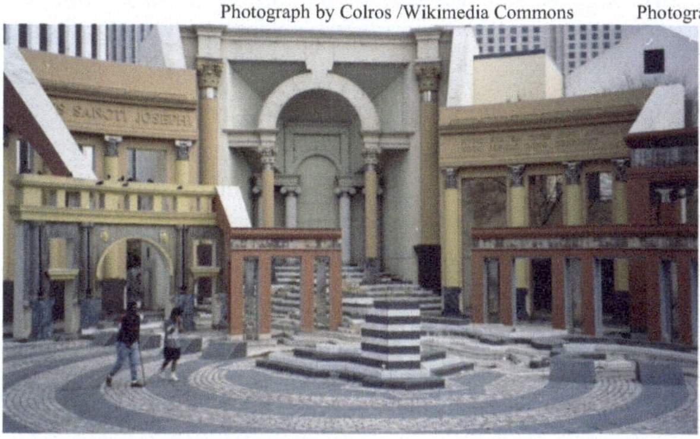

Piazza d'Italia, New Orleans (1975+)

Photograph by Gary Knight/Wikimedia Commons

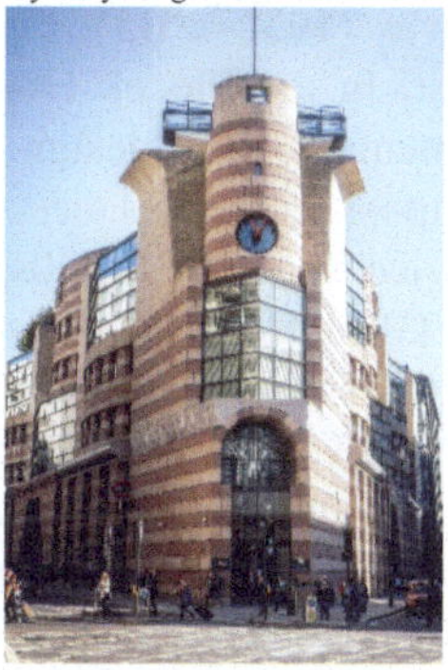

1 Poultry Street, London (1997)

Photograph by the author

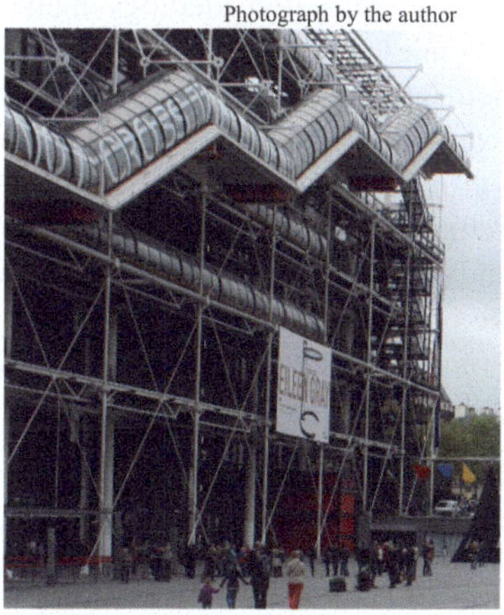

Pompidou Centre, Paris (1974+)

Photograph by Fredericus Magus /Wikimedia Commons

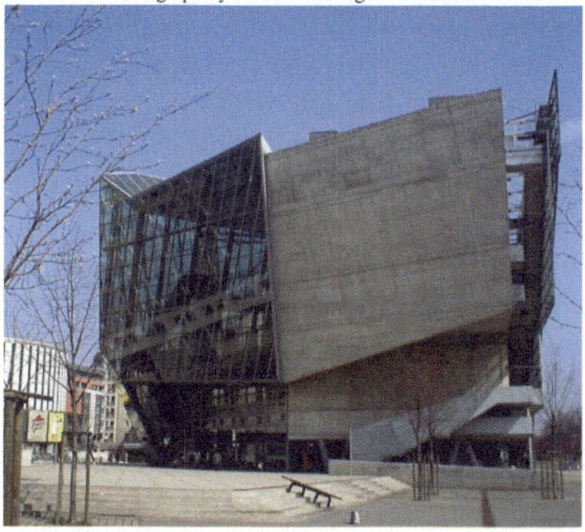

UFA-Kristallpalast, Dresden (1988)

Fig. 6.11 Post-modernist architectures

6.4.1 The Architects

The architects producing postmodern buildings were/are nonconformists uninhibited by the strictures of the Modernists. The firm of Venturi and Rauch (from 1989, Venturi, Scott Brown and Associates) led the way in forwarding an architecture based on abstracted patterns of precedents. Classical architecture influenced the explorations and work of several architects. Leon Krier in his Atlantis designs set the pace (Verbruggen 2017). The best-known Postmodernist in England is (John) Quinlan Terry (b. 1937), an architect both admired and derided for his love of traditional English architecture. His design of the Maitland Robinson Library (1992) in Cambridge is a Neoclassical design but his work elsewhere, such as for the Richmond Riverside in London (1984; see Fig. 3.9), draws on other historical forms and details. While imitative, it is not in the abstract manner advocated by Venturi.

The architects most closely aligned to Venturi's approach include Michael Graves (1934–2015). Graves found his early Rationalist work to be lacking in spirit. In the Plocek House (1977) in New Jersey, he drew on abstract classical elements for his design. He used colour and decorative elements to enrich the composition. He is best known, however, for his design of the Portland Public Service Building (1980s). Philip Johnson (1906–2005) followed a similar route. His Glass House in New Canaan, Connecticut (1944+) imitates the work of Mies van der Rohe but later, with John Burgee (b. 1933), he designed 550 Madison Ave in New York, a pragmatic design with a Chippendale cornice. The non-conformist designs in England by architects such as James Sterling (1926–1992) and Tony Farrell (b. 1938) draw on a variety of antecedent ideas and styles. Few of the architects designing Postmodern buildings show the intellectual rigour of Venturi and Scott Brown in their work.

Those architects whose work is aligned with deconstruction ideology include such luminaries as Bernard Tschumi in France, Frank Gehry in the United States, and, until her death, Zaha Hadid in England. Their influential designs reflect ideas occurring in the fine arts world. Frank Gehry's works, for instance, are juxtaposed collages of space and materials. Some are like crumpled bags. In contrast, Coop Himmelb(l)au (founded in 1968), a Viennese architectural firm noted for its flamboyant designs, was influenced by the Dadaist and Surrealist art movements of the 1920s and 1930s. The firm sought an architecture of "light and fluctuating clouds" (Mönninger 2010). While serving instrumental functional ends, their designs are, primarily, sculpturally exciting works of art.

6.4.2 The Clients

The clients for postmodern work were government corporations, companies, and wealthy families. They sought to get away from existing design expressions. Robert Venturi, himself, was the client for the Vana Venturi House that he designed for his mother. The well-educated Wislocki and Trubek families were the clients for the houses Venturi and Rauch designed in New England. A corporation, Best Products, was the client for the firm's design of the company's showroom in Langhorne, Pennsylvania. The City of Columbus in Indiana was the client for Fire Station No. 4 (1967) there. It had, however, a patron: The Cummins Foundation.

The foundation was established by the head of the Cummins Engine Company, J. Irwin Miller (1909–2004), a lover of architecture. It pays the design costs for public buildings designed in the city and surrounding county to be designed by internationally renowned architects. The buildings established Columbus as an outdoor museum of modern architecture. Miller wanted his hometown to have international recognition.

The goal of promoting a city was and is the end of many public clients who commissioned buildings displaying innovative architectural ideas. The Corporation/City of New Orleans was the client for the Piazza d'Italia (1978) designed by Charles Moore and Perez Associates in the city. The City of Groningen was the client for Bernard Tschumi's design of the Glass Video Gallery, a deconstructivist work. Much of the Deconstructivist work in Europe was paid for by city and national governmental agencies but it was often promoted by other people. In France, for instance, Presidents Valéry Giscard D' Estaing (1926–2022) and François Mitterrand (1916–1996) sought to make Paris the world's leading art centre.

Zaha Hadid's first built work was sponsored by another lover of advanced architectural ideas, Rolf Fehlbaum. In strong contrast to Fehlbaum's advocacy for Zaha Hadid's work was that of Hans-Jurgen and Helga Müller, art gallery owners in Cologne. They commissioned Leo Krier's Neoclassical Atlantis proposal (1986). Very differently, the communist-led council of Seine-Saint-Denis wanted a social housing scheme whose aesthetic did not stigmatize its inhabitants. The Taller de Arquitectura used classical elements at a large scale to achieve that end in the design of Les Espaces d'Abraxsas (1978–1983) in Noisy-le-Grand.

6.4.3 The Work

Five clear types of post-modern designs can be identified. The first is the organic architecture sought by Bruno Zevi. The second and third focus most clearly on the functions of building as communicators of identity. The first of the two used historical elements in an abstract manner and the second literally. A fourth approach is, explicitly or implicitly, the architecture of structural exposure and the fifth, the basing of the central idea of a design on the deconstructivist philosophy of Jacques Derrida and/or the Constructivist art of Russia. A sample of schemes demonstrates these prescriptions.

The Doolittle House designed by Kendrick Bangs Kellogg (1934–2024), is a 'sublime example' of organic architecture (Morris 2019). Built for artists Bev and Jay Doolittle in Joshua Tree, California, the house consists of twenty-six shell-like structures cantilevered out from the boulders of the site. It is a dramatic example of Kellogg's work and Bruno Zevi's advocacy. The intellectual underpinning of Venturi's designs differs.

The Vana Venturi house is Classical and Mannerist in its monumental, if at a much-reduced scale, façade (see Fig. 4.8). The façade has a broken pediment and applied ornamentation. The roof is a nineteenth-century shingle-style one while the interior is strictly modernist. The combination of ideas reflects the concept of a formal complexity and contradiction of ideas in a design. The Best Products Catalogue Showroom designed by the firm is a 'decorated shed'—a standard type with an avant-garde façade (Venturi et al. 1977). The box-like structure is clad in decorative panels painted with outsize flowers. It pays homage to folk culture.

In the third approach, the referents are more obvious. The easiest to understand is the work of Leon Krier in his Atlantis proposals (Verbruggen 2017). Atlantis with its agora and acropolis represents classical postmodernism. Other architects incorporated classical elements in a more playful manner. The design of the Piazza d'Italia in New Orleans is an example. Its central focus is a fountain organized as a relief map of Italy. The plaza is surrounded by a hemicycle colonnade and contains a clock tower, a campanile, and an abstracted temple façade (Lang and Marshall 2016).

Classical elements also appeared in a postmodern form, often abstracted, and changed in scale. Les Espaces d'Abraxsus designed by Ricardo Bofill and the Taller de Arquitectura is an example. On an even grander scale is the firm's contemporary design for the Quartier Antigone in Montpellier (1979+). It consists of grand neoclassic structures that have large pediments, entablatures, and classical pilasters. Its commissioning client, Georges Frêche (1938–2010), was a controversial, socialist, mayor of the city (Lang 2021, 243–244). He sought something grand.

Somewhere in between these approaches are the Municipal Services Building (1986) in Portland, Oregon designed by Michael Graves and the Secret Intelligence Service Building (1994) in London designed by Terry Farrell and Partners. The former has prominent decorative features and a variety of surface materials and colours. The latter has allusions to Mayan and Aztec temples. Similar approaches to building expression occurred around the world. In India, for instance, Gautam Bhatia (b. 1952) identified a series of domestic architectural types depending on the elements of past architectures on which they drew. Punjabi Baroque, Anglo-Indian Rococo, Bania Gothic were among them (Bhatia 1994). These approaches have in common a richness in visual character. They can also be interpreted as parodies of antecedents.

A different approach to design is shown in the architecture of structural exposure. The Centre nationale d'art et de culture Georges Pompidou (1977) in Paris designed by Renzo Piano (b. 1937) and Richard Rogers (1933–2021) has its structure, pipework, stairs, and elevators located on its façades. The stairs have treads of 'transparent' steel metal grids. The building gets added prominence from standing in strong contrast to Haussmann's Paris.

The Deconstructivist approach to architecture is displayed in Coop Himmelb(l)au's unbuilt design of the Merz-Schulz (1981). Its UFA Cinema Centre

(UFA-Kristallpalast) completed in 1998, in Dresden is a mature example of this genre of architecture. The Jewish Museum (1999+) in Berlin designed by Daniel Libeskind and the Vitra Firehouse (1993) by Zaha Hadid are others. A later scheme is the Walt Disney Concert Hall (2000) in Los Angeles designed by Frank Gehry. It is, as noted earlier, a work of sculpture with a mixed shoebox-vineyard type concert hall at its core. The Deconstructivist scheme most specifically related to Derrida's philosophies is, however, the Parc de la Villette (1984+) designed by Bernard Tschumi Architects (see Chap. 8).

6.5 A Postscript: Self-conscious Design in the Early Twenty-First Century

Today, much seems to be haphazardly designed. The justification is that it is in response to the perceived fragmentation of societies in an era when competitive, neo-liberal, 'anarcho-capitalist' politics dominate the world. As American architect Daniel Libeskind (b. 1946) noted "We are not living in [the era of] Haussmann's Paris. We have a pluralistic society" (cited in Lubell 2004, 47). His view echoes the calls to accept a radical heterogeneity in architecture today. Neoclassical designs co-exist with the Modernist, Neomodernist, Hypermodernist, and Neotraditional. The Neogothic is not extinct. Many buildings continue to be pragmatic responses to clients' basic demands.

Several large architectural practices today produce highly derivative but paradoxically innovative classical buildings. Their work is neither copy nor pastiche. The firms include those headed by South African born Allan Greenberg (b. 1938) and by John Blatteau (b. 1943) in the United States, John Simpson (b. 1954) in the United Kingdom, and Pierluigi Nicolin (b. 1941) in Italy. The Schermerhorn Symphony Center, a shoe box music hall (2006), in Nashville, Tennessee is the work of David M. Schwarz (b. 1951) Architects. Neoclassical architecture is seen as an architecture of enduring good taste in contrast to the ephemeral search for visually exciting building forms (Dowling 2004).

Many housing complexes in Europe and the Americas that were built in the early Post-World War Two years, either copying or imitating the Rationalist's design paradigm, have been demolished. In their stead Neotraditional schemes have been built that afford a wider range of functions than those offered by the Modernist buildings and urban environments that they replace. Many Modernist projects, nevertheless, continue to be built, particularly in East Asia. They function well enough in affording the ways of life and privacy needs of their inhabitants. Modernist architecture also thrives in a richer form. It has been called, appropriately, Neomodern.

Neomodern architecture retains the restraint and clarity of pure Modernist patterns but it is more varied in its geometric character, its interior spatial arrangement, and structural system (Glendening 2010). Much that is being built recognizes the potential role of buildings as space makers rather than as objects in space. This attitude is reflected in the work of architects such as the Venetian Carlo

Scarpa (1906–1978), who sought a regional modernism that respects an area's crafts tradition (Beltramini 2007), and the designs of 1996 Pritzker Prize winner (José) Rafael Moneo Vallé (b. 1937). Moneo's Neomodernist works include the Murcia City Hall extension (1990s) with its façade of broken symmetries, the Cathedral of Our Lady of the Angels in Los Angeles (2002), and the extension to the Prado Museum in Madrid (2007) (González de Canales and Ray 2015). Neomodernist design is more colourful too. Hypermodern architecture is much bolder in its geometries and, often, has greater structural ingenuity than pure Modernist or Neomodernist design.

Hypermodern architecture is, perhaps, *the* architecture of Neoliberal capitalism. Corporations seek buildings that impress and differ in appearance, if not in spirit, from others. They come in a variety of external geometrical shapes and colours and are built of steel and glass or free-flowing concrete. Each new building strives to outdo its competitors in its expression and recognizability. They are designed to be symbols of the future. Despite their novel forms they have generic interior layouts as they need to be pragmatically efficient and provide a sound return on the capital invested in them.

The rectangular, narrow, towers on small sites designed by architects for property developers, are pragmatic solutions to the need to increase the housing stock for the wealthy in cities such as New York. They represent the architecture of neo-liberalism as much as the hypermodern buildings on display in the United Arab Emirates and elsewhere. They test the limits of construction technologies and demonstrate how to maximize the profit that can be made on small sites by stacking apartments one above the other. They are anonymous structures of pragmatic capitalism.

Cities such as Doha have much-admired, bold, skylines of buildings in a variety of configurations of Hypermodernist buildings. Ashgabat in Turkmenistan is a mixture of awe-inspiring Revivalist, Modernist, and Postmodernist buildings. Designed to impress, the making of sound environment for pedestrians is neglected in their urban designs. Much Hypermodern architecture and urbanism has caught the imagination of the public and municipal governments whose wish is to be seen as up to date in an economically competitive world. It possesses a "wow" factor that amazes the beholder. If Anne Tyng's hypothesis holds, the world will see a return to a more restrained architecture in the future.

The Empiricists response to the perceived limitations of rationalist designs is equally bold. While garden cities and suburbs continue to be designed much as its advocates envisaged in the late nineteenth and early twentieth centuries, the prescriptions promulgated by the Neotraditionalists provide a radical rethinking of architecture and urbanism. Their goal is to ensure that buildings reflect their locale and their urban designs favour the pedestrian over vehicular traffic (Grant 2006; Congress for the New Urbanism 2013). The resulting work is less visually dramatic than that of the other contemporary advocacies. It is often considered to be bland.

The Neo-Traditionalists look for precedents in the local building stock and vernacular architecture. They seek to understand the principles on which their forms, fenestration patterns, ground floors, cornices, and roofs were based. They analyse how the materials available shaped building forms, structures, and details. They look at street patterns, the building-street relationship, and how vegetation was managed in urban areas and suburbs (Grant 2006). For New York's Battery Park City, it was the 1920s architecture that is much-admired by New Yorkers that set the standard for the formulation of the design guidelines for the precinct (Gordon 1997; Lang 2021). For Celebration in Florida, it was the architectural traditions of the locale that were the precedent. In Val d' Europa (part of the new town of Marne-la Valée in France), it was the street block and building configurations of Haussmann's Paris. A proposal for a new business district for Dammam in Saudi Arabia (2005), was based on the traditional Saudi settlement and building patterns (Lang 2021, 256). Neo-traditionalism is not a single style but a prescription for learning from local building types (Fig. 6.12).

Among the various recent architectural advocacies are those presenting prescriptions for a sustainable architecture (for example, Bergman 2012; Jodidio 2018). They show how buildings can be constructed of materials of low embodied energy and to be frugal in energy and water consumption when occupied. Fully sustainable buildings are also adaptable to the vagaries of climate change and dramatic weather events. If nothing else, its advocates believe that new developments should, at least, acknowledge the latitude of their locations and the movement of the sun through the sky over the course of the day and the seasons (Chandavarkar 2022).

6.6 Commentary

Architectural movements are generally defined by the common characteristics of their buildings' massing, façades, fenestration patterns and their decorative features. To fully understand them it is better to consider what each movement's proponents see as the problems worth addressing and the emphases they placed on the various potential functions of built form. Much of the thinking about architecture over the last century has been shaped by the problems recognized by members of two coexisting cultures, the societal and the professional. Each encourages explorations of specific lines of thought.

The history of architectural ends and means shows that one design paradigm succeeds another not in a neat chronologically order but in an overlapping sequence. The Neoclassical has persisted throughout many social changes, if not on a massive scale, for the past four hundred years. Each paradigm holds sway until architects and their clients find it does not fulfil their aspirations. The shifts from one paradigm to another only partially follow the experience in the sciences where a paradigm commands attention until it is proved wrong and abandoned. In architecture, however, shifts also occur as interest in one approach is saturated and a new approach is sought by architects and their clients for its own sake. They seek something new.

Photograph by D. Coventini/Wikimedia Commons Photograph by J. Schnake/Wikimedia Commons

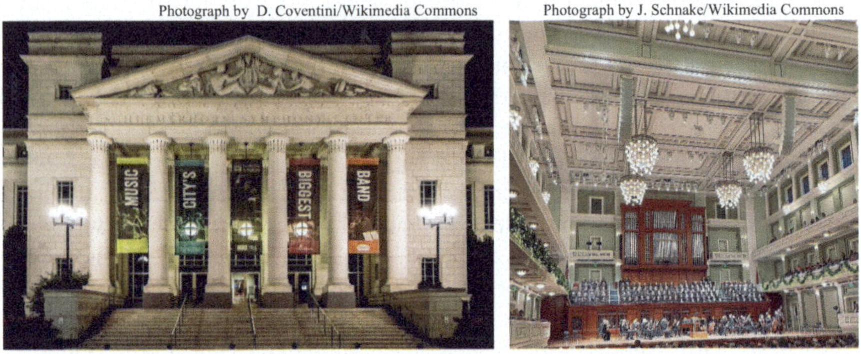

The Neoclassical: Schermerhorn Symphony Center, Nashville, Tennessee

Photograph by Alex Sergeev /Wikimedia Commons Photograph by John Pavelka/Wikimedia Commons

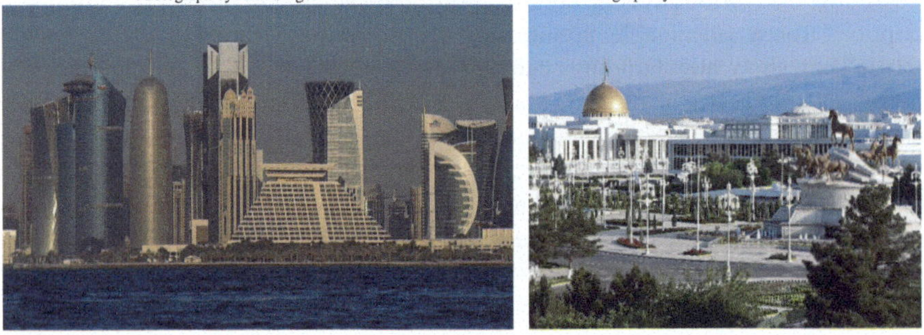

Hypermodern design and the architecture of radical heterogeneity: Central Doha (left), Ashgabat, Turkmenistan (right)

Photograph by the author Photograph by Elsabornfree /Wikimedia Commons

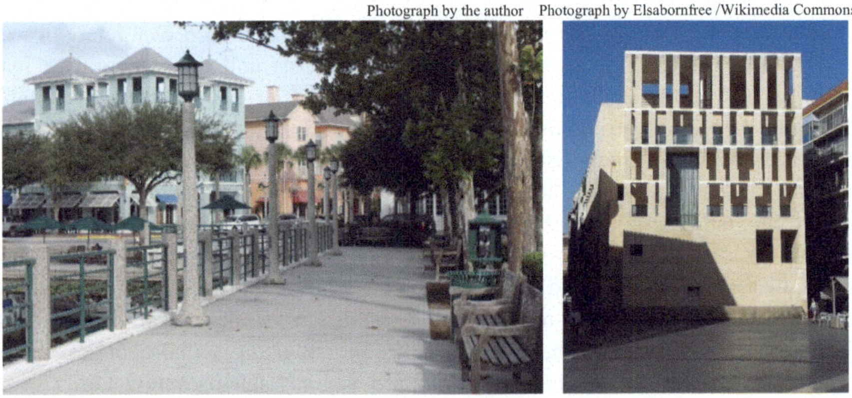

Neotraditional and Neomodern designs: Celebration, Florida (left) and Murcia City Hall (right)

Fig. 6.12 Co-existing architectural paradigms today

An overview of the history of architecture, such as the brief one supplied in this chapter supports (or, rather, does not negate) Anne Tyng's hypothesis that aesthetic tastes go from a desire for simple, undecorated geometries, to increasingly more complex forms after which there is a dramatic desire for the simple again. The ends sought are reinterpreted. The means for attaining them follow.

It is the Rationalists coming out of Europe, both eastern and western, with their bold ideas that attract the most attention in architectural circles. They keep breaking the boundaries of accepted architectural norms. The work of the Empiricists countered those ideas by following a more conservative approach in the search for liveable and enjoyable built environments and the behaviour settings they contain. The pragmatists are the realists. They get things done as quickly as possible.

Understanding the history of architectural paradigms enables practising architects to better understand their own work, the problems they are addressing, and the way they are addressing them. It also enables them to understand those concerns that, purposefully or subconsciously, they are avoiding. Understanding history helps students understand how design paradigms evolve and how what they are being taught and the designs they produce in the studio fit into the realm of contemporary ideas. The counter argument is that there is a utility in not knowing. Architects are then unhampered by an understanding of what they are doing and its impact on society. Luckily, people adapt, if not joyously, to the built environments offered them. They get on with their lives.

References

Borngässer B, Bednorz A (2008) History of architecture from classical to contemporary translated from the German by Maureen Bassedow, Mary Dobrian, Linda Marianiello and Sally Schreiber, edited by Rolf Toman. Parragon, Bath

Congress for the New Urbanism (2013) The charter of the new urbanism edited by Emily Talen. McGraw Hill, New York. https://www.cnu.org/who-we-are/charter-new-urbanism. Accessed 2 Mar 2024

Conway H, Roenisch R (2005) Understanding architecture: an introduction to architecture and architectural history. Routledge, London and New York. Aseel Rababah. Academia.edu. Accessed 24 Jan 2023

Ghirardo DY (1996) Architecture after modernism. Thames and Hudson, New York

Lang J (2021) The Routledge companion to twentieth and early twenty-first century urban design: a history of shifting manifestoes, paradigms, generic solutions and specific designs. Routledge, London and New York

Rudofsky B (1964) Architecture without architects. an introduction to non-pedigreed architecture. Museum of Modern Art, New York

Tafuri M (1968) Teorie e Storia dell'architettura translated from the Italian by Girogio Verrecchia as Theories and Histories of Architecture. Grenada, London, 1980

7 Designing Ends and Means: Prescriptions for the Design Process

Any history of architecture shows how embedded the typological and compositional approach to design is in the way architects work. It also shows how designing is driven by contemporary paradigms. Methodological changes are, however, occurring. Sketching of ideas is done on computers although the ability to rapidly sketch designs by hand remains fundamental to generating and testing ideas quickly. Working drawings, specifications, and product procurement are aided by computer-based algorithms that simulate such processes. Much continues to change. The application of artificial intelligence (AI) to architectural practice promises much. Parametric design may become *the* process for rapidly generating design possibilities in the future (Schumacher 2016; Çalişkan 2017; Leach 2021; Wainwright 2023). Conceptually, traditional approaches to conducting the design process, nevertheless, persist.

Colin Rowe's question about whether existing paradigms and generic solutions should guide design efforts or whether developing the program should dominate architects' thinking remains central to the discussion of how one should go about designing (Rowe 1983). His question can be elaborated into four issues that inspire different prescriptions for how architectural designing should take place.

The first addresses the question of whether one way or another of going through the design process should prevail. The second is whether an approach exists that offers more than the existing reliance on the adjustment/imitation of building types and generic solutions as a way of working. The third is whether computer-based design algorithms are more efficient and/or result in better designs than designers' internalized algorithms based on intuitive thinking. The fourth is over what and whose interests should guide the designing of buildings and, perhaps more importantly, urban designs.

J. Lang, *The Scope of Architectural Theory*, Design Science and Innovation,
https://doi.org/10.1007/978-981-96-3690-7_7

7.1 Operating the Design Process: Design-First and Program-Oriented Approaches Revisited

Implicit in the work of any architect is a prescriptive model of the design process. Architects differ in the amount of time and energy they put into generating designs in comparison to creatively understanding the problems at hand. Implicit in the difference are distinct attitudes towards architects' obligations and their views on the central concerns of professional practice.

7.1.1 The Design-First Approach

Some architects continue to argue for the design-first approach to architectural practice described in Chap. 2. They believe that a design should be developed as quickly as possible in response to a minimal program to understand the character of the concerns that need to be addressed. In doing so the energy spent on programming, they argue, should be minimized; the act of design helps in defining the program. Where necessary, the preliminary program is then fleshed out based on an analysis and evaluation of the design or altered to fit what the design affords. Advocates for this approach believe that historically major architects such as Frank Lloyd Wright, Louis Kahn, and Zaha Hadid worked that way and that leading architects today do so (Yaneva 2009).

Advocates for the design-first approach process believe that focusing attention on the creation of a rigorous program inhibits creative thinking by overwhelming the designers with issues requiring resolution. They argue that serious discussions about ends and means cannot be undertaken without a concrete proposal to discuss and evaluate (Walters and Brown 2004). Architects are always working under tight schedules and a design-first process eliminates endless circular arguments about ends and means in their view. The results are good enough.

Concentrating efforts on the design phase inevitably requires a relooking at the program. The advocates for the approach, however, believe that architects should avoid looking back unless it is essential to furthering the design. The counter argument continues to be that such approaches to design prematurely truncate the search for an understanding of what the scope of concern should be. It leads to an oversimplification of what are wicked problems. The evidence available suggests that the design-first approach leads to more creative solutions—ones that deviate in form from the norm. They, however, are of overall lower quality (Kruger and Cross 2006). They are not functional enough. The designs, nevertheless, suffice in the eyes of their creators and clients. Horst Rittel observed years ago that designers can get away without working from thorough programs. He was not arguing that they should (Rittel 1971). There are opportunity costs in doing so. Results could have been better.

7.1.1.1 Thoughts on Imitating Design Paradigms, Generic Solutions, and Building Types

Much designing, as noted in Chaps. 2 and 4, is carried out by imitating a building type or generic solution. Architects compare the problem they are assigned to other similar situations and the solutions developed to them (Colquhoun 1967). Using standard models representing a design paradigm is the basis for the explorations in the application of AI to the coding of design-generating algorithms (Harrell 2023). Today, Neoclassical, Modernist, Hypermodern, Neomodernist, and Neotraditional design paradigms are each meaningful to a group of architects and their clients. AI algorithms can produce such designs. The question is: How rich is the set of variables they address?

Zaha Hadid thought that the reliance on imitating existing types and/or imitating other architects' works is regressive design thinking. She abhorred the way architects rely on an existing design paradigm in producing their designs. She was, nevertheless, very much a design-first architect with the referents in her early work being the artistic output of Kazimir Malevich (see Chap. 8). By using Malevich's ideas, she sought to create a design paradigm suitable for the future (Hadid 1982; see Chap. 8). The implicit advocacy is that, within limits, the program has to fit the design rather than the other way around. Bernard Tschumi, basing his work on the deconstructive ideas of Jacques Derrida agrees but his answer differs (Tschumi 1987; see Chap. 8).

The observation is that architects not only use a design-first approach to design but that they prefer doing so. It is rewarding. It enables them to generate many ideas quickly. When following this approach, Colin Rowe argued that designers need to be thoroughly aware of existing paradigms and what they offer but that that knowledge is insufficient (Rowe 1983). Multifunctional designs need to be based on thorough programs.

7.1.2 Program/Problem-Oriented Design

The Rationalists among Modernists advocated a problem-solving approach to design that purposefully disregarded existing types and precedents. Using precedents would prematurely impose a formal solution on a design. They sought instead a function-based approach to architecture. The functions were based on rational thinking about the best way of carrying out activities, housing them in spaces and plan layouts that are safe, efficient, and well-lit. Efficiency in affording the movement patterns required for tasks was the end and the employment of appropriate technologies was the means.

In response to the shortcomings of the Modernists' model of functions, advocates for a problem-oriented approach to design argue that a thoughtful and detailed program is the key to creating a sound multi-functional design. It should be based on realities not on dreams (Horowitz 1967; Studer 1969; Cherry 1998; Kruger and Cross 2006). Increasingly, architectural firms argue for rigorously designed programs on which to base their work if only to avoid costly litigation if a design functions poorly. They advocate the use of refined program creation techniques to ascertain what they need to do when designing buildings, particularly sensitive buildings such as concert halls or juvenile psychiatric facilities (Peña and Parshall 2012; Cherry and Petronis 2016).

It is also recognized that programs are, indeed, designed. They are not simply a product of analyses of the situation being addressed. They are valued-laden prescriptions of what should be done. For some architects, programs developed for buildings should include specifications of the aesthetic qualities that they should strive to attain. Others believe that questions of aesthetic qualities should be left to those doing the design to answer. It is their prerogative (Peña and Parshall 2012). Doing so enables them to stamp their own brand on a design. The style is arbitrary.

The information sought to design programs, proponents for a problem-solving orientation to design believe, should draw on abstract functional theory, as presented in Chap. 3, as well as the critical analysis and evaluation of building types and precedents. The implication is that those architects, or consultants, who create programs need to keep abreast of current research on functional theory. In addition, the programs should be a 'performance' type. That puts considerable stress on the architect.

A 1970s advocacy for both enhancing architects' understanding of the functions that need to be addressed and their resolution based on evidence is the *Pattern Language*. Originally demonstrated as a method for designing multi-service centres (Alexander et al. 1968), the language has become a generic problem-defining first approach to designing. The structure of the language is generalizable to any design assignment that an architect undertakes (Steuteville 2024). AI approaches should be based on the language rather than the imitation of existing generic solutions.

7.1.2.1 The Pattern Language

All architects' heads are full of sets of patterns of built form that they believe resolve specific design problems and/or represent their brand of their work. These patterns are derived from their personal experiences in inhabiting places, analyses of other architects' work, and what they learnt in architectural school. The set serves as a vocabulary that an architect can draw on when designing and establishes and their brand of work. The set is not, however, open to public scrutiny. The Pattern Language as developed at the Center for Environmental Structure at the University of California at Berkeley by Christopher Alexander and his colleagues is (Alexander et al. 1977).

The pattern language consists of a hierarchy of patterns for designing places from settlements down to interior environments. Each statement describes a problem, provides a sketch of the pattern that resolves it, and an explanation based on evidence, for why it does so. The explanation contains an assessment of the quality of the evidence on which the pattern is based. Each statement follows the format: In situation A do B to solve problem C because of reason D. The argument for tailoring the language for a particular situation is that the clarity and logic of each pattern is open to the scrutiny of all the stakeholders engaged in the design of a scheme. The set of patterns proposed for the design of multi-service centres in the

1960s is shown as a cascade in Fig. 7.1 (Alexander et al. 1968). Some patterns inevitably conflict with others which is why the cascade is not a tree-like diagram but a web.

Several ways of organizing the flow of the patterns in such cascades are available to architects. One way is to have the patterns that constrain the design to the

Source: Alexander et al. (1968, 54): Courtesy of the Center for Environmental Structure, University of California at Berkeley.

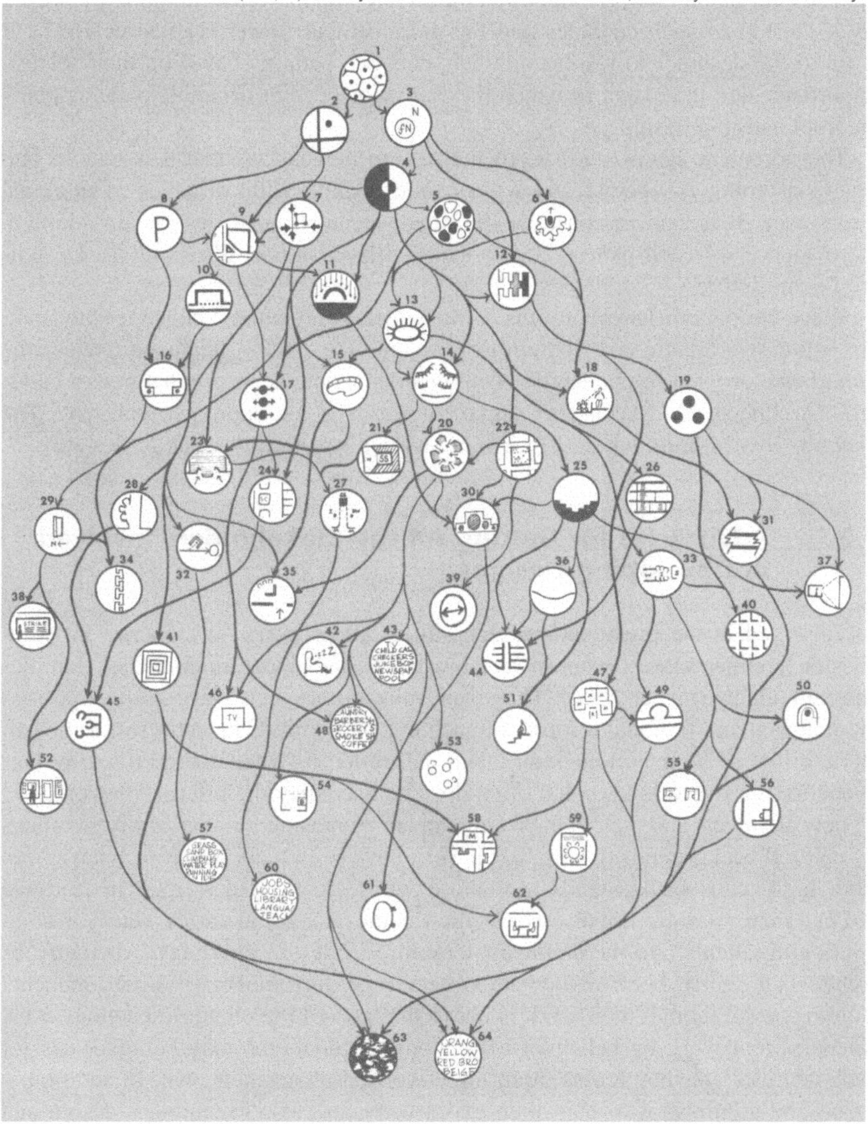

Fig. 7.1 Cascade of patterns for a multi-service centre

greatest extent at the top. It is efficient to do so as it reduces the potential number of lines of subsequent thought. Another is to begin with the patterns that simply must work well first and then proceed to the ones that have a lesser impact on what the building affords. A third is simply to deal with the large-scale decisions and then work the way down to the details. The problem is, as architect Mies van Rohe stated 'God is in the details' (a rephrasing of the German proverb 'The devil is in the detail'). Maybe architects should start from the details? René Descartes, in his third rule on method, suggested the patterns that are easiest to resolve should be at the top of such cascades and the more difficult lower (Descartes 1637). It is up to the designer to resolve conflicts between patterns based on their relative importance and the extent to which they allow or prevent desirable patterns being incorporated in a design.

The argument against the use of the pattern language is that it is tedious and time-consuming to operate and wastes time dealing with variables of marginal importance. It is said to constrain the timely generation of designs. In addition, the evidence for each pattern is, inevitably, drawn on past observations. Designs are for the future.

What the pattern language fails to recognize is architects' concern with making expressive artistic contribution to a design. The pattern language reduces the importance of what has generally been regarded as the heart of what makes buildings 'Architecture'. Alexander's argument was that his approach represents *The Timeless Way of Building* (1979). It is an applied science approach to design.

7.2 Advocacies for Getting All the Stakeholders Involved in Developing Designs

As observed in the previous chapters, almost all property developers who commission architects are corporations or wealthy and upper middle-class families and individuals (Fisher 2008). They want things done their way and they possess the financial and political resources to achieve their ends. The concern for the catalytic effect of their buildings on others is limited to what is prescribed by law. Dictators can override laws and often do. The users pay the price.

Few architects and their clients, although an increasing number, are from minority cultural backgrounds (Coles 2016). Most have been men; the women have been from high-status backgrounds (see those whose work is described in Hartman 2022). There is thus usually a congruence in values and tastes between commissioning clients and the architects they hire to design buildings. Alternatively, architects are hired because they are known to produce monetary value enhancing buildings even though their work is not to the taste of the client. Exceptions exist. Agencies that work on behalf of low-income populations, may be advocates for their interests but they are as often more worried about their own. In addition, a social and administrative gap often exists between those executing a design and those who will be inhabiting or using it. 'Most people don't want what architects want' is a recurring observation (Michelson 1968; Pelletier 2019).

Questions about for whom architects are designing arise constantly. Are they designing for the people who are going to inhabit a building or precinct of a city, or to meet the approval of the avant-garde among their peers. Commissioning clients almost always, if not always, have the ear of their architects. In many cases those clients are not the people who will be using the building when it is constructed. Strong advocates exist for involving the full range of stakeholders in the projects that affect them (Filippi and Hooper 2019). They argue that clients get more generous designs that way.

Walter Gropius's professed position, although not necessarily a practised one, is typically held by architects. It is that the public does not know enough to warrant consultation in making the decisions that affect them (Gropius 1955). Clients simply do not know what options are available to them. They have limited exposure to diverse built worlds. They, it is argued, do not understand what architecture is about or the possibilities that new design ideas afford them. All these observations about clients may be true, but they do know about their own lives and values although they may have difficulty in satisfactorily expressing them to architects. In response to this observation there have been calls for involving all the stakeholders in making the decisions that affect them. In doing so they have to resolve the conflicts that exist among the various perceptions of ends and means.

Architects have varied in their reaction to this call. Some believe that users get in their way; others argue for a more inclusive design process (Van der Linden et al. 2018). Many, however, listen to a range of views because they are obligated to do so, but have no intention of heeding them because it impedes the application of their brand of design. Advocates for the interests of clients, paying and user, argue that the design process should not be one of architects selling their ideas but be regarded as a process of mutual learning between the architects and the other stakeholders involved.

One of an architect's roles is surely to educate clients about the range of options available to them especially those of which they are unaware. In return, clients, particularly those of social groups different to that of the architect, have much to teach architects about their own aspirations. The design process becomes one of mutually supportive decision-making rather than one in which the architect seeks to impose his or her own ideas on the world in a heroic Howard Roark manner (Rand 1968).

7.2.1 A Note on Advocacy Design

Advocacy design is defined to mean that the interests of the un- or under-represented should be considered when designing the buildings and the districts that they will inhabit, work in, or visit. It has its roots in advocacy planning as proposed by Paul Davidoff (1930–1984), the founder of Planners for Equal Opportunity (Davidoff 1965). Its proponents strive to get architects to empathize with people who might not usually attract their attention, particularly when devising urban designs (Appleyard 1982; Jacobs and Appleyard 1987; Van der Linden et al.

2018; Jagtap 2019; Ornstein 2022). Such groups and architects, they argue, should be collaborators in the design process. The focus of attention when developing the program for any design should be on identifying and providing the supporting facilities required by marginalized groups to lead fulfilling lives.

Rationalists among architects make assumptions about what makes for a good healthy life for people in general and for the financially disadvantaged in particular. Empiricists focus on designing for the present or, at the best, predictions of evolving trends in the ways of life of people for whom they are designing. Advocacy design involves collaborating with people in a bottom-up manner. The aim is to understand what ends the disadvantaged wish to have met and the means for doing so. The focus is on more than meeting the basic shelter and communal needs but also on aesthetic concerns. The architect and client should together make conjectures about design possibilities (Jagtap 2019).

Co-designing methods are used in the attempt to understand the implications of having limited resources for operating and maintaining buildings. They are also used to understand and design for gender roles and how they might be changing, the relationships of adults to children, and the aspirations of the youth. The belief is that getting people involved in making decisions gives them a sense of dignity as well as the ownership of designs. Ownership of designs leads to people looking after them when implemented.

An alternative to advocacy design is to let people build for themselves through sites and services programs where the infrastructure—sewers and electric connection—are built and people build their own houses. For low-income people in resource poor situations, sites and services programs as advocated by John F. C. Turner (1927–2023) in *Housing by People* (1976) have been used with some success to provide residents with the possibility of owning their own homes. Turner's advocacies were based on his observations of what works in Latin American squatter settlements. Alternatively, the basic structure of buildings is provided for people to complete their own units 'as they will', as proposed by John Habrakan (1928–2023) in *Supports: Alternative to Mass Housing* (2021, originally 1962). Another possibility is for units to be partially built and the owners given the opportunity to complete them when they can (see the work of Alejandro Aravena described in Chap. 8).

Among the methods advocated for understanding and incorporating user opinions in shaping design programs and designs are focus groups and design charrettes. The former are widely used in public opinion surveys but can also help architect understand the issues of concern for those who will be affected by a design. The latter are more strictly concerned with producing designs.

7.2.1.1 Focus Groups and Design Charrettes

A focus group consists of a small set of people chosen to represent the cross section of the population that has a stake in a proposed design (Krueger and Casey 2014). The groups may include property developers, consent authorities, and potential users or surrogates advocating for their ends as well as the architects. The discussion is led by a moderator who has a series of questions on the issues of importance that need to be addressed and clarified. The information coming from the group, it is argued, provides more nuanced feedback than traditional methods involving individual interviews.

To yield valuable information, focus group meetings follow a rigorous outline. The questions raised aim at establishing the specific problems that need to be addressed. Whether or not the discussion incudes providing design ideas depends on the purpose of the meeting. Some meetings do attempt to come up with design ideas and, at least, partial designs based on a collaborative effort. Charettes more obviously do.

Charrettes are collaborative, problem identification and preliminary design creation gatherings (Pernice 2013). Strong advocates (for example, the Design Charrette Institute) argue that charrettes should be used as the means for getting diverse stakeholders involved in creating designs. The goal is to brainstorm what are the problems to be addressed and the opportunities to exploit. It is also to encourage the generation of design ideas. The process is coordinated by a leader whose task is to ensure everybody gets the opportunity to express their views (Roggema 2013; Lennertz and Lutzenhiser 2014; Todd and Lindsay 2016).

Charrettes must be brief in order to maintain participants' interest and ensure their attendance at follow-up meetings. The number of people participating can vary but once the group gets over twenty people in size it can easily become unwieldy and the less assertive participants get sidelined. Such gatherings are described as 'overmanned' (Barker 1968). If a charette has been conducted in an orderly manner, at its end architects will understand how best to proceed.

7.3 Advocacies for Systematizing the Designing Process

Developing and using pattern languages is one way of systematizing designing but other approaches also have their advocates. Several architects and even more critics consider the intuition charged generic design process, in both design-first and problem-oriented approaches to design, to be antiquated, inefficient, and unproductive. The disquiet was spurred by developments in engineering and cybernetics that were taking other design fields forward. The array of imaginative, new products devised by industrial designers such as Henry Dreyfus (1904–1972), author of the manifesto *Designing for People* (2003, originally 1955), led several architects to question their approaches to designing. Dreyfus created a vast array of products that function well both instrumentally and aesthetically (Flinchum and Meyer 2022). The products had to meet the needs of people of very different characteristics and competencies and they had to 'look good'. Dreyfus' success in achieving that end made some architects look at the design process anew.

Prescriptive models of the design process that examined the whole nature of creative thinking began to appear in the 1970s. They were a response to the black box approach to thinking about the nature of architectural designing. *The Universal Traveler* was one of the first books that offered a guide to design methodology and methods (Koberg and Bagnall 1981, originally 1974). Although descriptive of the whole problem-solving design process and of idea generating techniques, it implicitly advocated a systematic, transparent design process.

In England, practising engineers and theorists such as J. Christopher Jones (1927–2022) and L. Bruce Archer (1922–2005), a professor at the Royal College of Art, advocated the systematization of the design process and the integration of rationality and intuition in the thinking of designers. Jones' book *Design Methods: The Seeds of Human Futures* 1992, originally 1972) was the manifesto of the movement. Archer produced a 229-step prescriptive process for the design of industrial products. Implicit in his thinking was that the model also applied to how other design fields should conduct their design efforts. Later he became concerned with the visionary aspects of architecture and the need to think of the future using a more rationalist vein of thought and of incorporating random, chance methods of composition into design.

Other advocates for systematizing the architectural design process appeared among the members of the Design Research Society in the United Kingdom, the Design Methods Group in the United States, and at a few individual schools of architecture such as those at the University of California in Berkeley and at the University of Ulm. Their advocacies had a much greater impact, however, on industrial design where prototypes can be designed and assessed before being mass produced than on architecture where one-off designs are the norm.

The advocacy, instead, is for programs to be based on the extrapolation of evidence and for an evidence-based approach to the evaluation of potential designs (Kruger and Cross 2006; Zimring and Bosch 2008; Aksamija 2016). Adding rigorousness to methods employed in the intelligence phase of the design process means that the design of programs is now often outsourced by architectural firms to organizations that specialize in resolving politically charged debates over ends and means.

7.3.1 Computer-Based Design-Generating Algorithms

Computer-based design-generating algorithms have been used to assist architects produce designs for over half a century. In airport design, linear programming and queuing theory procedures (that is, those dealing with a succession of waiting lines) have been used to produce partial designs to assist the programming effort. The widespread introduction of AI algorithms to monitor and keep track of the design process and the procurement of materials has raised the question of whether similar algorithms can generate whole designs better than humans. Architects such as Patrik Schumacher believe so (Schumacher 2008, 2016; Wainwright 2023). He is not alone. Computer-aided design has already revolutionized workflow and reduced the time spent on drawing and repetitive tasks.

Explorations of how to generate floor plans of buildings based on Modernists' ideas of efficiency have been taking place since the 1970s and have become increasingly sophisticated. Proximity requirements of the spaces that serve instrumental functions are specified and automated algorithms generate plans based on mathematical optimization techniques. The spaces include both served and servant spaces in Louis Kahn's terms (Slunecko 2018). The algorithms replicate human problem-solving methods that they learn from existing types and precedents but they do so more efficiently and rapidly (Klawitter et al. 2021; Lobos and Donath 2021).

While traditional methods for creating designs persist, advocates for the use of advanced computer-based design generation techniques argue that AI can replicate architects' creative processes. Computer-based algorithms have already replaced much in dealing with structural and energy consumption design concerns. They can give architects further aid in generating complete designs (Leach 2021; Harrell 2023; Tovar 2023a). The research has focused on how best to generate three-dimensional designs using parametric methods. Advocates for taking these advances a step forward focus on the power of AI to rapidly produce designs. Algorithms have been developed to create designs based on a rich set of input variables including those that imitate the work of past and present architects of note.

In his manifestos, *Parametricism as Style—Parametricist Manifesto* (2008) and *Parametricism 2.0: Rethinking Architecture for the 21st Century* (2016) Schumacher, who has been identified to be a 'techno-evangelist', argues that parametricism is *the* new way for designing buildings and urban environments. Parametricism refers to the manipulation of variables whose relationships are described by mathematical equations based on the correlations among variables. The relationships are manipulated by scripts, or computer languages, that can be edited by the designer to generate different designs based on different input variables. The great advantage of such techniques is that the ramifications of changing one variable (for example, the height of a building) on all the other characteristics of a design becomes immediately apparent. It is thus possible to simulate how building forms change as design objectives are manipulated by the architect (Fig. 7.2).

Present computing capacity limits the number of variables that algorithms can handle. The danger is that definition of design ends become simplified. Streets, for example, are seen as channels for vehicular movement and not as the location of pedestrian activities and seams joining adjacent properties into a unit serving many functions. The variables that are included at present are those of importance to the neoliberal marketplace although in urban design they are like the key variables—streets, blocks, and buildings—identified by Jane Jacobs in the 1950s (Jacobs 1961). Parametricism promises much and as microchips are advanced so such techniques will come into common usage. Architects will need to retain control over them if their profession is to continue in its present form.

Courtesy of Zaha Hadid Architects

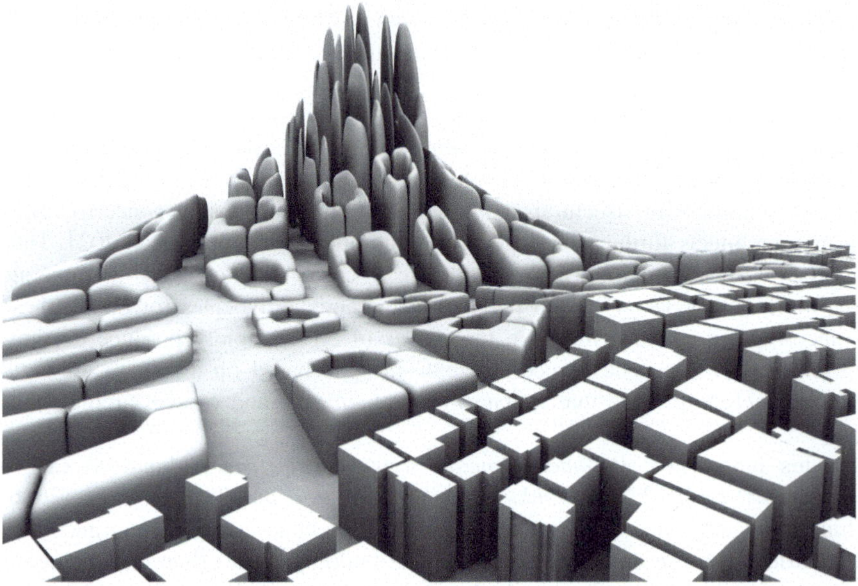

A design generated using parametric procedures: An illustrative design by Patrik Schumacher

Courtesy of Zaha Hadid Architects

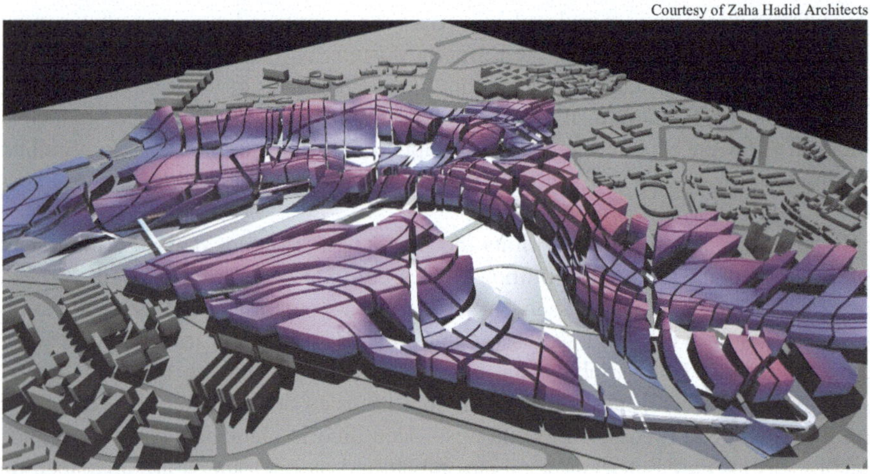

One North District proposal

Fig. 7.2 Parametric designing

7.4 Commentary

Considerable thought has gone into attempts to describe and prescribe design methods that enhance creative work. Professional practice has been enhanced through computer-aided design and the development of systematic management techniques. Much, however, remains unchanged both in architectural education and professional practice. The design process and how we should go about it remains a taboo topic in most architectural schools. AI will force a change.

The way design processes are conducted today remains based on centuries of experience in producing not only satisficing designs but also outstanding ones that have shaped the future. The processes architects use have served them well. Why should they be changed? Questions are, nevertheless, being asked about how best to conduct the design process efficiently in a digital age. It is not surprising that such concerns arose in the 1970s when the limitations of Modernist design thinking and the products it produced became clear. These perceptions generated prescriptions for an evidence-based design rather than one based on hopes and unsubstantiated self-serving claims (Zimring and Bosch 2008).

The generation of designs based on the imitation of generic solutions persists as the modus operandi for most professional practices. In the future no doubt automated processes of analysis, synthesis, and evaluation will be developed to ease the intellectual burden on architects (Aksamija 2016). Patrik Schumacher in Britain and Olgu Çalişkan in Türkiye argue that they are here already and should be widely incorporated into professional practice (Schumacher 2016; Çalişkan 2017). In the offing, is the automation of the entire design process from the generation of speculative design through to the construction phase (Leach 2021). Will it have levelling impact on architectural practice?

References

Aksamija A (2016) Integrating innovation in architecture: design, methods and technology for progressive practice and research. Wiley, Hoboken, NJ

Alexander C, Ishikawa S, Silverstein M (1977) A pattern language: towns, buildings, construction. Oxford University Press, New York

Çalişkan O (2017) Parametric design in urbanism: a critical reflection. Plan Pract Res 32(4):417–443

Coles RT (2016) Architecture + advocacy. In: Siener WH, Coles S. Buffalo Arts Publishing, Buffalo, NY

Filippi C, Hooper H (2019) The architect/the advocate: the role of advocacy and influence in architecture and design. Architecture Now. Accessed 10 Sept 2022

Leach N (2021) Architecture in the age of artificial intelligence. Bloomsbury Publishing, London

Schumacher P (2008) Parametricism as style—parametricist manifesto. Presented and discussed at the dark side club, 11th architecture biennale, Venice. http://www.patrikschumacher.com/Texts/Parametricism%20as%20Style.htm. Accessed 24 Feb 2017

Schumacher P (ed) (2016) Rethinking architecture for the 21st century architectural design. Wiley, Hoboken, NJ

Part III

Architects, Their Prescriptive Theories, and Their Designs

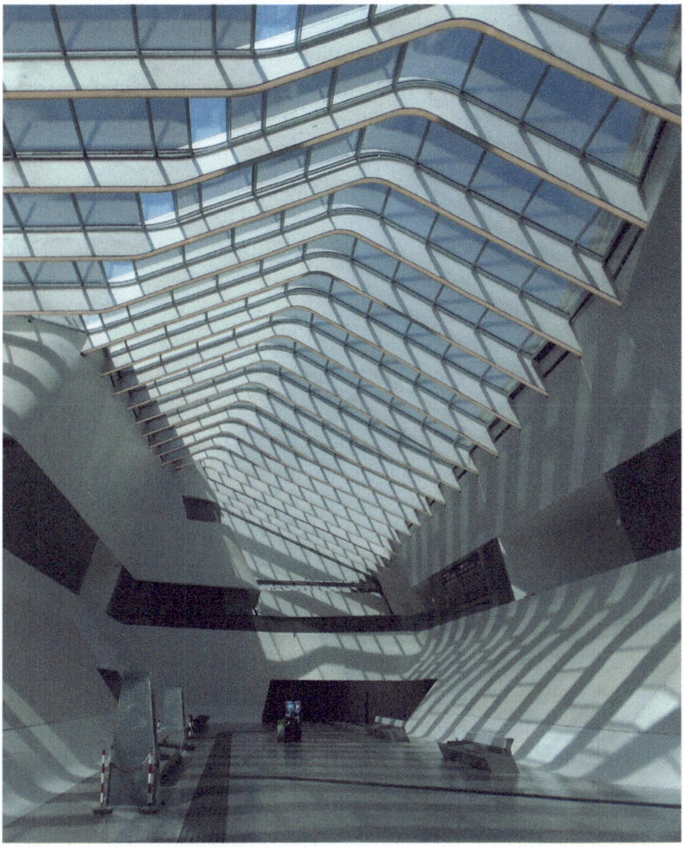

Stazione alta Velocita, Afragola Naples (2017)
Rete Ferroviaria Italiana developer and client, Zaha Hadid and Partners, Architects
AKT (Hanif Kara and Paul Scott) and INTERPROGETI (Giampiero Martuscelli) Structural engineers
Photograph by Pivari.com/Wikimedia Commons

Architectural theory has been traditionally focused on architects, their ideas, and designs. To better understand the strengths and limitations of the works of individual architectural firms, their clients and the context in which they are developed needs to be understood.

Architectural theory has long been concerned with describing the creative works of individual architects. The ones who have attracted the most attention have been those whose work differed or now differs substantially from contemporary norms and formed, or could form, the basis of an influential new design paradigm. It is their ideological positions that are included in anthologies on prescriptive architectural theories. Demonstrating the link between architects' prescriptions and the works that they have produced has less frequently been rigorously analysed. Indeed, who has actually been responsible for a design has often been glossed over. The collaborative nature of the authorship of designs is neglected. It has been assumed that a firm's design is entirely that of its principal.

In a sole chapter, the attempt is made in this outline of the scope of architectural theory to describe and explain the links among the philosophical orientations and ideological position of a sample of architects or architectural firms, the people who hire them, and the designs they jointly produce. The work of the nine architects selected for review is, as categorized earlier in the book, pragmatic, rationalist, or empiricist in nature. None of the nine fits entirely into the category they have been placed in here, but their works described do.

The descriptions of architects' ideological positions have been taken from their writings and/or their views as recorded by others. The discussion on each architect is brief. It is further restricted by only focussing on a part of an architect's firm's output. The nine selected, nevertheless, suffice to illustrate a key component of architectural theory—the relationship between architects' prescriptions and their work. The focus in the chapter is on showing how buildings reflect their designers' views of the world and the functions of built form that they emphasize in their work. Much depends on the characteristics of their clients and the ends they seek.

The chapter begins with a discussion of the relationship of the principal or principals of a firm, and an organization's work. The history of architectural practice shows that the design of buildings and urban designs is much a more collaborative effort than presented in the writings on key figures in the profession (Gutman 1988, Cuff 1992, Dickinson 2023). It is also clear that gaps often exist between the professed positions of architects and what they do in practice. The reasons are diverse but have to do with the realities of professional practice. In some cases, it is not clear what the architect's position really is because it is clouded in "architalk" (Johnson 1994).

The goal of the chapter is to bring together the arguments presented earlier in the book in order to understand how they can be used to analyse the works produced by architectural firms. It should also help individual architects understand where they fit in ideologically with their professional contemporaries who have different interests, beliefs and values than those they hold. The way the discussion in this chapter is organized should assist students not only to understand the implications of different positions on what architecture should be about but also the lines of thought open to them when in professional practice.

8 Understanding Architects' Prescriptions and Their Designs

The shelves of bookstores specializing in architecture are lined with well-illustrated, hagiographic works on architects and/or on architectural firms. The books are printed on high-quality paper. The photographs in them are taken by professional photographers when the sun is in the correct place to throw enhancing shadows on buildings' façades. Interiors have been spring-cleaned and every item of furniture is in its place. The photographs are usually devoid of people. Informative though these books are, they seldom tell the full story of the relationships among clients, architects, and the works.

The books describe buildings, what is innovative about their plans, their spatial qualities, and appearances, and why they are of intellectual interest. They often describe the evolution and increasing sophistication of an architect's ideas as reflected in the chronology of the designs portrayed. The architects written about are usually those held in high esteem by the architectural cognoscenti. The exceptions are architects who are highly prolific and/or have written about themselves or solicited books on their work. Clients might be mentioned in passing.

The buildings and their interiors are portrayed as objects to be admired. How they fulfil their instrumental functions and their multiplier and side-effects are infrequently evaluated. The books have seldom been about what the buildings afford specific people or organizations in terms of their ways of life or aesthetic values. The interests of the architectural cognoscenti are, in contrast, the central concern. The discussion's focus is on the relationship between one architect's ideas and those of earlier or contemporary design paradigms.

To inform architectural theory the nature of the people commissioning buildings and their demands need to be understood. The focus of attention of the architect in responding to those demands, and the outcome as displayed in plans and exteriors are fundamental concerns of architectural theory. An attempt is also made in this

J. Lang, *The Scope of Architectural Theory*, Design Science and Innovation,
https://doi.org/10.1007/978-981-96-3690-7_8

chapter to recognize who is responsible for the designs described and not simply the architects to whom the works are attributed.

8.1 Architectural Firms and Individual Architects

Almost all self-conscious building designs are the work of teams of people. The credit for the design of a building, particularly if it is visually innovative, is usually ascribed to the head of the architectural practice producing it. Even the junior employees of a firm completing working drawings may, however, make design decisions along the way. Senior designers certainly will be deeply involved even if their contributions are unheralded. Without the design input of structural engineers many of the world's best-known buildings never would have taken the form they have. Often celebrated architects have collaborators who ensure that the instrumental functions of a building are met well-enough. That participation frees time for leading architects to focus on their specific interests.

Philip Johnson is the architect of record for the Seagram's Building in New York. Mies van der Rohe, its designer, did not have a license to practise in New York State. Patrik Schumacher is listed as co-designer of several of Zaha Hadid's designs but those designs tend to be regarded as her work. Tony Henderson and George Marsh (1921–1998) were responsible for many of the designs attributed to Richard Siefert who, nevertheless, bore the brunt of the contemporary criticism of his firm's designs. He now receives recognition for the firm's contribution to British architecture. Le Corbusier is seen as the sole author of the Ville Savoye but (Arnold-André-)Pierre Jeanneret(-Gris) (1896–1967), his cousin, was the co-designer of record. Le Corbusier saw himself as Don Quixote and Pierre Jeanneret as Sancho Panza (Buggy 2012/13). Jeanneret's contribution to the development of Chandigarh has largely been ignored. Charlotte Perriand (1903–1999) worked with Le Corbusier for a decade. Her contributions to his designs have yet to be assessed (Hartman 2022). On occasions, employees of architectural firms break away to form new practices because they feel their contribution to much lauded designs was not recognized or was insufficiently acknowledged.

Le Corbusier was not alone among major architects in failing to credit the contributions of others to his work. Frank Lloyd Wright learnt much from his clients (Eaton 1969). His architectural ideas owe a debt to the Shingle style work of Joseph Silsbee. Wright, aged eighteen, a civil engineering student, contributed to the interior design of the Unity Chapel (1886) at Taliesin designed by Silsbee. His stained-glass window work builds on the designs of Marion Lucy Mahony (1871–1961). She, reputedly, was the one who developed the unique style of drawing for which Wright is renowned. She, along with Hermann von Holst, was the designer of Wright's Robert Mueller House in Decatur, Illinois (Gray 2022; Hartman 2022). Wright's early buildings also owe a debt to the ideas of Walter Burley Griffin (1876–1937). He did not give others any credit for the ideas they produced and the work they did.

> Such was the nature of Wright's studio … that those working for him were rarely acknowledged. To the outsider everything that came out of his studio was deemed to have been designed by himself. (Frank Lloyd Wright Home and Studio website)

Griffin left Wright's practice in a huff to establish his own office and Mahony soon joined him. They married shortly afterwards in 1911. Her contribution to Griffin's designs and drawings has only been belatedly acknowledged (Van Zanten 2011; Korporaal 2015).

The impact of wives on their husband's architectural development is little understood especially if they were also architects. Aino Aalto (1894–1949), an architect and carpenter, the first wife of Alvar Aalto was the cofounder of the design firm, Artek. Using her carpentry experience, she made contributions to her husband's interior and furniture designs and the domestic quality of his designs, although it is unclear how much (Hartman 2022). She has been regarded as the co-designer of the Paimio Sanatorium (1933), a major modernist design. Aalto's second wife Elissa (1922–1994), also an architect, worked on several of the firm's major designs and completed unfinished projects after his death (Wainwright 2021). Aalto, very much an autocrat, nevertheless, remained in total control. Similarly, the contribution of Anne Tyng to the geometric quality of Louis Khan's designs is yet to be fully comprehended (Hartman 2022).

Until recently, awards such as the Pritzker Prize for architecture, went to individual architects not firms. Robert Venturi received the award in 1991 but his professional and life partner, Denise Scott Brown, claimed that the award should have been a joint one as many of the ideas of the firm of Venturi, Rauch and Scott Brown were based on her input. Venturi acknowledged her contribution to his manifestos *Complexity and Contradiction in Modern Architecture*. The contribution of women to a firm's work has often been ignored (Desai 2017; Adajania 2018; Hartman 2022). In developing our understanding of architects' prescriptive theories these neglects are difficult to rectify.

8.1.1 Professed Positions and Practised Work

Each architectural firm is a business. It strives to develop a brand of architecture that can appeal to, or be sold to, a segment of the marketplace requiring design services. To obtain commissions firms state what their positions on architecture are and they show portfolios illustrating their work to potential clients. Several architects speak and write clearly about their intentions; others write in obscure language using words in unique ways to mean what they want them to mean as Humpty Dumpty does in *Alice in Wonderland*. Words such as 'dynamic' and 'organic' are used to describe buildings but mean different things to different people. They become meaningless (Orwell 1954).

All professed positions are developed within a socio-economic and cultural environment. Those of the architectural firms whose work is discussed in this chapter were developed in contexts ranging from the late Victorian era to the present. Changes in political attitudes, the two World Wars, the Spanish Flue pandemic, the Great Depression, the rise and fall in optimism about the future, and the battle between capitalism and socialism are all reflected in architects' attitudes. Those views are also affected by the perceived success or failure of specific design paradigms and the self-image of architects as artists or as applied social scientists. Today, in an era of neoliberal economics, uncertainty about the impact of climate change, the continuing developments in information technology and Artificial Intelligence, architectural attitudes continue to change.

Sometimes architects' designs closely reflect their professed statements on what the concerns of architecture should be and how those concerns should be addressed. At other times they do not. The reasons are diverse. Financial backers and clients may insist on certain ends being met, building codes designed to safeguard the public interest may dictate specific patterns of built form that can and cannot be used, and consent authorities may have their own criteria for deciding what is acceptable. In addition, architects' advocacies may change over time as the result of their professional experiences, their exposure to new ideas, and boredom with what they are doing.

The nature of architects' prescriptions and work is illustrated here by nine examples selected from a host of possibilities. The work of three is pragmatic, three rationalist, and three empiricist. Many firms, probably the vast majority, produce pragmatic work. Two that were highly productive are discussed here. The work of the one headed by Howard Shaw at the beginning of the twentieth century was empiricist in nature but influenced by Arts and Crafts ideas. The work of another, Richard Siefert and Partners, in the post-World War Two years, owes much to imitating rationalist models. The third, Alejandro Aravena and Elemental Consultores, in other than their social housing work, is very much Neomodernist in tone. The three Rationalists—Le Corbusier, Bernard Tschumi and Zaha Hadid—are highly lauded in architectural circles. The three Empiricists—Dashan Kumar Bubbar, Abdel-Wahed El-Wakil and DPZ CoDesign—produced ideas and designs that fall outside the domain of interest of most of the international architectural cognoscenti whose tastes are more allied with the avant-garde art world. Their works are, however, increasingly being recognized as products of highly creative minds. What the nine firms have in common is that their principals were fine advocates for their prescriptions. They were outspoken salespeople.

8.2 Three Pragmatists

All buildings require some pragmatic thought if they are to proceed from the computer screen to reality. Most of the buildings that adorn cities are the result of pragmatic thinking but they are, as noted earlier, touched by the ideologies and works of leading architects. Although disparaged by the contemporary cognoscenti many such buildings and the logic of their architects are now being re-evaluated. The work of Howard Van Doren Shaw and that of the firm Richard Seifert and Partners are two of them. That of Alejandro Aravena and Elemental Consultores differs from those two. Its work has always been held in high esteem, although it seems that it was the firm's Neomodernist work that led to Aravena's Pritzker Prize in 2016. All three, in common, were/are highly successful practices.

8.2.1 Howard Van Doren Shaw

Howard Shaw, as he was known to his contemporaries, was born in Chicago, the son of Theodore Andrews Shaw, a prosperous businessman and Sarah Van Doren, an artist. His father served on the committee for the 1893 World's Columbian Exposition. The family lived in the wealthy and socially prestigious Prairie Avenue district of the city that was full of what were regarded as modern homes. He attended a prestigious school in Hyde Park and then Yale University and Massachusetts Institute of Technology where he studied architecture. MIT, like other contemporary architectural schools in the United States, based its educational philosophy on the methods of the École des Beaux Arts in Paris. Shaw's studies in classical and Georgian architecture there set the tone for his career (Fig. 8.1).

After graduating from MIT in 1890, Shaw worked for Jenney and Mundie, a firm that served as a finishing school for many later renowned architects. There he came under the tutelage of William Le Baron Jenney (1832–1907), designer of the first skyscraper in Chicago. Shaw was, however, more influenced by what he saw on his 1890 travels in Europe and, particularly, in the United Kingdom. He admired the work of contemporaries such as Edwin Lutyens and the Arts and Crafts designs of Charles Rennie Mackintosh in Scotland. After his travels, Shaw established his own firm. He never took a partner, "He wanted to do every detail on his own" (Shaw 1926). His work received grudging praise at the time. He received the American Institute of Architect's gold medal on 6th May 1926 in recognition of his contribution to architecture. He died on his birthday the next day.

8.2.1.1 Shaw's Views on Architecture

Shaw's wrote little about his work. It was, however, clearly inspired by the architecture of diverse earlier American and English architectural paradigms.

> In keeping with his essentially pragmatic approach, Shaw developed an architecture almost devoid of [explicit] ideological content. … Shaw simply never gave much thought to the architectural implications of the American political system. … He likewise made no attempt to express a Midwestern regionalism [or] the realities of the twentieth-century industrial civilization. His houses could have been set down in any pleasant upper-class suburb. (Eaton 1969, 148–149)

Prairie School architect Thomas Tallmadge (1876–1940) who held Shaw in high esteem, called him "the most rebellious of conservatives and the most conservative of rebels" (cited by Moser 2015).

Shaw was a member of The Eighteen Group of architects who gathered at lunch time in Chicago to discuss architectural concerns. The group included Frank Lloyd Wright. Shaw, unlike the other members of the group, did not follow a

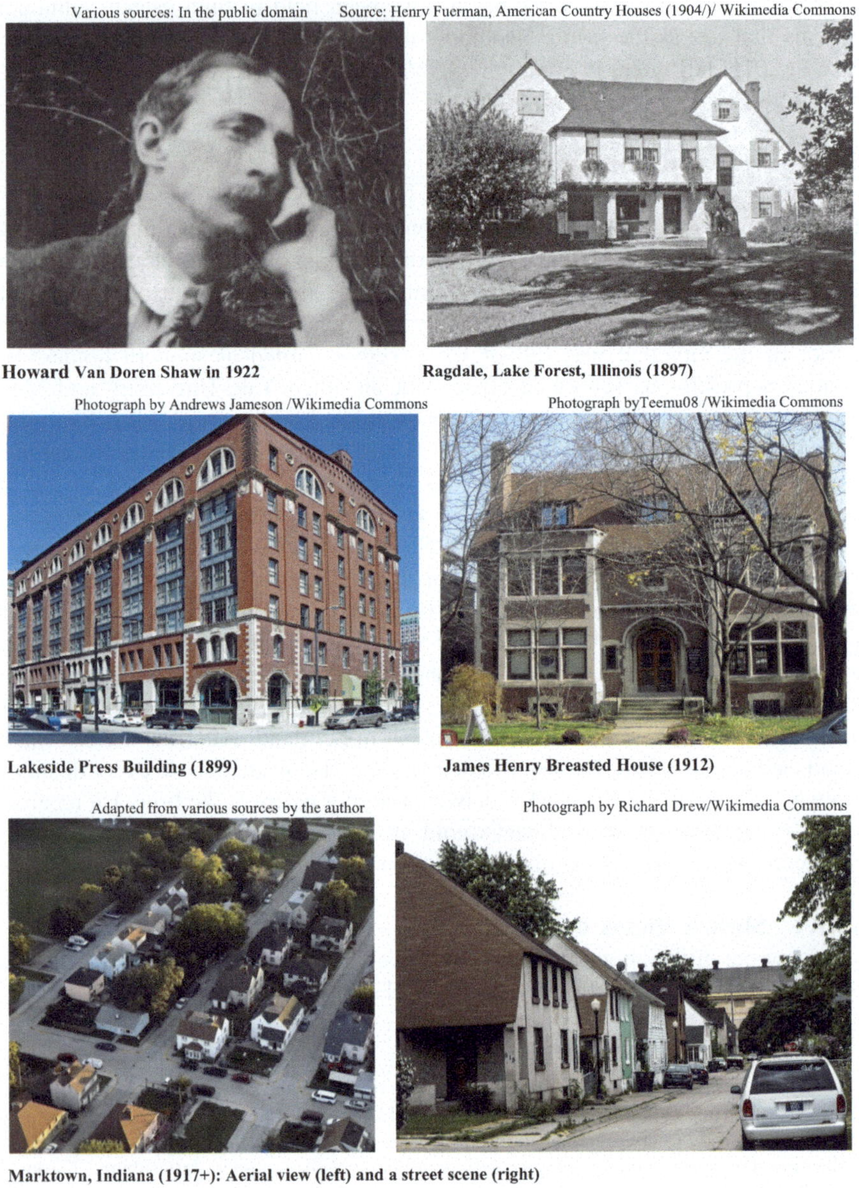

Fig. 8.1 Architecture of Howard Shaw

progressive line of architectural thought. His work was more pragmatically aligned with the tastes of the established patrician elite of Chicago who valued the types of European antecedents that the other members of The Eighteen decried. Shaw was known for his magisterial but quiet, reticent architecture and for being a leader in American craftsman approaches. He was highly skilled in amalgamating a variety of acceptable architectural precedents in the design of his houses for the wealthy.

8.2.1.2 The Clients

Shaw's clients for the first buildings he designed were members of his family—his father-in-law and then his sister-in-law and her husband (Shaw 1926). His first commercial building, the Lakeside Press Building (1899), was designed for Richard R. Donnelly (1838–1899), the father of Thomas Donelly, one of his Yale friends. Much later, Thomas Donnelly hired him to design a printing works. The clients for his mansions were members of the educated elite of Chicago. Their wealth was generally built on inheritances rather than their entrepreneurial skills (Eaton 1969). His client, for the Breasted House, for instance, was Yale and Berlin educated James Henry Breasted (1865–1935), an archaeologist and historian. The house, Camp-Woods, that he designed in Villanova, Pennsylvania was for a prominent banker, James A. Wilcox (1852–1935), who later hired Howe and Lescaze to design the ultra-modern and 'practical' Philadelphia Savings Fund Society Building (1934) in the city. Although best known for his suburban mansions, Shaw's designs also included buildings such as the Mentor Building (1906) for a commercial client and buildings for universities.

The client for his one urban design, Marktown, Indiana was Clayton Mark (1858–1936). Mark differed from Shaw's other clients. He, the owner of the Mark Manufacturing Company, a steel works, was public school educated and then only up to the seventh grade. Like most of Frank Lloyd Wright's clients Mark was a self-made man and not one born as part of the Ivy League educated social leaders of Chicago (Eaton 1969). He established himself among the city's elite. He was President of The Civic Federation of Chicago, where he worked with Jane Addams (1860–1935), the social reformer, and he served influentially on the Chicago School Board.

8.2.1.3 The Work

Shaw's office was prolific. Although he collaborated with other architects on some projects his designs were entirely his own. His most admired works are The Quadrangle Club at the University of Chicago (1893), the Lakeside Press Building (1899), and the Fourth Presbyterian Church (1914–19) designed with Ralph Adam Cram (1863–1942), a Gothic Revival architect (Cohen 2015). Shaw is, however, best known for the houses he designed (Eaton 1969; Green 1999).

The houses dot the prestigious suburbs of Chicago. They include several built contemporaneously with Frank Lloyd Wright's very different designs in Oak Park (see Fig. 5.2). One, Ragdale (1897), that he built for himself, is very much an Arts and Crafts building. Another house is, however, more highly regarded. Today a fraternity house, the Breasted House (1905) illustrates Shaw's architectural values.

It combines a classical organization with an Arts and Crafts touch. His portfolio includes a variety of building types other than houses.

Shaw's first major design was the Lake Side Press Building. It was carefully laid out for its instrumental functions. Its fine brickwork and eight storey façades are Beaux Arts in organization with a clear base, tall central portion, and a clearly delineated top storey. The later brown brick and terracotta Mentor Building has the same Beaux Arts organization. It was built at a time when other architects in Chicago were beginning to eschew the Beaux Arts as the design paradigm to follow. The building has a clear four-storey base, middle, and classically inspired cornice. Its windows grouped in bands, however, owe a debt to the Prairie style buildings beginning to appear in the Chicago area. Market Square (1916), which he designed with Edward H. Bennett (1874–1954), the coauthor of the 1909 City Beautiful *Plan of Chicago*, is often considered to be the first shopping centre in the United States to accommodate automobile parking. A U-shaped mall with a central courtyard, the building is surrounded by parking spaces.

Shaw's predilection for English precedents is displayed in his design of Marktown, Indiana. The development was an effort to recreate a pastoral English village in the heart of industrial East Chicago. It was built at the tail end of the Progressive Era in America that saw socially conscious towns, such as Pullman (1888) in South Chicago, being built by industrialists for factory workers. Begun in 1917, Marktown was designed to house the families of 8000 workers in 28 sub-districts. Only four were built. World War One and the sale of the company stopped its development.

Marktown was designed to possess all the amenities of a small town (Smith and Mark 2011). There were to be schools, recreation facilities, including tennis courts, a movie theatre, and shops. While its layout is a pragmatic gridiron one, the town had narrow roads designed for both pedestrian and vehicular circulation. Cars were parked on the sidewalk. The pattern is like the much later *woonerf* in the Netherlands. The streets, as in many English villages, were treeless but trees were planted in the houses' yards. They give the development a leafy appearance. An unusual feature for a company town was that workers could purchase their houses. The houses themselves were sturdily built in a Tudor revival style and fireproofed. Although now in decay and surrounded by industries, all the houses built in Marktown still stood in 2008. The apartments over shops did not.

8.2.1.4 The Functions of Central Attention

Howard Shaw's buildings met their basic functions of shelter and security and provided the affordances for their instrumental activity patterns well enough. They still do. The house plans followed standard rectangular types but on a generous scale that allowed for adaptations to be made easily. The siting of houses on their lots was standard too. They contributed to the unity and quality of the streets they faced and the neighbourhood in which they were located. The siting of the Lakeside Press Building functions similarly, but it is clearly urban being built to the property line.

While Shaw was concerned with meeting basic safety needs through fireproofing of his designs his focus of attention was on creating prestigious buildings for prestigious clients. The architectural references he chose contribute to the associational meaning of the building for his clients and the public. For the mansions, Shaw's own background and architectural studies led to his predilection for the values of the Beaux Arts and the Arts and Crafts movements. In landscape design, whether on his own or in collaboration with others, he looked to naturalistic rural layouts as referents in the manner of the picturesque English Landscape Garden Movement. Shaw's focus of concern was ultimately on meeting his clients' aesthetic values. The characteristics of his designs reinforced their sense of belonging to the contemporary high social status group. His clients selected him because they belonged to the same taste culture and social class as he did. Clayton Mark saw himself as a member of the group. His home, designed by Shaw, is Italianate in nature and his mausoleum in Illinois' Lake Forest Cemetery is a Neoclassical temple.

8.2.2 R. Seifert and Partners and Richard Seifert

Richard (originally Reubin) Siefert was born to Jewish parents in Zurich. The family came to England when he was young. He attended the Central Foundation School in Islington, London. The churches designed by Christopher Wren (1632–1723) that he saw on school walking tours in the City impressed him. They led him to study architecture, aided by a scholarship, at the Bartlett School. As a student his work was informed by the contemporary writings and work of Walter Gropius and the work of CIAM. He graduated in 1933. During World War Two Seifert was a colonel in the Royal Engineers in India and Burma (now Myanmar). That experience, he said, gave him his leadership skills.

After graduation he worked for several firms. In 1958, he formed and headed the highly productive and commercially successful firm of R. Seifert and Partners (later R. Seifert Company and Partners). Unlike many of his contemporaries he did not dress as a bohemian but as a businessman in pinstriped suits, regimental ties, and glasses. He always seemed to have a smoking pipe in his hands. *Private Eye*, an English satirical current affairs magazine, described him as an "ugly-looking little fellow" which smacks of anti-Semitism (Harrison 2019). During the 1960s and 1970s the firm had over 300 employees and designed over 500 buildings in the United Kingdom and around the world. It is, however, the London buildings that have attracted the most attention. The firm designed more of them in the City of London than any firm since Wren did after the Great Fire of 1666 (Harrison 2019; Bradbury 2020). He was a designer of eclectic talents (Pawley 2001).

8.2.2.1 Seifert's Views on Architecture

Like Howard Shaw, Seifert never provided a clear statement on what he considered to be good architecture. He simply got on with his work. His practice, unlike those of his contemporary luminaries, was not known for a few highly prestigious projects, but for the volume of its work. Their numerousness had a considerable catalytic impact on property values and the morphologies of cities in which they were located. Seifert saw himself as a creator, a technologist, a lawyer, a cost-controller, and a sociologist. He attributed his success to his affability (Bradbury 2020). His was a nonsense-free and professional approach to architecture.

The firm focused on providing its clients with profitable buildings. In doing so it learnt how to use planning laws and building codes to maximize a return on the capital invested in new buildings. Seifert was fully conversant with the regulations controlling building forms and their legal interpretations. He often knew more about the controls than the officials in charge of implementing them. His recognition of loopholes enabled him to maximize building envelopes, heights, and site coverage. His philosophy was simple (Fig. 8.2):

> You work, you work very hard and you work all the time and produce cost effective, program specific buildings. A satisfied client briefs another client … up to now, that's the only kind of promotion we've used. I've never chased after work or 'sold' the practice. What people want is a building worth the money paid for it. (cited in Harrison 2019)

This position is reflected in the firm's productivity. Seifert had the reputation of getting jobs done speedily.

The commercial success of the firm did not endear Seifert to architectural critics. His close relationship with property developers saw him dismissed as a property developer's architect. In the contemporary political climate of socialist Britain, his work was seen as opposed to the public good. He was regarded as a traitor to welfare state ideals. Attitudes have changed. The firm's work is now seen in a different light. Like many once critically disparaged works, Seifert's firm's efforts are now being evaluated for what they achieved both in structural innovations and in expressive design qualities. Centre Point is now said to have "the elegance of a Wren steeple" (Harrison 2019).

8.2.2.2 The Clients

R. Seifert and Partners' clients were primarily property developers. They were not patrons of the arts. Most were self-made men who were prima donnas anxious to get things done *their* way. At the outset they were Jewish speculative property entrepreneurs such as Felix Donovan Fenston (né Fenstein, 1915–1975) and, through serendipitous connections, Harry Hyams (1928–2015), a young property developer born in Handon, Middlesex. Both Fenston and Hyams entered property development from working in real estate agencies. Hyams was the client for Centre Point, the firm's most politically notorious tower. Other buildings were also designed for commercial clients. The National Westminster Bank was the client for Tower 42 in the City of London. Seifert was the developer and designer for his own house in Mill Hill, north London that he bought in 1946 and extended over the next half-century.

© National Portrait Gallery, London:
Courtesy of the National Portrait Gallery

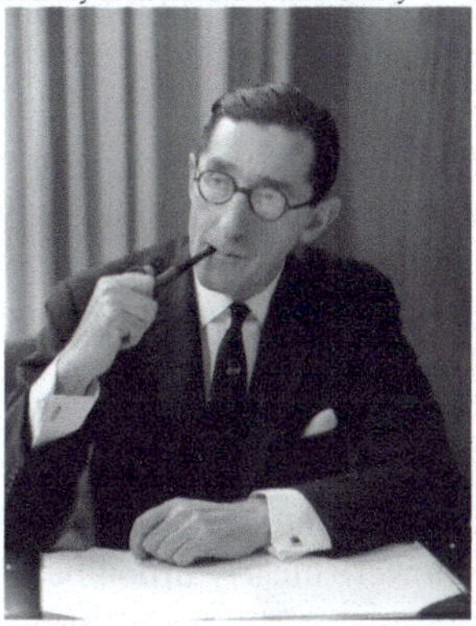

Richard Seifert

Photograph by Philafrenzy /Wikimedia Commons

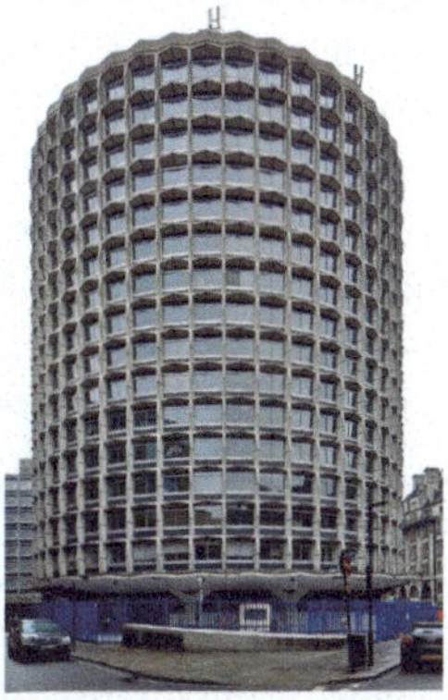

Space House, London (1968)

Photograph by the author

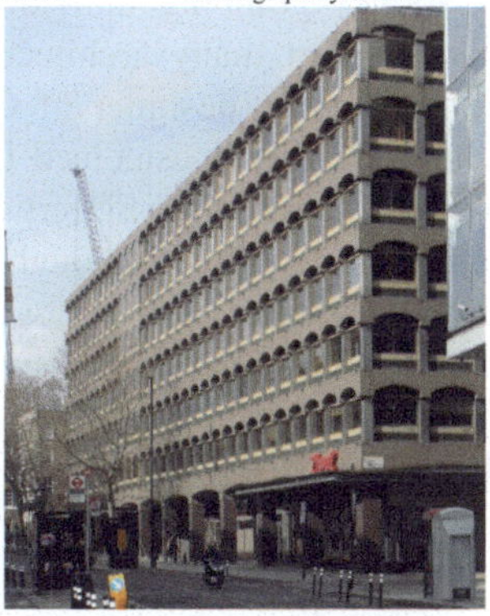

236 Gray's Inn Road, London (1976)

Photograph by Colin/Wikimedia Commons

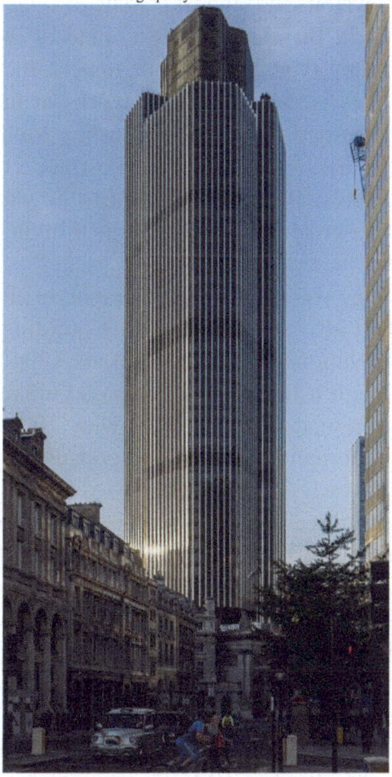

Tower 42 (NatWest Tower), London (1971-1981)

Photograph by the author

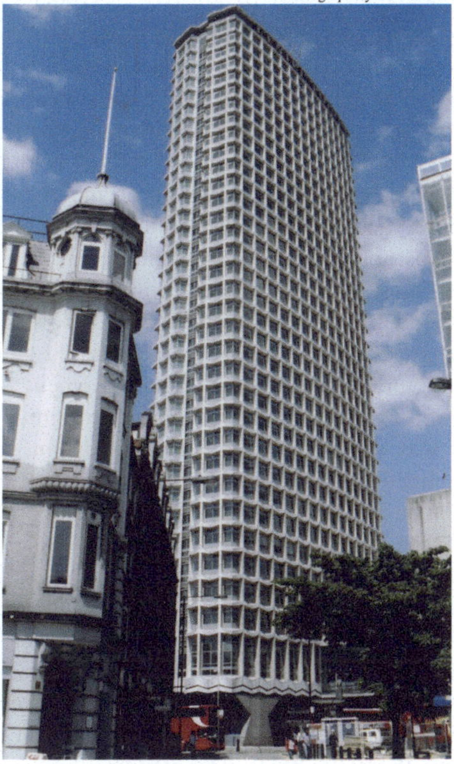

Centre Point, London (1960s)

Fig. 8.2 Work of Richard Seifert and Partners

During his working life, many of the public considered property developers to be money grabbing individuals. Working for them was seen as degrading. The irony is that many architects, such as Sir Basil Spence (1907–76), the competition winning designer of Coventry Cathedral, and Alison and Peter Smithson, who were highly esteemed by the cognoscenti, also executed designs for similar property developers. Seifert's work, in contrast, was regarded at that time to be an architectural failure although financially successful. He himself, with his pinstriped suits and regimental ties, was seen to be, as he was, a member of the capitalist establishment that progressive commentators deplored.

8.2.2.3 The Work

The firm's designs evolved from the Neoclassical to a Corporate Modernism that was popular among property developers. Its office towers were clearly urban buildings contributing to the maintenance of the quality of the streets they faced. Seifert regarded the NatWest Tower (now Tower 42; 1970–80) as his firm's major work along with Centre Point (1972–75) and Space House (1964–68), both designed by George Marsh, and Alpha Tower (1970–72) in Birmingham. The brand of the firm's work was not so much a style as a building type—the slab or tower office block. Many were structurally innovative. The stylistic nature of the works of the firm, like that of many other commercial firms, varied from production-line Modernist to Brutalist (described by his contemporaries as bombastic), Postmodern, and Neotraditional depending on the client's tastes (Bradbury 2020). In this way Siefert and the co-designers in his firm showed a great flexibility in their thinking. It was not driven by a single aesthetic position.

Tower 42, a 182-m-tall commercial building, was the first skyscraper in the City of London. As such it was controversial. It also displaced existing buildings of interest. The building's plan is based on the National Westminster Bank's logo. Structurally, it is a technical achievement although it took so long to pass bureaucratic scrutiny that the electronic revolution soon made it obsolete (Pawley 2001). It was engineered by Pell Frischmann, an international technical consultancy founded in 1920.

The building consists of a concrete core with three chevrons cantilevered from it. The cantilevers were designed to take advantage of the air rights granted to the site to protect a banking hall of note. The two lower floors consist of a single attached 'leaf' while the two floors above it have two leaves, a pattern repeated at the top of the building. Double decker lifts speed passengers from the lobby and mezzanine to sky lobbies at levels 23 and 24. Both concepts were novel at the time. The building was clad in vertically striped metal said to resemble the suits that Seifert wore. Completed in 1980, it was opened by Queen Elizabeth II (1926–2022). The cladding was replaced after the 1993 Provisional Irish Republican Army's bomb attack that devasted the area (Pawley 2001).

Centre Point is a 34 storey, 117 m (385 ft.) tall tower. Its design became a precedent for much of the commercial work that followed it in England. The building's external columns, load-bearing façade and floors were prefabricated off site. Doing so enabled the building to be erected at the rate of a floor a week. The beehive-like grid of glass and concrete is seen as an example of Brutalist architecture (Bradbury 2020). The concrete was highly polished to give it a luxurious appearance. In 2015 the Concrete Society gave the building a Mature Structures Award. During the same year it was converted to a mixed-use building containing luxury apartments. Elegant though the tower is, it is the building's financial history that attracted attention.

Hyams wanted the building to house a single tenant. He could not attract one. Given the tax breaks available at the time, the building was allowed to stand empty for fifteen years. Its vacancy had the catalytic impact of driving up rental prices elsewhere. Standing unoccupied, the building soon became the rallying point for protests against capitalism. Architecturally, it was thought to be a badly designed, coarse building by the cognoscenti (Harrison 2019; Bradbury 2020). If one is looking for precedents, the building reflects the design vocabulary of Modernist architects such as Marcel Breuer (1902–1961). Another building, Tolworth House, shows more an influence of Oscar Niemeyer's work in Brazil and that of Gio Ponti (1891–1979) in Italy (Pawley 2001).

8.2.2.4 The Functions of Central Attention

Most of the towers designed by R. Seifert and Partners were based on standard speculative, commercial office floor plan types. They have a central core of services and floor plates as large as legally possible yet providing natural light to their interiors. The designs were strictly Modernist in terms of functionality. They were efficient in layout and the floors were open to various internal configurations. They were easy to clean. The design objective was to get the buildings constructed and operating as quickly as possible and thus returning interest on capital invested in them—their primary function.

The firm paid much attention to the efficiency of the construction methods employed but it often used innovative techniques. It pushed the boundaries of technical knowledge and once proposed a 150-storey high tower for Liverpool. In doing so, it was as much concerned with the 'firmness' of its work as its efficiency and its presentation as modern.

While the contemporary blend of the modernist aesthetic of the firm's work symbolically appealed to its clients, to the contemporary cognoscenti the firm's early work was architecturally uninteresting and drew little comment. Its aesthetic quality was bland and failed to break new ground. The cognoscenti, imbued with the progressive, political spirit of contemporary Britain, deplored the commercialism of the firm's work. It represented unbridled capitalism (Bradbury 2020).

8.2.3 Alejandro Aravena Mori and Elemental Consultores S.A

Alejandro Aravena was born to a middle-class family in Santiago de Chile in 1967. His parents were teachers. They ensured that Aravena attended fine private schools and that he studied at the Universidad Cátolica de Chile, the home of Chile's outstanding architectural school. He graduated in 1992. His studies occurred during the austere years of the dictatorship of Agusto Pinochet (1915–2006). Access to international architectural design concepts was limited. Aravena regards this lack of exposure to emerging architectural prescriptive theories while he was an undergraduate to have been an asset (Kimmelman 2016). Buildings, instead, needed to fulfil basic functions.

The architectural school's professors were pragmatists not theoreticians advancing their personal prescriptions for the aesthetics qualities of buildings. They taught students how to get buildings built without much fuss. Aravena's graduate studies (1992–1993) differed. He attended the Università Iuav di Venezia, in Italy where he studied art theory and history. In Venice he was exposed to international contemporary architectural ideas as well as those of classical Italy.

Aravena started his own practice in 1992 and later founded Elemental with Andres Iacobelli del Rio (b. 1969), a civil engineer. Iacobelli had studied economic development at Harvard's Kennedy School. Their joint goal was to improve social housing in Chile. Elemental is referred to as a 'do-tank' rather than a 'think-tank'. Iacobelli subsequently made a career in academia and public service and Aravena continued in his role as executive director of Elemental. He is now in partnership with Gonzalo Artega, Juan Cerda, Victor Oddó, and Diego Tomes. Contemporaneously, Aravena is a professor at the Universidad Católica de Chile. He had also taught at architectural schools in the United States.

8.2.3.1 Aravena's Views on Architecture

Aravena is regarded as the most socially conscious of the Pritzker prize winners. He believes architects must be explicitly, rather than implicitly, concerned with contemporary political, economic, and environmental problems and issues of their societies (Chatel 2019). His stated goal is to improve people's ways of life by responding to their social needs and cherished hopes. His views reflect those of architects such as James Stewart Polshek.

Aravena's position on architecture is expressed in *Los Hechos de la Arquitectura* (Pérez Oyarzún et al. 1999) and in Diego Hernandez's monograph *ELEMENTAL: Incremental Housing and Participatory Design Manual* (2012).

> We don't think of ourselves as artists. Architects like to build things that are unique. But if something is unique it can't be repeated, so in terms of it serving many people in many places, the value is close to zero. (cited in Kimmelman 2016)

He adds:

> If architecture is about giving form to the places where we live, and life ranges from basic needs to artistic desires, the task of architecture and the difficulty of producing architecture is that it's not about choosing one or the other but integrating the two.
>
> Not in one single project that we have done, not even social housing, did we forget that life cannot just be about mere satisfaction of physical needs. And the other way around; if

> you just concentrate on the artistic cultural aspect, but you haven't satisfied basic needs, you don't even have the possibility to have a life beyond mere survival. So, the task of architecture should be that of integrating. (cited in Kimmelman 2016)

Elemental's position on social housing is that it is an investment in society not an expense to it (Elemental 2008). Much of the firm's work has been in response to each client's pragmatic needs (Fig. 8.3).

8.2.3.2 The Clients

Elemental's clients have been as varied as the commissions the firm has secured. Several have been Chilean mining and forestry companies interested in providing their workers with good housing but his clients include universities, government agencies (Chilean and foreign), and private individuals. A university administration was the client for the Siamese Towers and for the Faculty of Medicine at Universidad Católica. Angeli Group donated the funds for the Universidad Católica's Innovation Centre. The Chilean Government was the sponsor for the social housing project at Quinta Monroy and a timber works for that at Villa Verde. In these projects, the future inhabitants were considered to be significant clients. In developing the programs for social housing, many requirements were gleaned via the participatory techniques advocated by Advocacy architects as described in Chap. 7. The goal was to understand how to improve the lot of people struggling financially and socially (Hernandez 2012).

Only a few clients have been closely associated with the art world. Most have shared his concern for developing architectural ideas that serve people well and for buildings that are sustainable in the materials and construction techniques employed. The client at St. Edward's University, a private Catholic institution, in Austin, Texas, for instance, supported his use of rammed earth walls in the dormitories he designed for them. The residents in his half-house designs have embraced Aravena's design philosophy and state that they are happy with their homes. The commissioning clients also say they are pleased with the results of Aravena's work.

8.2.3.3 The Work

In addition to the Chilean projects Aravena has designed buildings in the United States, Germany, Mexico, Switzerland, and Inner Mongolia in China. Much of it has focused on meeting a building's pragmatic instrumental functions but most buildings are also works of art. This duality of goals is evident in his design of the Torres Siamesas, a workshop building for the School of Architecture at Pontificio Universidad Cátolica de Chile (UC). In the design of Edificio Anacieto—the U.C. Innovations Centre—Aravena avoided using the standard symbols of innovativeness—glass walls. Among Elemental's designs are the Colegio Huelquén Montessori, houses for individual clients, the Johnson residential halls of St. Edward's University, and an art workshop on the Vitra campus at Weil am Rhein where it sits alongside works designed by avant-garde architects such as Zaha

Photograph by Centro de Politcas UC/ Wikimedia Commons

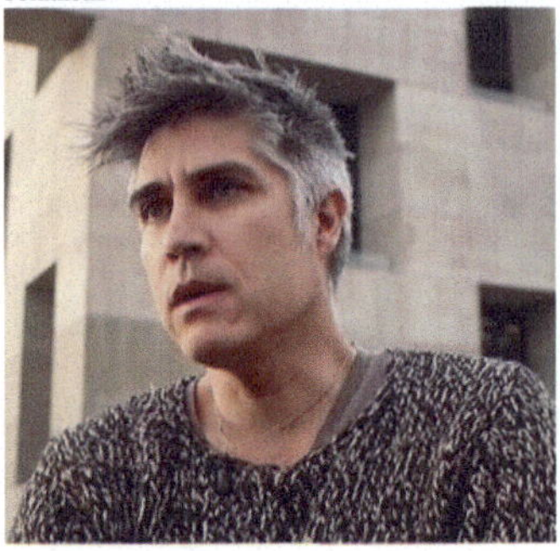

Alejandro Aravena in 2015

Photograph by Gabriela Alcanta/ Shutterstock.com

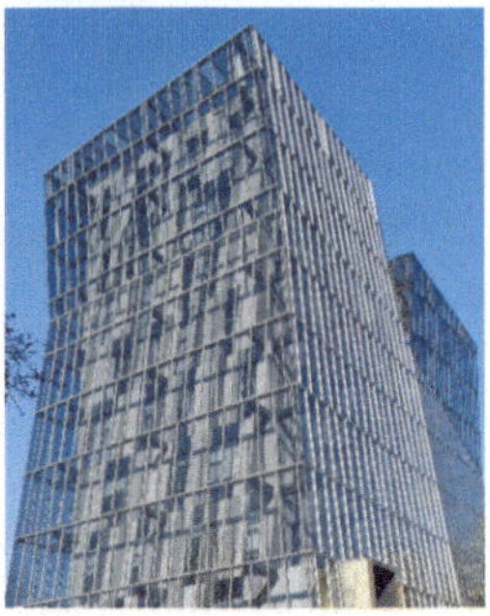

Las Torres Siamesas (2003)

Photograph by V.Morande/Wikimedia Commons

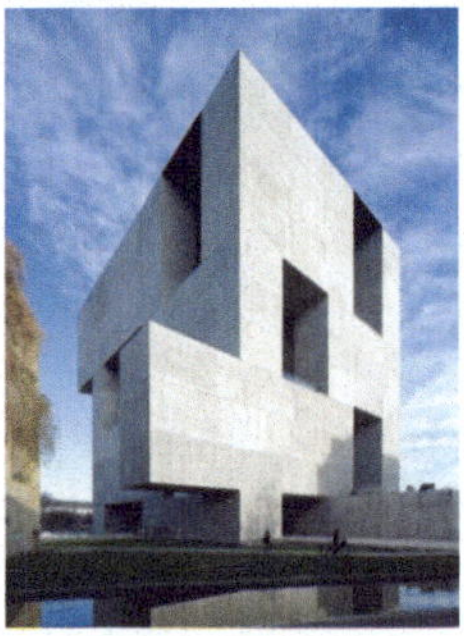

U.C. Innovations Centre (2014)

Photograph adapted from various sources by the author

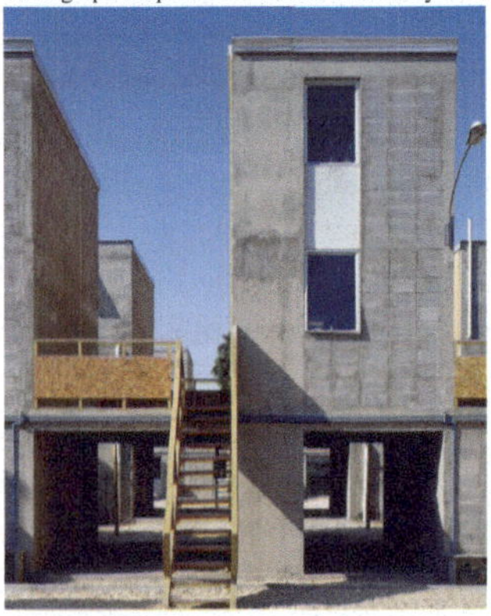

Photograph adapted from various sources by the author

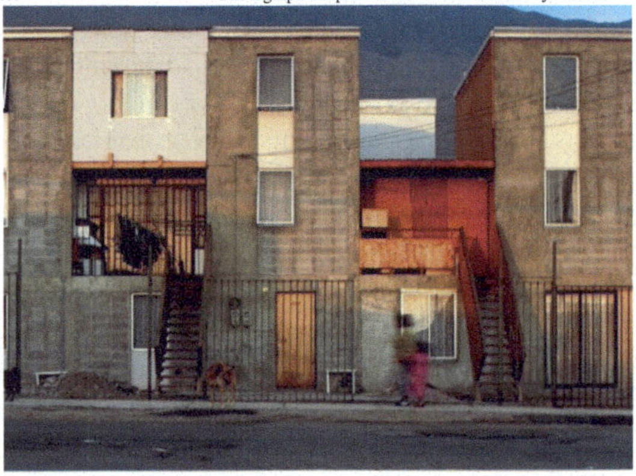

Quinta Monroy, Iquique, Tarapacá (2003); as built (left) and as in-filled (right)

Photograph adapted from various sources by the author

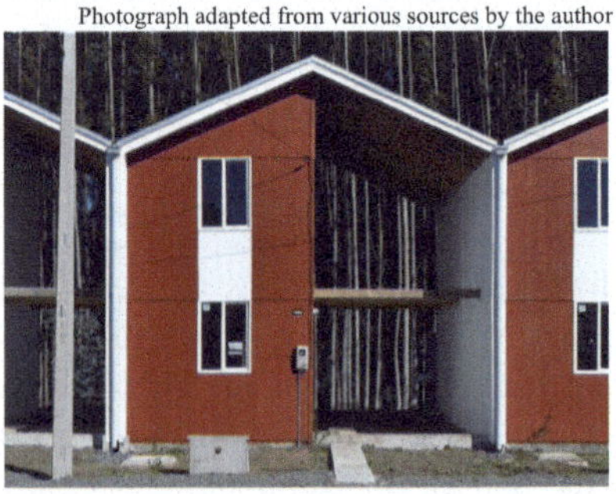

Photograph adapted from various sources by the author

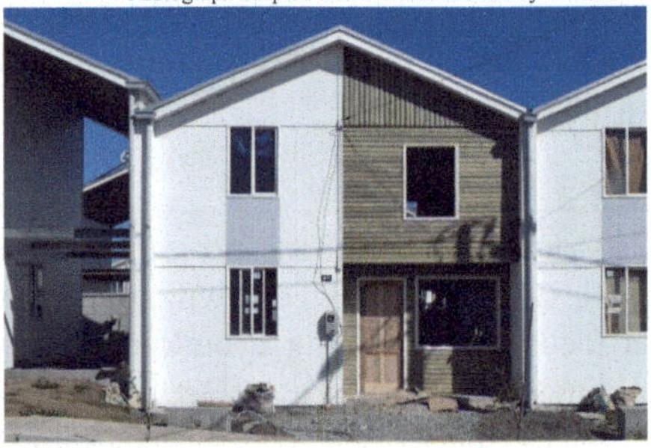

Villa Verde, Constitución, Talca (2010): As built (left) and as filled-in (right)

Fig. 8.3 Architecture of Alejandro Aravena

Hadid and Frank Gehry. He has also designed public spaces and a children's playground in Santiago's Metropolitan Park. This body of work contributed to Aravena receiving the Pritzker Prize.

Aravena's engagement in the rebuilding of the port city of Constitución in central Chile following the 2010 earthquake and tsunami brought his attention to the political and social nature of architecture and, particularly, urban design. His design of a cultural centre and the waterfront as a public park rather than a seawall required considerable discussion about ends and means with government agencies, private developers, and the public.

Designing housing while recognizing its potential residents' low-income status is a major challenge everywhere. Elemental has designed hundreds of two-storey, two-bedroom homes with kitchens and bathrooms, but with half of each unit left unfinished. Once settled in their completed half of a house, residents can assess their needs and resources and decide how to complete the house's other half when they are financially able to do so (Marinovic and Baek 2016). It is pragmatic solution.

The half-houses of Quinta Monroy in Iquique (2003) in northern Chile and Villa Verde in Constitución (2010) south of Santiago are a step up from the sites and services work developed by Christopher Benninger (b. 1942) in India as advocated by John Turner in *Housing by People: Towards Autonomy in Building Environments* (1976). They are also a more complete solution than that promoted by John Habrakan in *Supports: An Alternative to Mass Housing* (2021, originally 1962).

The 5000 m^2 site of Quinta Monroy in the centre of Iquique was illegally occupied by one hundred families for fifty years. The lot was bought by the Chilean government to enable the householders to be close to potential places of work rather than relocated on the periphery of the city. The Chilean Department of Housing provided a subsidy equivalent to US $7500 per family for a house on the site. Building detached house was financially and spatially impossible; a high-rise building would have been culturally inappropriate. The need was to provide high density housing without their inhabitants feeling overcrowded. The solution was to build half a two-storey 72 m^2 row house. The portion built was the part that a family would have difficulty in building on their own and be financially incapable of hiring a contractor to build it for them. Since occupancy many families have built in the open parts of their housing units. In design the complex special attention was paid to the nature of the outdoor space. A hierarchy of spaces from private to collective semi-public to public was provided on the site (Elemental 2008).

The philosophy behind the design of Villa Verde was similar but it was privately financed (Elemental 2013). Villa Verde was commissioned and sponsored by Arauco Forest Company to house its workers and contractors. The goal was to provide them with home ownership at a cost that they could afford. Elemental developed a set of types of various size housing within the funding guidelines of the Fondo Solidario de Viviendas of the Chilean government. Villa Verde's half-built design follows the type developed at Quinta Monroy but for a higher income group. Its design is less spartan.

8.2.3.4 The Functions of Central Concern

The fundamental objective in the half-housing designs is to provide families with shelter. It is also to provide them with a sense of financial security by providing them with the tenure of their dwellings. As a result, families achieve some control over their own lives. They thus enhance their feelings of dignity and self-worth. Given their common problems, they possess a sense of belonging to a group. They obtain a sense of community by having a common housing type and a shared communal open space. The population was also largely at the same stage in their lives.

For people to not feel overcrowded in high density environments, privacy between units and within is essential (Alexander and Chermayeff 1964; Altman 1975). In the residents' designs for the unbuilt half of their houses, the possible configurations are restricted. The purpose is to prevent them creating layouts where individuals must pass through one private space to reach another. The fear was that reduced privacy would lead to internal conflicts and consequential social problems. The same concern for privacy and control over space was applied to the outdoor areas. The hierarchy of spaces provided in each housing complex approximates, intuitively or self-consciously, that prescribed by Oscar Newman in *Defensible Space* (Newman 1972). Some families have colonized the open space in front of their houses and thus have added semi-private territories to the hierarchy.

Involving users in the design process not only gives them pride as co-designers of the final product but it is also necessary to get worthwhile design input. In planning for the rebuilding of Constitución, Aravena noted:

> We knew from the start that the people had to participate in the reconstruction process. In effect, we needed to create the right client. So, we devised a consortium: Arauco, the government, the public and us. We were taking an intuitive approach because we actually knew nothing about planning. In the end, ignorance helped. (cited in Kimmelman 2016)

The low-income residents also get an added sense of self-worth by getting what are essentially middle-income dwellings within their limited budgets.

The choice of materials for building not only in the half-houses but also as demonstrated in the dorms at St. Edward's in Austin shows Elemental's concern for sustainable architecture. The dorms' walls not only provide fine insulation but have the plasticity of form sought by many architects. At the Innovation Centre, the concrete frame serves as an energy saving device while the atrium gives exposure to daylight on the interior. The evidence of the outcomes of the employment of such measures is, however, limited. Elemental's concern for environment sustainability, nevertheless, is recognized by Aravena receiving the Gothenburg Prize for sustainable development in 2017.

8.3 Three Rationalists

The three Rationalists chosen for discussion demonstrate approaches to design that reflect different socio-economic, technological and political eras. In common they saw themselves as artists. The first, Le Corbusier, the giant among the early twentieth-century architectural philosophers, has had much written about him and his architectural attitudes. The thoughts and work of the second, Bernard Tschumi, presents one postmodernist reaction to the work of Le Corbusier, the Bauhaus, and the International Style. The prescriptions of the third, Zaha Hadid and her colleagues, are among those of recent Rationalists. All three drew on empirical experiences in developing their ideas so none can be regarded as a pure Rationalist. Many of the design decisions they made were simply pragmatic.

8.3.1 The Le Corbusier and Pierre Jeanneret Partnership and Le Corbusier

Le Corbusier, as noted earlier, was raised imbued with Calvinism's puritanical values (Brooks 1997). His mother favoured his brother, Albert, over him which may account for his fanatical drive. He fought for his ideas with what some analysts believe to be a messianic, autistic determination (Sussman and Chen 2017). He received a mixed architectural education. He attended the École de Arts Decoratifs in La Chaux-de-Fonds from the age of 13. He studied art history under the direction of Charles L'Eplattenier (1874–1976) and learnt about architecture from René Chapallaz (1881–1976). L'Eplattenier's ideas and the work of Viollet-le-Duc were highly influential in Le Corbusier's intellectual development. He also learnt from stints as an apprentice with Auguste Perret (1874–1954), the French structural rationalist and pioneer in the modern use of concrete, and from Peter Behrens (1868–1940), a leading German architect and industrial designer.

Le Corbusier was a voracious reader with an extensive library of much-annotated books, a frequent visitor to museums, and a traveller visiting places around the Mediterranean (Turner 1977). He was fully aware of the history of architecture and the prescriptions of architects such as Tony Garnier. The monastic simplicity of forms in places such as L'abbeye du Thoronet in Provence, reinforced his views on architecture (Sternberg 2012). Perhaps it was cubist art that left the greatest mark on his work. Later, the plasticity of his work was influenced by the vernacular architectural forms in France's African colonies that were beginning to be known (Steyn 2014). He saw himself as a revolutionary saving the world through architecture (Hewitt 2023). To him it was architecture or revolution.

Le Corbusier's twenty-year on-and-off partnership with his cousin Pierre Jeanneret allowed him to devote his mornings to painting and the exploration of architectural ideas. Jeanneret was born in Geneva and, like Le Corbusier, he was raised as a Calvinist. He studied architecture at the École des Beaux Arts in Geneva. His professional relationship with Le Corbusier ended during World War Two when Le Corbusier worked with the Vichy government while Pierre joined the French underground. It resumed when they worked on and in Chandigarh. Jeanneret was long Le Corbusier's chief protagonist. His own work was more critically regionalist than Le Corbusier's. He is best known for his contribution to the design of housing in Chandigarh, the design of the Panjab University campus and Gandhi Bhawan—a sculptural building—on the campus. B. F. Mathur, an architect and engineer, was co-designer of the Bhawan but Le Corbusier's hand is clear in it too (Fig. 8.4).

8.3.1.1 Le Corbusier's Views on Architecture

The focus here is on Le Corbusier's early career and thoughts on architecture as presented in *Vers une Architecture* (1923). The book argues for a total rejection of all past prescriptions, the embrace of advanced technologies, and the application of Cubist artistic ideas to building design. His ideas on architecture were developed when the austerity of World War One and the devastating impact of the Spanish Flu were fresh in people's minds. Polluted industrial cities still filled the valleys in many European countries. Urban areas, le Corbusier believed, had to be made for healthy lives.

Le Corbusier's well-known aphorism on the nature of architecture as Cubist art was expressed in the book.

> Architecture is the masterly, correct, and magnificent play of masses brought together in light and our eyes are made to see forms in light; light and shade reveal these forms; cubes, cones, spheres, cylinders, or pyramids are the great primary forms which light reveals to the best advantage; the image of these is tangible ... these ... are the most beautiful forms. (Le Corbusier 1923, 31)

Le Corbusier gave a clear prescription for the composition of buildings. It has been summarized as follows:

> *Pilotis* – a grid of slim reinforced concrete pylons that assume the structural weight of a building. They are the foundations for aesthetic agility, allowing for free ground floor air circulation to prevent surface dampness, as well as enabling the garden to extend beneath the residence. *Free design of the ground plan*, with its construction dictating new architectural frameworks. The absence of load-bearing partition walls affords greater flexibility in design and use of living spaces; the house is unrestrained in its internal use. *Free design of the façade* – separated exterior of the building is free from conventional structural restriction, allowing the façade to be unrestrained, lighter, more open. *Horizontal windows* – ribboned windows run alongside the façade's length, lighting rooms equally, while increasing the sense of space and seclusion and providing interior spaces with better light and a view of the surroundings. *Roof garden* – flat roofs with garden terraces serve both harmonic and domestic utility, providing natural layers of insulation to the concrete roof and create space. (Le Corbusier's five points of architecture, Undated)

His position on urban design was presented in *Urbanisme* (1925a) and on both architecture and urban design in CIAM's declarations especially in *Charte d'Athènes* (1933). His early urban design ideas are, however, better represented

Photograph by Jeet Mehrotra/ Wikimedia Commons Photograph by the author Photograph by Shakespeare/ Wikimedia Commons

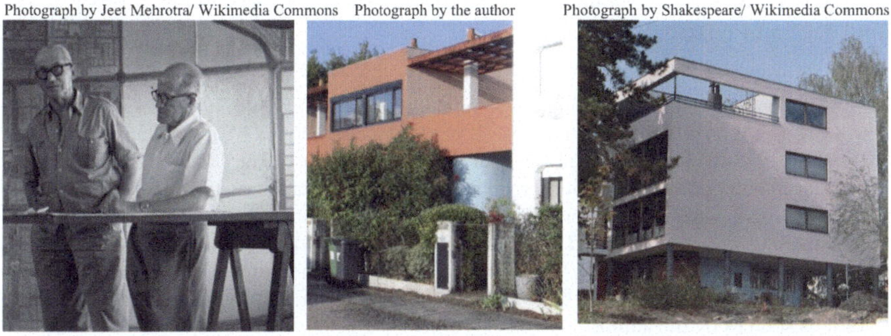

Le Corbusier and Pierre Jeanneret, Les Quartiers Modernes de Fruges, (1924), and a Maison Citröhan (1925)

Photograph by Dijana Alic

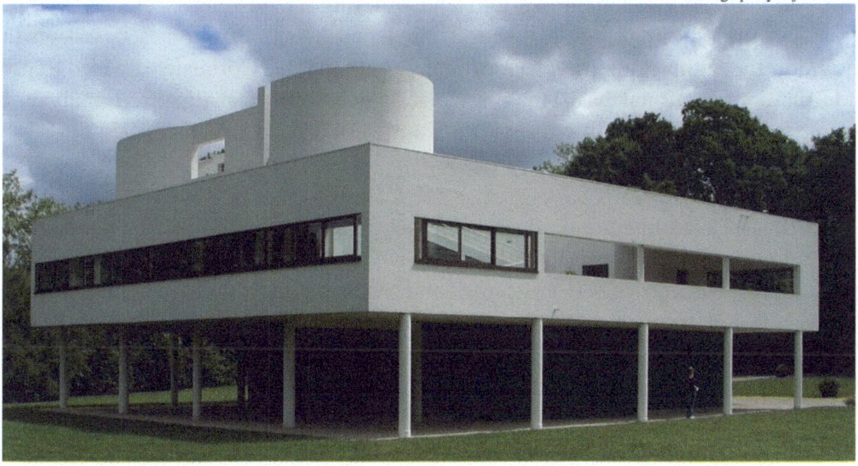

The Villa Savoye as rehabilitated (originally 1928)

Source: Mumford (1954): Collection of the author. Adapted by the author from a drawing in the British Library/Wikimedia Commons

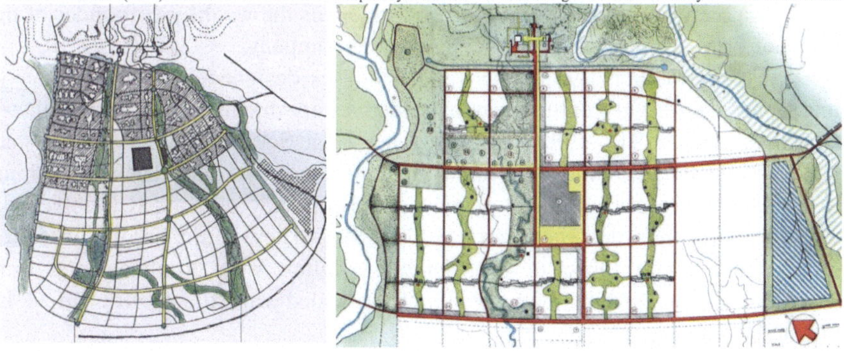

Chandigarh (1956 +): The Mayer- Nowicki plan (left) and the plan "de-Anglo-Saxonized" by Le Corbusier (right)

Fig. 8.4 Work of Le Corbusier and Pierre Jeanneret

in CIAM's La Serraz Declaration (1928) and shown in his generic designs (see Figs. 4.12 and 6.9).

> Town planning is the organization of the functions of collective life; it applies just as well to rural places as urban agglomerations. It cannot be conditioned by the pretensions of an established aestheticism; its essence is of a functional nature. The functions it embraces are four in number: dwelling, work, recreation, and transportation (which connects the first three functions with one another). The chaotic subdivision of land as a result of real estate speculation should be corrected. Present technical means, which multiply ceaselessly, are the very key of town planning. They imply and propose a complete change in existing legislation; this change should be commensurate with technical progress. (compiled from various sources by the author)

This rationalization, standardization, and ordering of designs is clearly demonstrated in the buildings he designed. Their façades' organization is based on the Modulor, the anthropometric proportional system that Le Corbusier devised to provide the compositional basis for his work. It is a somewhat distorted analysis of his ideal 6 ft. (1.83 m) tall man with his arm raised (Le Corbusier 1951).

8.3.1.2 The Clients

Le Corbusier's architectural ideas were supported by progressive clients (Soth 1983). Leonard Eaton describes those for whom he designed houses after World War One as "eccentrics, art collectors and avant-gardists" (Eaton 1969, 55). His clients differed from those earlier of Howard Shaw and later of Richard Seifert. They acted as patrons for his work. The type of people who were Le Corbusier's clients did not exist in the United States or in Britain.

Henri Frugès, a sugar refinery owner in Bordeaux gave Le Corbusier a free hand in designing Le Quartier Moderne, in Pessac (1924). Frugès had been inspired by an article written by Le Corbusier that he had read in *L'Esprit Nouveau.* Pierre Savoye and, even more so his wife, Eugène, were patrons of Le Corbusier's design for the Villa Savoye (1928–1931). Pierre Savoye was the wealthy cofounder of the still extant Gras Savoye, an insurance brokering company.

The post-World War Two clients for the houses designed by Le Corbusier in Ahmedabad were highly cultured industrialists. Some members of the families—Gautam Sarabhai (1917–1995) and his sister Gira Sarabhai (1923–2021)—had studied architecture at Taliesin West with Frank Lloyd Wright. They were Jains whose precepts are not unlike those of Calvinism. They demanded humble, climatically appropriate designs from Le Corbusier. They could hardly be called eccentrics nor could those in the Catholic Church who commissioned him. The experience in working with the latter coincided with the freeing up of Le Corbusier's designs.

His urban design explorations, as already noted, were sponsored by Gabriel Voisin, hence the centrality of the automobile in his designs (and aeroplanes in the skies of his drawings). Businesspeople rejected Le Corbusier's approach. His clients for the design of Chandigarh, which largely falls outside the timeline of this analysis, were very different people. Their mention here is simply to show the changing nature of Le Corbusier's clients. They were Oxbridge educated Indian civil servants who had worked for the Government of India prior to Independence in 1947 and continued doing so uninterrupted afterwards.

Le Corbusier straightened out the Mayer-Nowicki design for Chandigarh that he inherited to 'de-Anglo-Saxonize' and rationalize it. That design may have originally been directed by Indian Civil Service officer Anthony Leocadia Fletcher (1909–1974), a Keralan Christian, on behalf of the Government of India. Le Corbusier's client group was, however, headed by Prem Nath Thapar (1903–1969), an Oxford educated Indian Civil Service officer, and P(armeshwari) L(al) Varma (1920–*ca* 2000), Punjab's chief engineer. Jawaharlal Nehru (1889–1964), independent India's first prime minister, hovered in the background as a quasi-patron. His charge to Thapar, Varma, and Le Corbusier was for Chandigarh 'to be expressive, experimental and not be hindered by tradition' (Seguin 2014). Nobody involved seems to have been aware of India's own urban design canons as expressed in the *Shilpa Shastras* (see the discussion of the work of D. K. Bubbar later in the chapter).

8.3.1.3 The Work

Le Corbusier's designs for the Citröhan houses (1920s) illustrate his concern for the spatial qualities he sought in his architecture. They show a reinterpretation and evocation of historical types such as Palladio's Villa Foscari la Malcontenta (*ca* 1555) (Rowe 1976; Etlin 1987). It is, however, the Villa Savoye (1928–1931) that clearly demonstrates his prescription for buildings. Applying his five points of good architecture to an actual building overrode his consideration for other programming necessities. He had to make some changes in applying those principles to meet financial constraints on what could be accomplished. His ideal structural grid size had to be compromised. Pierre Jeanneret worked out a new grid and how to relocate some spaces (Buggy 2012/13). More seriously, when completed the building suffered structural, leakage, and overheating problems. The Savoyes were infrequent visitors.

Villa Savoye was damaged by the German and American troops who occupied it during World War Two. As a result, the Savoyes abandoned it. Funded by the French government it was renovated under the direction of Jean-Louis Véret (1927–2011) and is on display as a work of art expressing the philosophy of its creator. It is the twentieth century equivalent of Palladio's Villa Rotunda (Rowe 1976). Both are much admired for their innovative designs that established intellectually new approaches to architecture. Neither afforded highly liveable spaces.

Le Corbusier's earlier design with Pierre Jeanneret for Les Quartiers Modernes de Fruges (1924), a small suburban development in Pessac, went through much the same fate as the Ville Savoye. The intended inhabitants declined to move in because of its inconvenient location and unusual design. (The location problem was something the Chilean government and Elemental avoided at Quinta Monroy.) The houses at Les Quartiers were sold on the open market and bought by people who took advantage of their flexible structural characteristics to make changes that would meet their requirements. The structural system that Le Corbusier devised allowed walls to be moved, windows to be punched into the exterior walls or to be closed. The complex was allowed to deteriorate as it changed owners and maintenance costs rose (Boudon 1972). It has now been restored to its original state by new owners and exists as an exemplar of Le Corbusier's suburban design ideals. Unlike the Villa Savoye the houses are inhabited. The people who live in them appreciate them for what they afford and they enjoy the association of the complex with a world-renowned architect. It matches their aspirations well (Lang 2021, 155–158).

The ambient temperature problems at Ville Savoye may have led to Le Corbusier's post-World War Two collaboration with André Missenard (1901–89), a leader in studies on thermo-physiological comfort in the built environment (Requena-Ruiz 2014). The designs of the Ahmedabad buildings illustrate Le Corbusier's later preoccupation with climate and design. The spatial layout of the capitol complex in Chandigarh that imitates that of the Mayer-Nowicki plan on a grander scale is, however, blistering hot on a summer's day, drenched during the monsoon, and freezing in winter. It remains suitable for grand gatherings, although not for watching them, when political conditions allow.

Le Corbusier's generic urban designs prescriptions are best illustrated in *Ville Contemporaine* (1920s), and *Ville Radieuse* (1930). In them he presented design patterns to be imitated in the rebuilding of bomb damaged, to him outmoded European cities. The plans were never implemented but some elements were further explored in his design for the Unité d'habitation in Marseilles (1947–1952), by Sert in Barcelona (Fig. 4.12) and by Lúcio Costa in the design of the Pilot Plan for Brasília (1956).

The plan of the Ville Radieuse was based on the shape of the human body with each part serving a separate instrumental purpose. Designed for high-speed automobile traffic, it possessed an orthogonal grid road system with largely independent pedestrian paths. The buildings were either identical high-rise towers or slab blocks set in a zig-zag pattern. The residential buildings had roof terraces and jogging tracks. The open space among the buildings was devoted to athletic facilities and parkland. The design was as much a blueprint for social reform as for a physical city. A good environment, it was assumed, would make good people. What Le Corbusier failed to consider were the behaviour settings that offer joy to people's lives.

The rational segregation of building uses and the repetitive pattern of building geometries set in parklands resulted in sanitary but uninteresting environments when applied to large scale mass housing projects. The streets in the designs were channels for moving vehicles and not seams joining their two sides into a unit making public spaces for daily life. The model of the human being that he chose as the basis for his designs was himself. He said little about the everyday requirements of men, women, and children of different cultures and beliefs. His was an imagined ideal world for an imagined ideal people.

8.3.1.4 The Functions of Central Attention

The functions Le Corbusier emphasized in his works differed considerably from project to project. He was clearly concerned with creating salubrious environments full of sunshine and fresh air. The spacing of buildings, their orientation to the southern sun and their settings in open green areas, show his concern for the health of the inhabitants of his work. He did not always get matters right because he had to make trade-offs between what was needed by the potential residents of his buildings and his own intellectual and aesthetic ends. His generic housing schemes were, nevertheless, full of opportunities for what he considered to be healthy pursuits.

His house designs for wealthy clients were clearly artistic displays that reinforced their owners' social status. They are much admired to this day by the cognoscenti for their ingenuity and departure from aesthetic norms. The link between contemporary artistic movements such as Cubism, and later Purism, and Le Corbusier's design patterns continues to attract attention. More generally, however, his mass housing was designed to eliminate opportunities for the visual display of social status and individual tastes. The function of buildings as individual displays was limited to his detached houses and even there it was his brand of architecture that was really on display. The structural integrity of his housing in Pessac, certainly allowed for substantial personalization by its inhabitants. The original residents took advantage of it. Those of today admire the houses as works of art.

In his generic urban designs, the program that Le Corbusier set was limited in scope. The designs, thinking broadly, served too few functions. The designs fail to address the everyday concerns of people and the aspects of life that give them enjoyment. Enjoyment was not something held in high regard by Calvinists. To them an austere, Spartan life is the appropriate one. It is what God expects.

8.3.2 Bernard Tschumi Architects and Bernard Tschumi

Bernard Tschumi was born in Lausanne, Switzerland, the son of a prominent architect and professor, Jean André Tschumi (1904–1962). He received his architectural education at the Federal Institute of Technology in Zurich where he found the conservative and pragmatic approach to architecture limiting. Instead, he was inspired by the hypothetical designs of architects such as Cedric Price (1934–2003) and the Archigram group. After a stint working for Candilis-Josic-Woods in Paris he took a teaching position at the Architectural Association (AA) in London. Several radical avant-gardists—Rem Koolhaas (b. 1944), Zaha Hadid, Daniel Liebeskind (b. 1946), Nigel Coates (b. 1937), and Elia Zenghelis (b. 1937)—were colleagues there. He found the AA atmosphere congenial (Astbury 2022a).

Tschumi first made his reputation as a theorist offering prescriptions for how buildings should express intellectual ideas. Among his writings are *The Architectural Paradox* published in 1975 and republished in 1994, *The Manhattan*

Transcripts originally published in 1985, in which he presented a series of hypothetical designs for New York, and *Notes towards a Theory of Architectural Disjunction* (1988). Tschumi has taught at major schools of architecture in France, the United Kingdom, and the United States where he was Dean of the Graduate School of Architecture, Planning, and Preservation at Columbia University from 1998 to 2003. As a practitioner he runs offices in New York and Paris. They have designed buildings for diverse clients in Europe, the United States, and the Dominican Republic among other countries (Walker and Tschumi 2006).

In developing his design manifestoes, Tschumi has been much influenced by film and literary theory. The thoughts and ideas of Jacques Derrida in challenging established ways of interpreting literary texts have had a major impact on his architectural and urban design ideas. His reputation among the cognoscenti stems from the matching of his words and deeds.

8.3.2.1 Tschumi's Views on Architecture

Tschumi opposes conventionally accepted ways of considering the built environment and the activities and interpretations afforded by its patterns. He challenges traditional concepts of harmony, unity, and stability. He believes that there is no fixed relationship between the instrumental function of buildings and their forms. A building may, after all, serve a series of purposes over its life span depending on what its layout affords (Migayrou 2014). At the same time, it must accommodate an original set of activities well enough.

Tschumi seeks to make people pay attention to new buildings by using revolutionary architectural forms. The clearest statement of his prescriptive theory appears in *The Pleasure of Architecture* (1977).

> The architecture of pleasure lies where concept and experience of space abruptly coincide, where architectural fragments collide and merge in delight, where the culture of architecture is endlessly deconstructed and rules are transgressed. No metaphorical paradise here, but discomfort and the unbalancing of expectations. Such architecture questions academic (and popular) assumptions, disturbs acquired tastes and fond architectural memories. Typologies, morphologies, spatial compressions, logical constructions – all dissolve. Such architecture is perverse because its real significance lies outside utility or purpose and ultimately is not even necessarily aimed at giving pleasure. (cited in Jencks and Kropf 1997, 268)

Another statement is also important.

> Architecture is as much about the event that takes place in a space as about the space itself. (cited in Charitonidou 2020, 5)

For Tschumi architecture is a tool for questioning fixed social values. He challenges standard programs, the role movement plays in experiencing spaces, and the relationship between milieu and events. He reforms standard building types and expectations. His basic tenet is that space, movement, and events are ultimately independent, yet form a relationship. The result is that the conventional components of architecture and urban designs are all dissolved and rebuilt along new lines. His views foreshadowed the more widespread consideration of deconstruction philosophy in architecture.

8.3.2.2 The Clients

The clients of many of Tschumi's major projects have been public institutions seeking eye-catching buildings. Their goal has been to establish their statuses in the avant-garde, international commercial and intellectual world. The Japan Arts Council was the client for the Tokyo National Theatre and the City of Groningen in The Netherlands for the Glass Video Gallery. Both were promoting their cities as progressive centres. The Parc de la Villette (1984–1987) in Paris was no exception. The sponsor was the Établissement Public du Parc de la Villette (EPPV), a public property developer, headed by Paul Delouvrier (1914–95). In the background for that design was, like Nehru for Le Corbusier in Chandigarh, a major patron, President Valéry Giscard d'Estaing (1926–2022).

The Parc de la Villette was part of Giscard d'Estaing's effort to displace New York as the world's pre-eminent art centre. His initiating of major projects in the Paris region had that goal. They include the Grande Arche (1989) at La Defénse, the Musée d'Orsay (1980s), and the Arab World Institute (1987). President Francois Mitterrand (1916–1996) later initiated the *Grands Projets* (Grandes Opérations d'Architecture et d'Urbanisme) to take the program forward. Giscard d'Estaing sought unconventional designs and commissioned architects who could produce them (Fig. 8.5).

EPPV established the program of facilities to be included in the Parc de la Villette but others emerged from analyzing the design competition entries. It was a design first approach to creating the program. EPPV footed the 1990s €1.4 billion bill for the project. Other stakeholders included the operators of the various cultural institutions in the park, potential users—adults and children—and those Parisians who simply demand 'more open space'. Despite, or because of, these diverse interests Tschumi had almost total sway over the design.

A property developer, Angelo Consentini and John Carson Company, was the client for the Blue Condominium, a visually distinctive residential tower in New York (completed in 2007). It was built by their construction company, On the Level Enterprises. They sought something commercially viable but visually different from the mainstream of New York apartment buildings. That the building towers over its surroundings seems to have been the source of more local ire than its aesthetic character. Gentrification was already well underway in the area and residents wished for no more (Mele 2000). The worry in the neighbourhood was that the growth in luxury condominium buildings would threaten the Lower East Side's physical and thus demographic character.

8.3.2.3 The Work

Tschumi's firm has produced many widely applauded designs. The Tokyo National Theatre and Opera House (1986), Glass Video Gallery (commonly known as the Tschumi Gallery, 1990) in Groningen, the architectural school (1998) at Marne la Vallée, and the Blue Condominium are among them. The Video Gallery in

Photograph by Columbia GAAPP/Wikimedia Commons

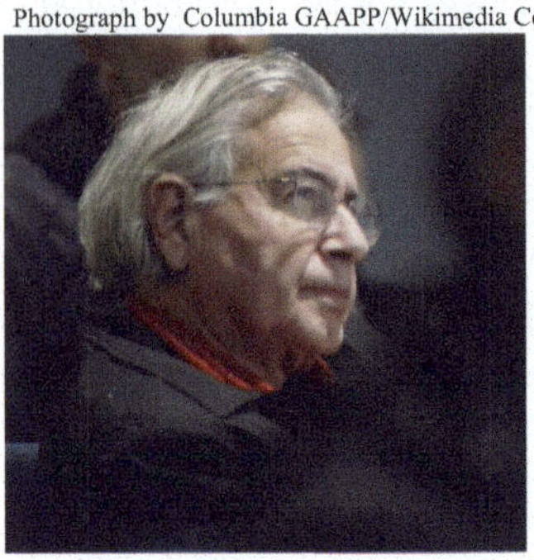

Bernard Tschumi

Courtesy of Bernard Tschumi Architects

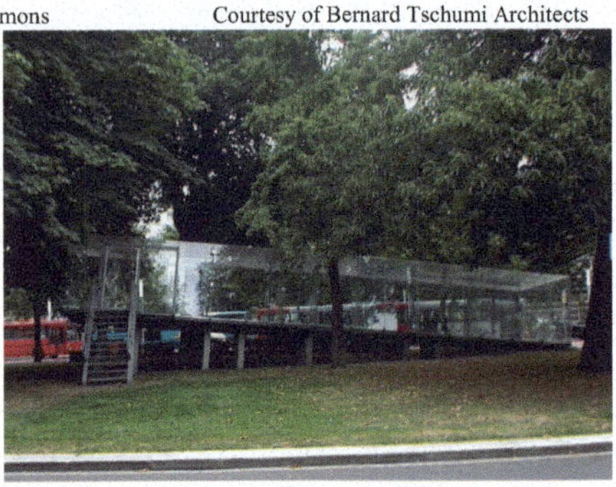

Glass Video Gallery (1990), Groningen

Drawing by Yin Yin: Courtesy of Yin Yin

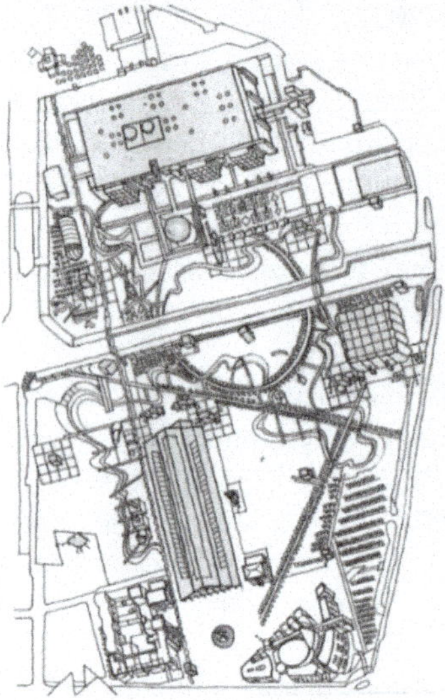

Courtesy of Bernard Tschumi Architects

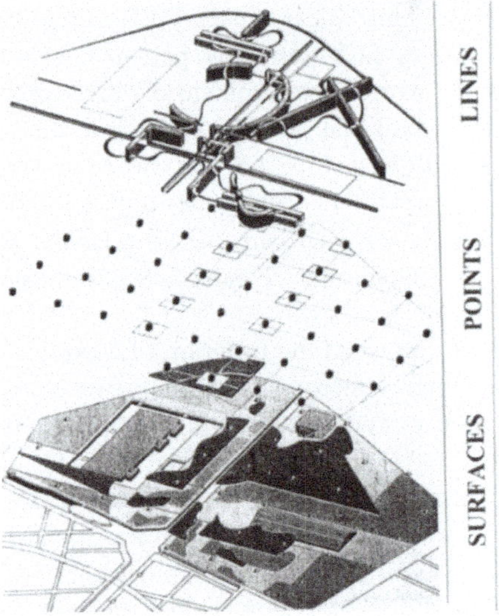

Parc de la Villette (1979): The plan (left), and the design deconstructed into three patterns (right)

Photograph by the author

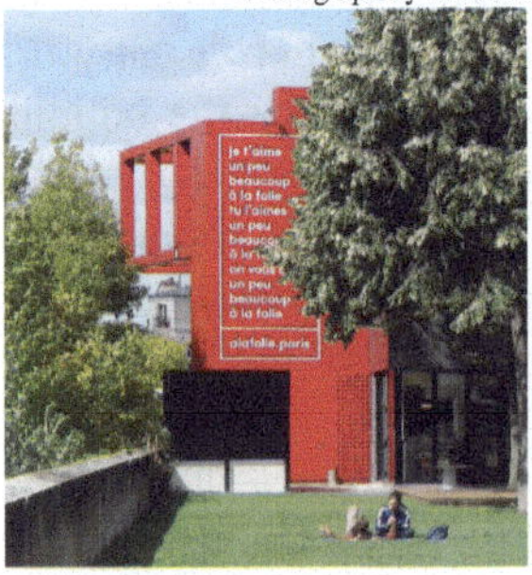

Photograph by the author

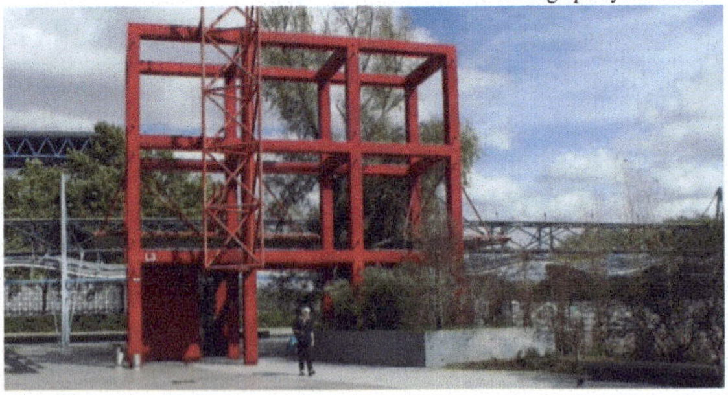

Parc de la Villette: Follies

Fig. 8.5 Work of Bernard Tschumi architects

Groningen has transparent walls and tilted floors that reject the norms of walls, interior spaces, and the relationship of building interiors to their exterior settings. In the School of Architecture at Marne la Vallée the spaces challenge the norms of expected uses and sequences.

Tschumi's professed position on architecture is most clearly demonstrated in an earlier design, that of Parc de la Villette. It was designed in partnership with Colin Fournier, an architect highly influenced by Buckminster Fuller. Fournier was a founding member of the Archigram Group and, with Peter Cook (b. 1936), he was the architect of the Kunsthaus Graz (2006). The Parc de la Villette established Tschumi's reputation and his brand of work (Walker and Tschumi 2006; Astbury 2022a).

The Parc's design resulted from a 1982 competition organized by EPPV for the redevelopment of a 55 ha (136 acre) site of semi-abandoned industrial land in Paris' 19th Arrondissement. The instrumental program included a museum of science and industry, a *cité* of music, a major hall for exhibitions, and a hall for popular concerts. It required two existing structures to be reused (Tschumi 1987; Migayrou 2014; Lang 2017). The individual buildings were to be designed by several leading architects. In creating their design proposal, Tschumi and Fournier followed Le Corbusier's maxim that a city should be in a park rather than a park in the city. Most major cultural institutions are landmark buildings located in the fabric of a city but in the Parc de la Villette, they are in a park. Although serviced from streets, they are connected primarily by pedestrian paths. As an urban design, the complex is an anti-urban gesture.

The design is composed of three logically consistent but independent patterns superimposed on each other. The first of the three consists of a series of follies at the intersections of a 120 m (390 ft.) grid. Each folly is a 10.78 m (36 ft.) cube "divided three dimensionally into 12 ft. cubes forming red-enamelled 'cases' that "can be decomposed into fragments … or extended through the addition of other elements" (Tschumi 1987). Each was intended to be a garden decoration. The second system is comprised of lines that form pedestrian paths. One set consists of cross axes of covered galleries; another is a meandering 'cinematic' promenade". The third system is formed by the surfaces of the park. Avenues link the major activity sites of the design (Tschumi 1987; Lang 2017, 75–77). Although Tschumi intended the design to have no inherent referents, the three systems evoke the clashing of surfaces, grids, and forms of the Russian Constructivists. This connexion predates Zaha Hadid's preoccupation with Kazimir Malevich's work.

The reactions to Parc de la Villette vary. The cognoscenti hold the intellectual rigour of Tschumi's work in high regard but it has been criticized for sacrificing human needs for intellectual purposes (Salingros 2007). Some critics consider its design to be inhuman. Others see it aesthetically reflecting the past industrial use of the site. Locals care little about the logic behind the Parc's design. They appreciate what it affords them for recreation, its instrumental function that, to Tschumi, was a secondary consideration in its design. The views of his design of the Acropolis Museum in Athens (2009) are similar. Architectural critics love it, but other intellectuals—the contextualists—think that its design has nothing to do with its location. It could be anywhere.

The Blue Condominium (2007) designed with New York based SLCE Architects, is a more conventional design. A high-rise residential building, it consists of 32 apartments, a gallery on the ground floor, and a roof garden at the third-floor podium level. The design is a pragmatic response to the constraints of the site and the demands of the marketplace. The apartments' designs closely follow standard types in their price range but they are said to have an 'edgy' character. The construction method of cast-in-place reinforced concrete is also typical of New York. The curtain wall cladding system is conventional although the blue panels make the building visually distinctive. The building is clearly a developer driven project in which Tschumi's written prescription for architecture and the nature of the design task are at odds.

8.3.2.4 The Functions of Central Attention

Tschumi and Fournier's professed goal for Parc de la Villette, was to create a new form of artistic expression. Most of the Tschumi's firms' work, such as the Glass Video Gallery in Groningen and the architectural school at Marne la Vallée, demonstrate a means for attaining that end. The buildings are a departure from the public's expectations. One of the prime functions of such designs is to make people pay attention and think. Liveability can become a secondary consideration.

The Parc de la Villette affords many functions. It allows for the development of a variety of behaviour settings and aesthetic experiences. It is meaningful to both lay people and the cognoscenti although their interpretations differ widely. Lay people consider the park in terms of what it affords them. It is rich in offerings for activities, both formally organized and informal. The lawned areas are used for impromptu football games, exercising, and experiencing nature within the city. The paths function not only as pedestrian ways but as potential play surfaces for children. Their curving lines provide ever-shifting vistas for the people walking along them. The follies function as landmarks much as the sculptures and statues in tradition urban parks. The institutions function as destinations and drawcards as well as reinforcing the picture of Parisians as a cultured people. In winter, the Parc can be a miserable place but the attractions and special events draw people.

Above all, Tschumi's designs function as food for thought for the cognoscenti. The Parc de la Villette's intellectual association with the contemporary deconstructivist literary analysis of Jacques Derrida is what matters to them (Benjamin 1988; Wigley 1993). Derrida presented a radical view of the ways of interpreting the meaning of phenomena. Tschumi seeks to achieve the same end in his firms' designs. As for most architects, the catalytic effect of the firms' designs is regarded as somebody else's responsibility. The impact of buildings such as the Blue Condominium and the Parc on their surroundings is the government's concern. The architect's focus of attention is on creating the best designed object and living environment possible.

8.3.3 Zaha Hadid Architects (ZHA) and Zaha Hadid

Zaha Mohammad Hadid, a Pritzker prize winner in 2004, was an Iraqi-British architect. She was born in Baghdad and died in Miami, Florida. The daughter of a wealthy industrialist and politician and an artist mother, she attended boarding schools in England and Switzerland and studied mathematics at the American University in Beirut before enrolling at the Architectural Association's school in London. Bernard Tschumi was one of her instructors there. After a spell working for the Office for Metropolitan Architecture in Rotterdam she branched out on her own. Her firm, Zaha Hadid Architects continues its work after her death in 2016.

Like Tschumi, Hadid first established her brand of architecture through conceptual drawings, oil paintings, and teaching. Her illustrations of imaginary free-form, sculptural buildings influenced by the art of Kazimir Malevich caught the attention of the cognoscenti. They displayed a sense of fragmentation and instability. Perhaps her work was also, consciously or subconsciously, influenced by Arabic calligraphy (Haider 2011; Sabastian et al. 2018). Hadid's reputation was boosted when her art works were displayed in an exhibition on Deconstruction in Architecture held at the Museum of Modern Art in New York in 1988. The display alongside the work of influential late twentieth century architects gave her enormous publicity. It not only enhanced the brand of her architecture but led to commissions for buildings. Her first commissions were never built and gave her the reputation of being a paper architect. She persevered. In common with other creative architects, she possessed an enormous inner drive.

Zaha Hadid designed furniture, jewellery, and handbags, but it is the buildings she authored that have attracted most attention. Getting all her designs built involved an array of consultants to ensure that they fulfilled their instrumental functions and to allow their free-flowing geometries to be constructed. Peter Rice (1935–1992) at Arup Associates assisted in the structural design of several of her early buildings. Studio Mouton, civil engineers, was the consultant for Havenhuis in Antwerp.

8.3.3.1 Hadid's Views on Architecture

Hadid was contemptuous of most contemporary architectural designs. No intellectual rigour, she observed, underpinned much that was being built. Most designs she recognized are based on adapting standard types. She considered their designs to be arbitrary.

> Arbitrariness has to do with a generation [of architects] that has been brought up on shopping for ideas. A catalogue exists from which they freely copy anything and apply it with little relevance to any situation. (Hadid 1982)

Their designs were pastiches. Her key goal was to 'reinvent modernity'.

> … in architecture … we must create new dynamics … We must understand the basic principles of liberation. (Hadid 1982)

> The twentieth-century triumph of technology and our accelerating and ever-changing lifestyles have created a totally new condition. These changes, despite the difficulties, have a certain exhilaration which is yet to be matched in architecture. It is … the absolute need for inventiveness, imagination and interpretation that makes our role in architecture more valid. We can no longer fulfil our obligations as architects if we carry on as cake decorators. Our role is far greater than that. We, the authors of architecture, have to take on the task of reinventing modernity. (Hadid 1983)

She sought this reinvention by using sharp angles or curving façades of free-flowing, smooth surfaces in the buildings she designed. She found ninety-degree angles boring and avoided them as much as possible (Jodido 2020; Hartman 2022). She is seen as a postmodern deconstructivist (Fig. 8.6).

8.3.3.2 The Clients

Zaha Hadid Architects' clients have included public agencies, institutions, dictators, and other private individuals. They have sought, in common, visually exciting buildings by an internationally renowned designer. Many of her works resulted from winning design competitions, open and closed, where the jurors were drawn from the avant-garde architectural world. The client for her first completed building, the Vitra Fire Station was, as noted, Rolf Fehlbaum, the president of Vitra, a family-owned furniture company. He was an enthusiastic supporter of avant-garde architecture and an early patron of the then unknown but now internationally renowned American architect Frank Gehry. Gehry designed the Vitra Design Museum. Fehlbaum was much involved in architectural circles and served as a jury member for the Pritzker Prize awards. He acted as the patron for Hadid's firehouse design.

The Havenhuis (Port House) in Antwerp was designed for a governmental agency and One North in Singapore, for another public agency, the city-state's Redevelopment Authority. The Authority has been responsible for turning Singapore from a colonial backwater into the world's most up-to-date urban environment. A university, relying on public funding, was the client for the Library and Learning Centre (2019) of the University of Economics in Vienna. The client for Heydar Cultural Centre (2012) in Baku, Azerbaijan was Ilham Heydar oglu Aliyev (b. 1961), the autocratic ruler of Azerbaijan and son of another dictator known for his human rights abuses. Working for dictators and in situations where the construction workers' lives are at risk is a contentious issue. Hadid's position, when working in Oman, was that such concerns are those of governments and not architects.

The audience for her designs clearly included the avant-garde among architectural critics and the cognoscenti. They seek the innovations that her works display. The buildings she designed were seldom delivered on time and on budget (Fairs 2015). They are also tourist destinations not only for architects but the public. They have had commercial effects on the cities in which they are located.

Photograph by Demetri Ternovoy/ /Wikimedia Commons

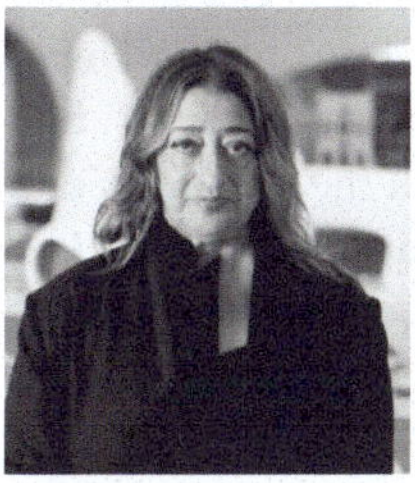

Zaha Hadid

Photograph by Yale/Wikimedia Wikimedia Commons

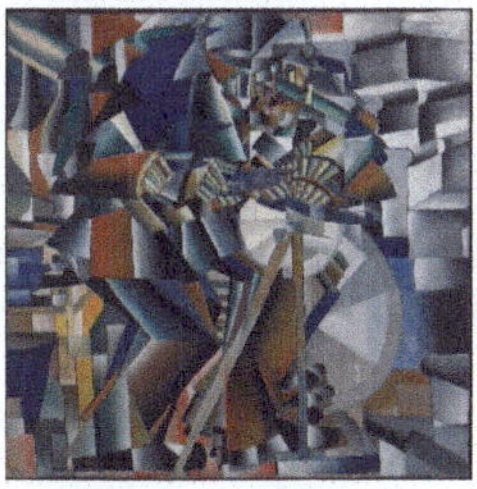

Malevich's *Knife Grinder*

Photograph by Bob Boom/ Wikimedia Commons

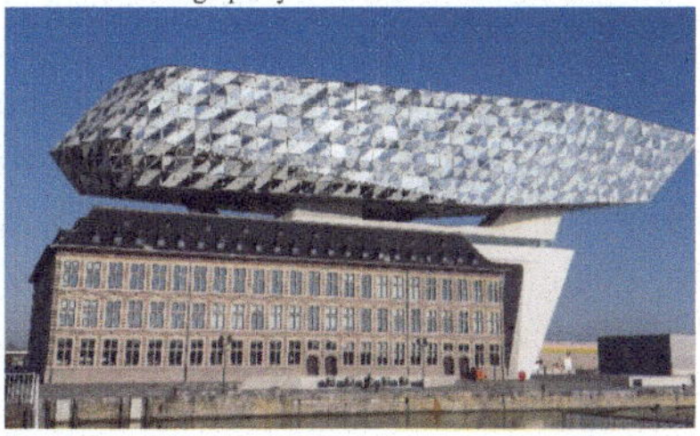

Havenhuis, Antwerp (2009)

Photograph by Brenac/Wikimedia Commons

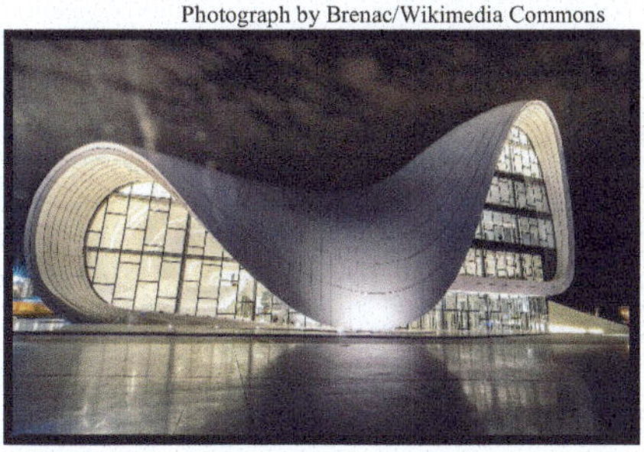

Photograph by Khalilov/Wikimedia Commons

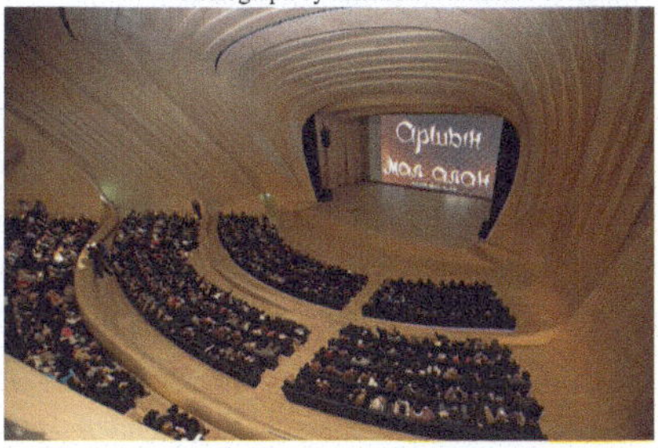

Heydar Aliyev Centre, Baku, Azerbaijan (2012)

Image by Flying Architecture: Courtesy of Zaha Hadid Architects

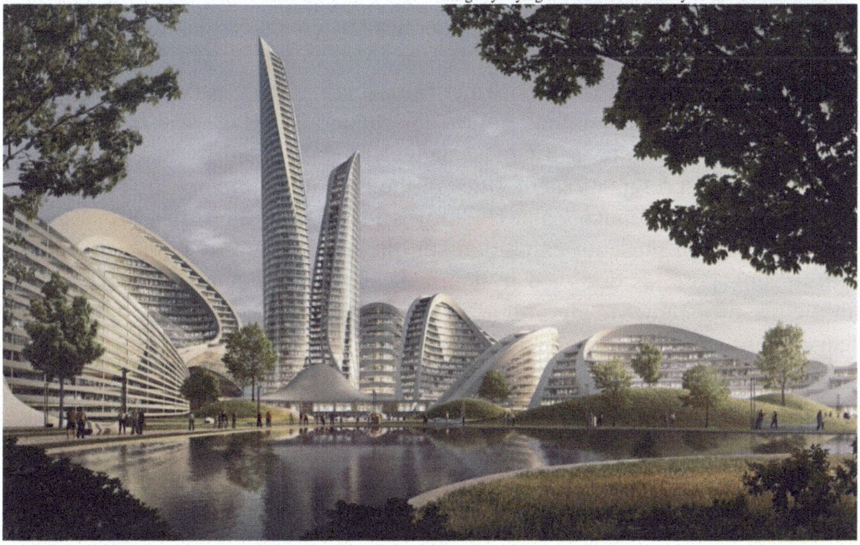

Proposal for the smart city of Rublyovo-Arkhangelskoye, near Moscow, Russia (2020)

Fig. 8.6 Work of Zaha Hadid and Partners

8.3.3.3 The Work

The diversity of ZHA's work can be seen in a sample of its designs. The Vitra Fire Station (1993), Hadid's first built work, differs in form from those of the Heydar Cultural Centre (2012) and the Havenhuis (2000). Many of the firm's designs, however, have in common a fluidity of curvilinear forms as seen in the Heydar Centre and in the proposed Rublyovo Arkhangelkoye (Moscow International Finance Centre) design. Rublyovo Arkhangelkoye is being taken forward posthumously by the firm under the leadership of Patrik Schumacher.

The Fire Station of the Vitra Company in Weil-am-Rhein, Germany consists of obliquely intersecting concrete planes (see Fig. 3.10). It showed that Hadid's Malkovich-like drawings of buildings can be turned into three-dimensional structures that serve instrumental functions. The Heydar Aliyer Centre is very different. The influence of Malkovich is gone. The building's conference hall, gallery, and museum are each represented by a concrete fold that give each's entrance an identity. Havenhuis is a structurally innovative building. The brief required reusing a derelict building and adding new facilities on to it to serve as the port authority's headquarters. A new ship-like structure of glass and steel floats above the original building. This extension has a skin of glazed panels that reflects the shifting light of day.

Most of the buildings that ZHA has designed sit as sculptures on their sites. The Lois and Richard Rosenthal Museum in Cincinnati (completed in 2003) is an exception. It is an urban building built to the property line to form part of a continuous street edge. Few of ZHA's other designs are place-making structures or relate to their context (Fairs 2015). If Howard Van Doren Shaw's buildings could be in any an upper middle income American suburb, ZHA's buildings could be in any city anywhere. The buildings follow the modernist siting ideal but the forms have a hyper-modern, deconstructivist nature. They are said to be futuristic like the comic book representations of cities of the next millennium.

It is not clear how much Zaha Hadid was involved in the firm's urban design work other than in museum exhibitions on the architecture of future cities. One North in Singapore, as designed by Patrik Schumacher was an early experiment with the use of parametric algorithms as a design tool. Turning ZHA's conceptual urban designs into built form without severely compromising their overall form has proven to be difficult. One North, as built, barely resembles the free-flowing lines shown in Fig. 7.2. It takes a little imagination to perceive anything of the original design in what has been erected. Its components have been built by several property developers, public and private. A variety of architectural firms have been involved. It is no total urban design by one firm (Lang 2021, 286–288). The architectural forms of Rublyovo-Arkhangelskoye are clearly those of Zaha Hadid's brand. Whether the design retains its brand flavour when implemented remains to be seen.

8.3.3.4 The Functions of Central Attention

Hadid's design ideas were applied universally to buildings as sculptures without reference to the characteristics of their context. She was no contextualist. Her clients bought a product to put on display. The works established a new sense of place—a location in the international art world rather than one in its environmental context. Some of the patterns used in the Galaxy SOHO (2012), a shopping centre in Beijing were said by Hadid to reflect traditional Chinese forms but the Beijing Cultural Heritage Preservation Society disagrees (Staff Writer 2013). It is certainly difficult to see the connection.

Zaha Hadid's primary concern, and the one that established her brand of architecture, was with the self-conscious formal aesthetics of buildings. Her focus was on bringing her imaginative fragmented or flowing geometric forms into fruition in concrete, steel, and glass. In several designs, the circulation patterns were of central concern not only from a narrow functional view but as a sequence of visual experiences (Sabastian et al. 2018). The innovativeness of the geometric patterns that the firm uses have, on occasion, resulted in safety concerns such as at the Guangzhou Opera House (2010) where cladding elements came loose soon after construction (Waite 2011). The problems have been variously attributed to poor workmanship or to poor design.

All Hadid's works provide satisficing solutions to the basic functions of the programs that ZHA receives. Sometimes, it has been her collaborators who have dealt with meeting the interior spatial requirements of the building forms she designed. Events Communications, a museum design firm, were consultants for the interior of the Glasgow Riverside Museum (2011).

On occasions how well a building serves its instrumental function gets less attention than critics expect. The Vitra Fire Station was not fully suited to its primary function and disliked by the firemen who worked there (Astbury 2022b). The reasons appear to be that the layout did not help response times to emergencies and that there was a lot of unusable space. The building is now an events hall and exhibition space reflecting the change of the district in Weil-am-Rhein from an industrial production zone to a cultural centre. The building was easily adapted to serve an instrumental function more in tune with its character.

Some of the observations on ZHA's work is harsh but they raise questions about the nature of Rationalist thinking today. Critic Stephen Bailey (b. 1951) noted: "Her latest buildings always win approval from the supine architecture and design media, so work very well as salvation-via-design for repressive regimes" (cited in Fairs 2015). The cognoscenti, particularly those in the fine arts community, have warmed to Hadid's designs and their Postmodern Deconstructivist nature. As a woman she is seen as 'rocking the world' (Hartman 2022). During her life, she received numerous awards. She was made a Dame Commander of the British Empire by Queen Elizabeth II in 2012.

8.4 Three Empiricists

Empiricists may rely on traditions but they consider themselves to be producing responsible designs for real, not imagined, people for today and tomorrow. Frank Lloyd Wright drew on many antecedent ideas to create an innovative new architecture. Robert A. M. Stern (b. 1939), whose work is more regressive utopian in character than Wright's argues that he is creating a modern architecture (Stern 1975). The three empiricist whose prescriptive positions are presented here are imitators not copiers.

The work of The Angles Architects in India and its principal, Darshan Bubbar (1937–2021) was Modernist in appearance but it drew on design principles that are described in several ancient Indian treatises—*Shastras*—on architecture and urban design. Bubbar's search for a canonical basis for design is an exemplar of the genre. The work of Abdel-Wahed El-Wakil (b. 1943) is traditional in form and building techniques although its simplicity of form echoes that of Le Corbusier's works and Romanesque architecture. The work of the third firm, DPZ CoDesign (DPZ) led by Andrés Duany (b. 1949) and Elizabeth Plater-Zyberk (b. 1950), is known as Neotraditional although its urban design position is labelled New Urbanist. The firm's directors, unlike many architects, savour urban life.

8.4.1 The Angles Architects and Darshan Kumar Bubbar

Darshan Bubbar was born to a middle-class family in Quetta, Baluchistan in what is now Pakistan. After the Partition of India his family moved to Delhi where he received his schooling that had a Dayanand Anglo-Vedic orientation. It blended Western scientific and Vedic spiritual and cultural traditions. That background shines through in his architectural designs. From an early age he showed a talent for drawing that drew him to the arts. He started his working life in Delhi's Methodist Engineering Office but he wanted to be an architect. He moved to Mumbai to study at the Academy of Architecture after his application for admission to the Delhi Polytechnic was rejected. The Polytechnic did not warm to his background in engineering at a time when architects in India were trying to forge an identity separate from engineers. After graduation in 1965 Bubbar opened his own professional practice, The Angles Architects, in Bandra, Mumbai.

Bubbar was neither a historian nor an academic but a practising architect and the sole proprietor of an architectural office although he cooperated with artists such as Satish Gujral (1925–2020) and with a variety of consultants in implementing his designs. He taught at the Academy of Architecture in Mumbai and conducted design workshops at universities in India, Australia, and Singapore. Throughout his career he was concerned with pragmatically 'doing' in the present and sought the basis for good designs in the rules specified in the *shastras*. He is not alone. Other architects see the shastras as *the* architectural design paradigm to apply in India (Sachdev 2011). Bubbar's magnum opus, *The Spirit of India Architecture*

(2005) laying out his design philosophy, led to his invitation to discuss architectural matters with the eleventh president of India, Avul Pakir Jainulabdeen Abul Kalam (1931–2015). He remained a sole practitioner throughout his career.

8.4.1.1 Bubbar's Views on Architecture

Bubbar's design approach harkens back to the teachings of the nineteenth century Hindu mystic Ramakrishna Paramahamsa (1836–1886), his wife and spiritual consort Sarada Devi (1853–1926), and Swami Vivekananda (1863–1902), and to the prescriptions presented in *Manasara*, a text on architectural proportional systems. When applied, the *Manasara*'s edicts are said to lead to harmonious buildings, something Palladio sought but neither Tschumi nor Hadid did. Bubbar did not advocate copying the edicts of the *Vastu Shastra*, the book on which many Indian architectural gurus rely, but imitating them. The book predicts dire outcomes (such as, in modern times, electrocution) if their edicts are ignored by those who occupy spaces not designed in accordance with them. Rather, Bubbar sought an understanding of the principles on which the texts are based. He reinterpreted them to deal with today's world.

Bubbar observed:

> The ancients elevated architecture to such an art where life blended seamlessly with the sublime – the spiritual. … However, todays' architecture is a far cry from the glorious expression of culture, art, functionality, and appropriateness that it once used to be. …

He asked:

> Why did our education system put its overwhelming trust in western techniques and style? (Bubbar 2005)

Many Indian building designs based on the *Manasara* and *Shilpa Shastras* are revivalist in form and decoration. They copy the past. Such a design prescription is advocated by the Government College of Sculpture and Architecture in Mamallapuram, Tamil Nadu. Bubbar's prescription was different. In *The Spirit of India Architecture*, he presented the design principles that architects should follow in creating harmonious, peaceful environments for daily life. Unlike the Rationalists, he did not seek to jar people into looking at the world anew. It is the *spirit*, not the style, he argued, that is important to consider when designing.

> Space is enclosed by length, width, and height. All dimensions must marry in harmony. The site and the edifice must be compatible … To work out the plan the mandala system is applied. The mandala is a celestial graph. … It includes all the cosmic elements without which life cannot exist. (Bubbar 2005)

Mandalas come in a variety of geometric forms. The square *Vastu-Purusha Mandala* is fundamental to the science of building. It is the unequivocal geometric form representing the absolute. Its relationship to the cosmos is clear in its orientation to the cardinal points of the compass. The spirit of the site is said to be pressed prone to the ground within the bounds of the square as can be seen in Fig. 8.7.

The square mandala can be divided into sub-squares. The subdivision that is used as the basis for a design must be appropriate for a proposed building's location and the program given. Nine-by nine squares is typical; seven-by-seven is less so. Each sub-square is related to a god with the central square being the abode of the god Brahma, the creator. The location of the limbs specifies the use to apply and those to avoid in each part of a building. Bubbar recognized that life exists in a changing world and what is appropriate will change as time passes. What remains constant is the proportion of parts to the whole of a building. The proportional system he advocated is based on the *hasta*, the distance between the elbow and the fingers of the principal client for a building.

High Renaissance architects sought mathematical proportional systems based on the Golden Mean Ratio to organize and give a feeling of stability to their designs. Le Corbusier developed the Modulor to give him a tool for organizing the overall form and the façades of the buildings he designed. Bubbar's use of proportions differs. He sought a meticulously worked out logic that his clients could understand and appreciate. He also sought to be modern. This was his goal in urban design as well.

Bubbar believed that the City Beautiful, Garden City, and Radiant City urban design paradigms have severe limitations (Bubbar 2005, 267). The mandala to use as the basis for the layout of the road system and the distribution of activities in a new settlement depend on the climatic zone in which it is to be located. For a semi-rural settlement in a warm temperate zone, for instance, he considered the 27×27 *Vipra-Kanta Mandala* to be the appropriate basis of the plan. The logic of the choice is not made clear. The centre square, Brahma's domain, should be the administrative and business heart of the settlement and the other uses should be distributed in accordance with each sub-square's designation.

8.4.1.2 The Clients

Bubbar's clients included families, institutions, and commercial corporations such as the First National City Bank and Berger Paints. They were part of his extended list of personal connections dating back to his time with the Methodist Engineering Department in Delhi. His clients were people who knew him or heard of his work through word of mouth. None were major art or corporate organizations seeking a new architectural expression for their buildings to establish or enhance their identities as avant-garde members of society. They sought well-functioning, peaceful, modern environments. If anything, they were pragmatic people with little understanding of architecture.

Generally conservative in outlook, his clients, nevertheless, sought an aesthetic comparable to the contemporary buildings being produced by architectural practices in India whose work followed international norms. They were drawn to Bubbar because of the detailed quality of his work, his knowledge of the shastras, and his flexibility in applying them within the acceptable mores of Hinduism.

Courtesy of Chandrika Bubbar

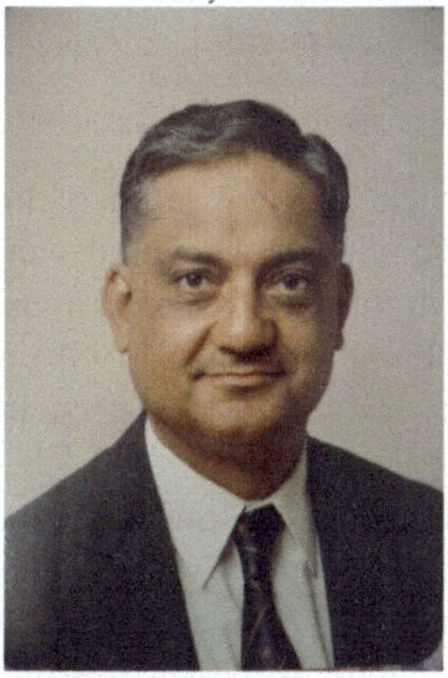

Darshan Bubbar

Source: Lang, Desai and Desai (2022, 30): Courtesy of CEPTU Press

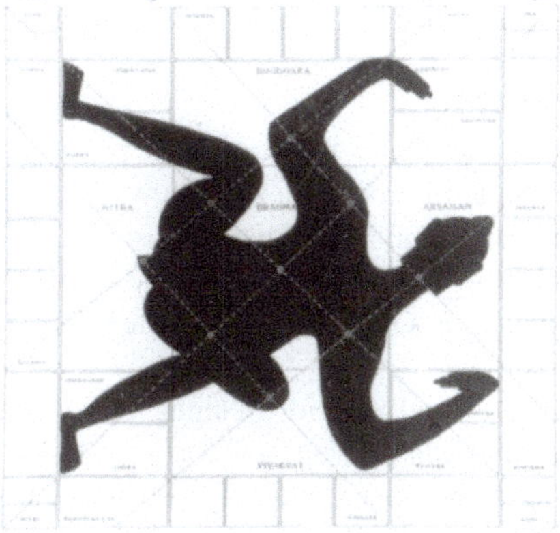

The Vastu Purusha Mandala

Source: Bubbar (2005, 164): Courtesy of Chandrika Bubbar

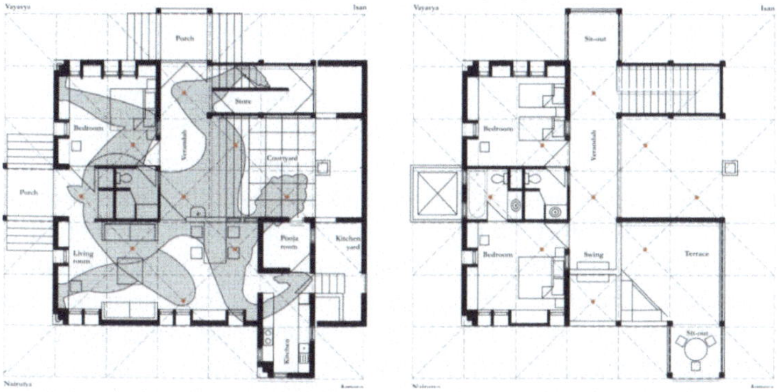

Shashital House, Pune: The plans according to the Sthandila Mandala (7x7)

Source: Bubbar (2005, 118): Courtesy of Chandrika Bubbar

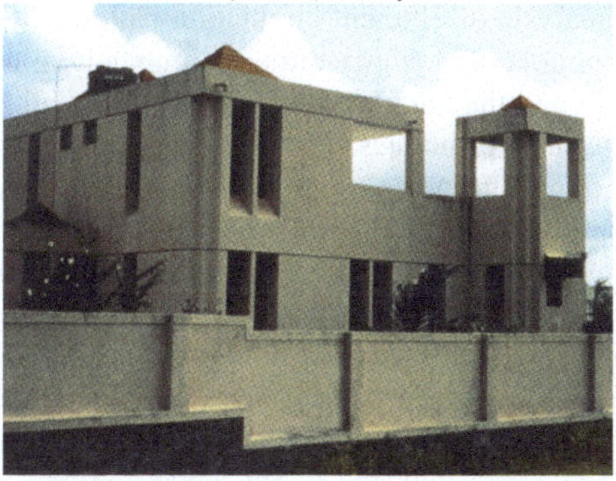

Shashital House, Pune

Source: Bubbar (2005, 180) Courtesy of Chandrika Bubbar

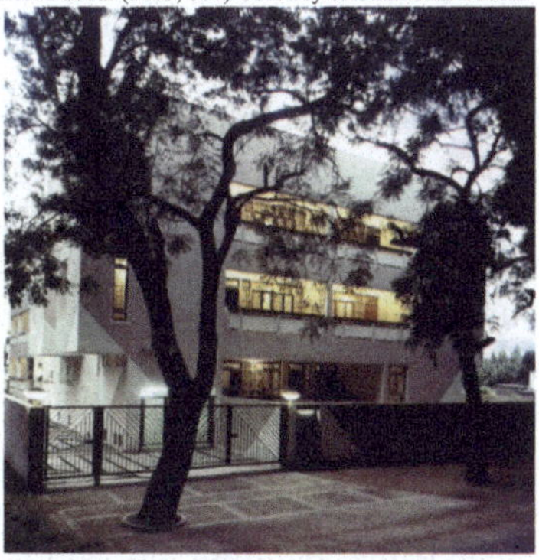

Berger Paints building, Vallabha Vidya Nagar

Fig. 8.7 Work of the Angles Architects

Those clients who were sceptical of his logic were happy to see the mandala system applied to the design of their buildings 'just in case' it is valid.

8.4.1.3 The Work

The application of Bubbar's philosophical position is clear in the design of his buildings. They range in type from bungalows to housing complexes, hospitals, and commercial buildings. The Methodist Centre (1967), in Mumbai, a weekend house for the Shashital family in Pune (1990), and the Berger Paints Building (1988) in Vallabh Vidya Nagar in Gujarat are examples. His urban design proposals remain on paper.

Bubbar's building and interior designs for corporate clients follow his interpretation of the mandala appropriate for a building's location and for the client. In the case of the Methodist Centre, for instance, it was an 8 × 8 mandala for the structural grid and the general *hasta* measurement—18 in. in the Imperial measurement system—since there was no one client for the project. Eight hastas formed the basic grid dimension.

The Shashitals wanted to have a building that housed and symbolized their well-knit family's way of life and social stratus. As Mr Shashital was the head of the family, his forearm was the basis of the proportional system applied to the design. The height of the Shashital House is 25 such units. For the house's plan, Bubbar considered the *Sthandila Mandala* (7 × 7) with seven hastas forming the size of the grid to be the appropriate one to use in establishing the distribution of rooms. That mandala would protect the family from the torments of everyday life.

In the design of the Berger Paints Building, the client wanted to have a building that served its instrumental function well and have a modern corporate image that promoted the company. In the design of its building, Bubbar considered the *Asana Mandala* (10 × 10) the best to use for a corporation. As there was no obvious individual on whom to base the hasta, Bubbar applied the standard dimension.

The distribution of rooms in both the Shashital House and the Berger building was based on the mandala not on an analysis of efficient circulation patterns. On the ground floor of the house, the bedroom is in *Vayavya* corner, living and dining spaces in *Nairutya* through *Yama* quarter and the kitchen is in *Agreja.* The location of rooms on the second floor follows the same principles. The east facing courtyard receives the purifying effects of the morning sun. At its centre is a *tulsi* plant sacred to Hindus. The pyramid roofs of the building concentrate energy so are conducive to the mental, calmness, and alertness of the building's inhabitants. They also give the house a distinct appearance.

According to Bubbar, the Shashital family told him that however stressed and exhausted they are when they arrive at the house, entering it gives them a feeling of calmness. They feel satisfied with life and are happy. The people using the Berger Paints building are said to feel at peace with the world while working in the building (Bubbar 2005). To critics, such observations are deterministic and fail to recognize a host of possible intervening variables that affect people's moods. Maybe simply getting away from the daily grind relaxed the Shashitals.

8.4.1.4 The Functions of Central Attention

The goal of meeting basic functional needs, both in traditional Modernist terms and in terms of Maslow's hierarchy of human motivations are met well-enough to solicit few complaints. The focus of attention in Bubbar's designs has been elsewhere. As in many societies that are seeking an identity other than to be part of the neo-liberal, avant-garde, global, social world, some local precedent is often sought as a basis for a design. In India, particularly in the south of the country, several architects rely on one or other of the *shastras* as the basis for their designs. To some observers, however, the design principles prescribed in such texts are merely superstitions ungrounded in any evidence.

Two main concerns formed the basis of Bubbar's designs. The first was to meet security concerns and the second was to provide the owners and users of buildings with a sense of belonging to a society and to a tradition. The means used to achieve these ends do imply an acceptance of the canons of the *shastras* as the empirical basis for making decisions about the plans of buildings and the proportions of rooms and other internal spaces. At the same time, the exterior appearance of his designs, although based on a proportional system, shows that their inhabitants are modern within the generally accepted values of the Indian middle-class taste culture. The buildings are set back from the street in walled compounds giving their inhabitants symbolic control over the exterior space.

Bubbar strove to give his clients' buildings an individual identity so they stand out from their surroundings, not in the way they are sited, but by their appearance. While the compositions of the plans and elevations may be based on the shastras, the façades reinforce corporate or familial identities. They provide little for the cognoscenti to savour but the logic behind the design is there for them to analyse and admire, or not.

Neither academic institutions nor the cognoscenti much heed the work of firms such as The Angles Architects. It is too simple visually and geometrically unexciting. The increasing consideration of the shastras as a mechanism for establishing an Indian identity, as a sense of Hindu nationalism grows may lead to an enhanced interest in *The Spirit of India Architecture.* Similar functions are served in China by *Feng Shui*, in Japan by *Fusui*, in Vietnam by *Phong Thuy*, and by *Pung-Jiri* in Korea.

8.4.2 Abdel-Wahed El-Wakil, Architect

Abdel-Wahed El-Wakil is an Egyptian architect renowned for his knowledge of Islamic architecture. He was born into an upper middle-class family in Cairo. He received his schooling at British schools—Victoria College and the English School—in the city. He obtained a distinction in applied mathematics. Subsequently, he was an outstanding architectural student at Ain Shams University graduating in 1965. It was there that he was exposed to John Ruskin's *arborescent* concept—the idea that the patterns of a building should be visible in finer and finer detail as one approaches it (Ruskin 1849, 1853). After graduation he taught at the university. He has long been based in England but he has offices in Cairo and Jeddah.

The character of El-Wakil's early work was International Modernist reflecting his education at Ain Shams. It possessed a simplification of form, layouts designed to efficiently afford a building's instrumental function, and ornamentation based only on structural necessities. His architectural attitudes changed after meeting Hassan Fathy and working with him for five years. The influence of his European Modernist education at Ain Shams on El-Wakil's work remains subtle. Fathy's influence is obvious (Fig. 8.8).

Fathy's work focused on meeting the ways of life and values of those for whom he was designing as inexpensively as possible. He did so by using the traditional methods and tools of indigenous house builders. In his manifesto *Architecture for the Poor*, Fathy echoes Frank Lloyd Wright's observation on vernacular architectures. Fathy wrote:

> When the full power of human imagination is backed by the weight of a living tradition, the resulting work of art is far greater than any an artist can achieve when he has no tradition to work in or when he wilfully abandons his tradition. … (Fathy 1969, 25)

Al-Wakil's designs were shaped by Fathy's architectural attitudes. He added:

> And I do believe it is the lack of sacred attributes that's the cause of so many problems today. (cited in El Sayed Foundation 2009)

> The cathedral, the temples in Egypt – they all have messages to give. That is what I attempt in my work – something universal. (cited in El Sayed Foundation 2009)

> To abandon tradition, to disregard the achievements and models of the past and to be caught up in the trauma of change, means being incapable of handling the new. I seek the essence within the natural order to guide my work. God Almighty has requested man to seek inspiration in the natural world. (School of Architecture 2023)

El-Wakil seeks to balance continuity and change in his work. He believes in fostering a sense of community and minimizing the catalytic impact of his work on the biogenic environment. In this way he is allied with the canonical classicists in the Western World, such as John Blatteau, Alan Greenberg, and John Simpson rather than his Islamic contemporaries (al-Asad 1992).

El-Wakil's work has received international recognition. He has twice received Aga Khan awards for his designs. His research on Islamic architecture brought him a King Faud Award and in 1996 he was made an honorary member of the American Institute of Architects. In 2009 he received the Richard H. Driehaus Prize for classical architecture from the University of Notre Dame. His designs have also been criticized for their lack of modernization and being too eclectic in the precedents on which they are based (Sakr 1987).

Photographer unknown: Collection of the author

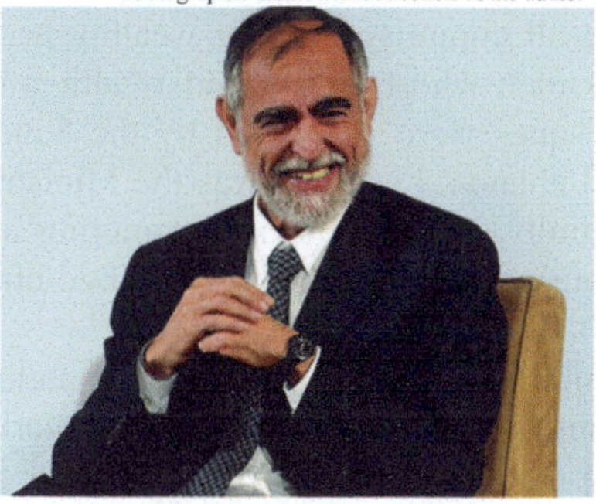

Photograph by Najah Alsalami/Wikimedia Commons

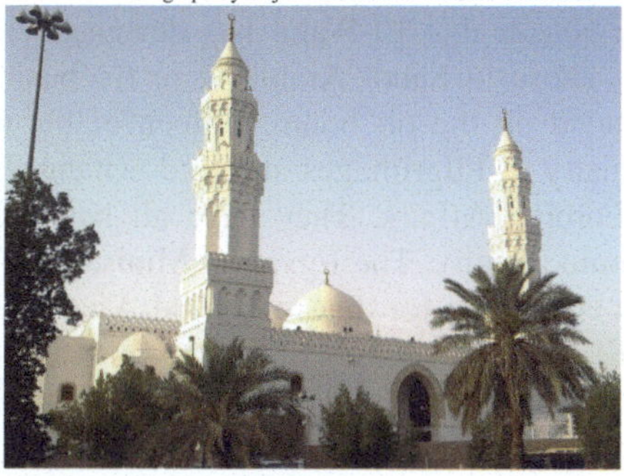

Abdel-Wahed El-Wakil

Masjid al-Qiblatayn, Medina

Photograph by Leonardo Martin/Shutterstock.com

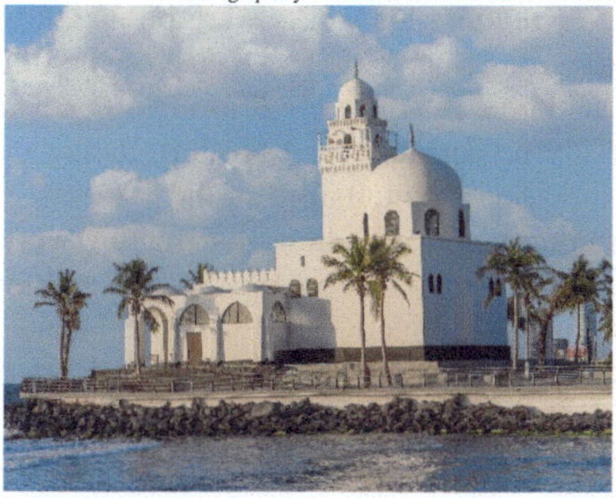

Photographer unknown: Collection of the author

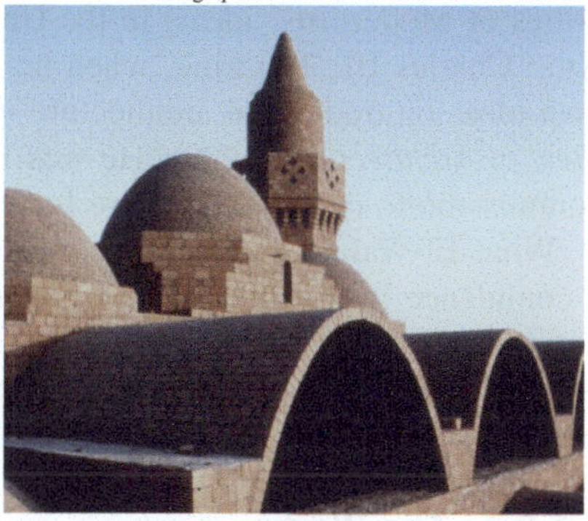

Al-Khobar Corniche Mosque, Jeddah (left) and an example of the brickwork of mosques (right)

Photograph by H5TAYF/Alamy

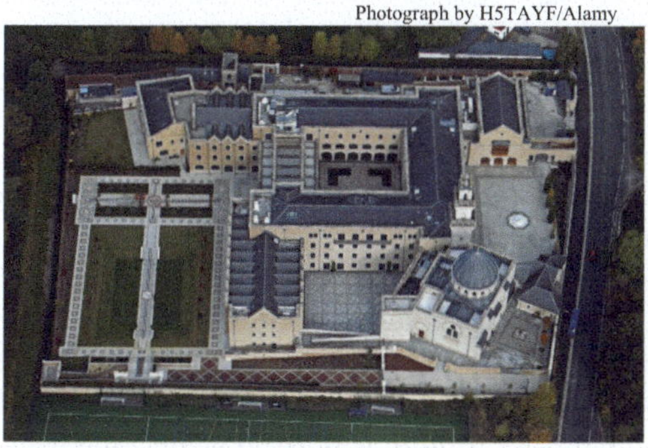

Photograph by Motacilla/Wikimedia Commons

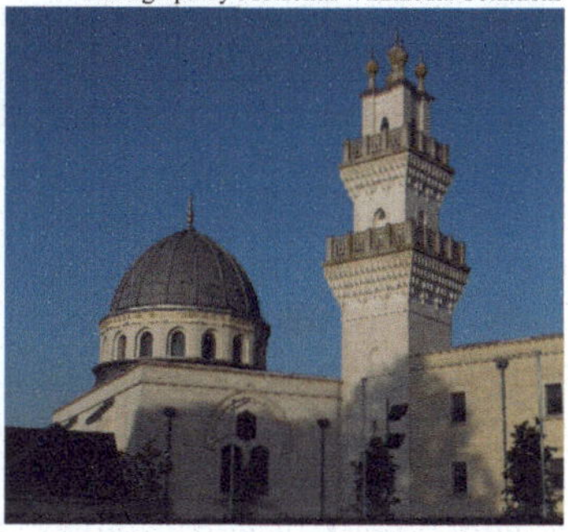

Oxford Centre for Islamic Studies

Fig. 8.8 Work of Abdel-Wahed El-Wakil, Architects

8.4.2.1 The Clients

The houses that El-Wakil has designed were all commissioned by wealthy people. Those in Saudi Arabia were for businessmen whose new-found wealth was founded on the oil boom. Fifteen of his mosques were sponsored by the Saudi Ministry of Pilgrimages and Endowment. In Lebanon, his client was the Ministry of Foreign Affairs. They have all been centrally concerned with promoting an Islamic identity. The Egyptian Ministry of Tourism and UNESCO have hired him as a consultant for that reason (al-Asad 1992).

El-Wakil is seen as a non-threatening competitor by the mainstream of architects so he has received strong support from designers who admire his work. They have helped him to obtain architectural commissions. For instance, Andrés Duany (b. 1949) invited him to Florida where he completed a design for property developer and venture capitalist Thomas Kramer. Other supporters have been critics of Modernists' ideas. In the United Kingdom his works were promoted by King Charles III. The king, when he was still Prince Charles, had presented his own ideas on traditional architecture in his book, *A Vision for Britain: A Personal View of Architecture* (1989). He was largely responsible for El-Wakil getting the commission to design the Centre for Islamic Studies in Oxford in the 1990s.

What El-Wakil's clients have had in common is an admiration for his work, a confidence in his approach to design, and an appreciation for the construction methods he employs. His clients seek something modern but distinct from International Modernism and certainly from International Hypermodernism. His work has a popular appeal.

8.4.2.2 The Work

El-Wakil's professed position and work closely coincide. His designs for houses, public buildings, and mosques are imbued with traditional forms amalgamated with a Modernist spirit. They are not copies but imitations of past architectural ideas. His design of the Halawa house (1975) in Agamy, near Alexandria, Egypt initiated his practice and established his brand of architecture in the eyes of potential clients and the profession. The house's design was solely his, although some details were left for local workmen to complete in their own way.

The house, while luxurious, possesses an indigenous character. It has the traditional Egyptian elements of vaults, domes, and arches (Mirza 2016). A courtyard and loggia, fountains, and wind-catchers for drawing air down into the interior, alcoves, and masonry benches are other traditional features. Sunlight filters through the lattice work of *mashrabiyas*. The built-in furniture is pressed against the walls. Members of the household and visitors can squat on the seating without having their legs dangling. Great care was shown in choosing paving materials. That on the inside is marble; on the outside it is Muqattam stone on which people can walk barefooted in summer. Much of the construction was completed by unskilled Bedouin workers but professional plasterers and carpenters did the work requiring high manual skills. The house's success and the ensuing publicity led to commissions for far grander houses in Saudi Arabia. El-Wakil reputation is, however, based largely on the mosques he has designed (al-Asad 1992; Mirza 2016).

He has designed more than fifty mosques. Fifteen in a decade were for the Ministry of Pilgrimage and Endowment in Saudi Arabia. That commission represents the effort of a nation to reinforce its own identity in the face of globalizing cultural trends. The mosque designs demonstrate El-Wakil's effort to marry traditional Arab architectural forms and construction techniques with modern plans. The mosques were constructed of load bearing brick walls, vaults, and domes whose exteriors are covered with white plaster. Their interior surfaces have been left unplastered but sealed with brown paint so that their brickwork and mode of construction are visible. The Corniche Mosque (1984) in Jeddah, is typical of his work. Larger mosques, such as al-Qiblatayn in Modena, came later. The construction approach employed in them differed from that used in most contemporary mosques whose domes are made, as in the Pantheon, of concrete (El Sayed Foundation 2009).

The enthusiastic reception of El-Wakil's work led to commissions in countries as diverse as Bahrain and South Africa where he designed Johannesburg's Kerk Street Mosque (1990). In the United Kingdom, he designed the Oxford Centre for Islamic Studies (completed 2007). The Centre is composed of buildings that form quadrangles typical of the colleges of Oxford University (although the Centre is not part of the university) while having Islamic forms such as a dome and minaret that gives the centre a clear identity. The Centre was built without the use of concrete or steel. In Qatar he is designing an urban precinct combining low-energy design practices with climate specific design.

8.4.2.3 The Functions of Central Attention

El-Wakil's central concern is with designing culturally appropriate buildings based on observations of the ways of life of his clients. He focuses on buildings as communicators of meaning by integrating local forms with modern ideas. His designs draw inspiration from the architecture of the whole Islamic world: the traditions of the Ottomans of Turkey, the Mamluks from Egypt, the Rasulids from Yemen, and the Seljuqs from Iran (Sakr 1987; al-Asad 1992). A casual observer may not recognize the specific referents but will certainly understand their Islamic roots. The way structural and constructional techniques function to conserve resources is an essential component of his work. These approaches, and the quality of workmanship, not only serve utilitarian ends, but are also highly symbolic gestures.

His focus on the symbolic function of built forms is intertwined with his environmental concerns and, particularly, for designing with the climate of a locale in mind. His design of courtyard houses and how they deal with the sun, provide shade, enhance air movement, and incorporate vegetation, show his concern for the comfort level of a building's occupants while minimizing energy consumption costs. The interiors are also culturally sensitive in the way their fixed feature furniture layouts are organized. Given the family privacy values of the Islamic world, the houses are closed off from the street by a perimeter wall. Courtyards provide outdoor space.

There is much in El-Wakil's work for the cognoscenti to ponder. They are able to understand the individual referents on which he draws. Some of them, the regressive utopians, have lauded and awarded his work; others are less sympathetic and dismiss his work as romantic (Sakr 1987).

8.4.3 DPZ CoDesign and Andres Duany and Elizabeth Plater-Zyberk

Andrés Duany and Elizabeth Plater-Zyberk are a husband-and-wife architectural team. Duany was born in New York City but lived with his family in Cuba until he was eleven. He attended the prestigious Choate school in New England and Aiglon College in Switzerland. Plater-Zyberk was born in Bryn Mawr, Pennsylvania where her father, Count Jozafat Plater-Zyberk (1906–1994) was an architect and her mother, Maria Meysztowicz, an academic. She attended the selective Sacred Heart Academy in Bryn Mawr. Both Duany and Plater-Zyberk studied architecture at Princeton and Yale. Duany also spent a year at the École des Beaux-Arts in Paris. The two, as a couple and individually, are much awarded architects.

In 1977, the pair along with Bernardo Fort-Brescia (b. 1950) and Laurinda Spear (b. 1950), founded the architectural group Arquitectonica in Miami. Arquitectonica is known for its lively and idiosyncratic Latin-American, Postmodernist work. Choosing a different approach to design, Duany and Plater-Zyberk broke away to form their own firm, Duany Plater Zyberk and Company, that has evolved into DPZ CoDesign (DPZ). A highly prolific organization, its output is that of several partners—Galina Tachieva, Marina Khoury, Senen Antonio, and Mathew Lambert. Duany and Plater-Zyberk were among the founders of the Congress for the New Urbanism, an advocacy group for Neotraditional urban design and architecture. Duany has been the primary spokesperson for the firm's positions while Plater-Zyberk shares her professional commitments with an academic career at the University of Miami.

DPZ has received two American Institute of Architecture awards, and a Vincent Scully Prize for exemplary practices. Its urban design work has been much lauded. Its first major project, Seaside, was selected by *Time* as one of the ten outstanding projects of the 1980s.

8.4.3.1 DPZ CoDesign's View of Architecture

DPZ CoDesign's views are best expressed in the manifesto *Suburban Nation: The Rise of Sprawl and the Decline of the American Dream* (2010) co-authored by the Duany, Plater-Zyberk and Jeff Speck. The book's message is simple (Fig. 8.9).

> In terms of the exterior of homes and the layouts of communities, Americans build the worst public forms in the world with the least walkable communities. … Developers need to learn to design open space with a square or plaza and walkable, charming streets. (Duany 2021, originally 1991)

Photographer unknown: Collection of the author

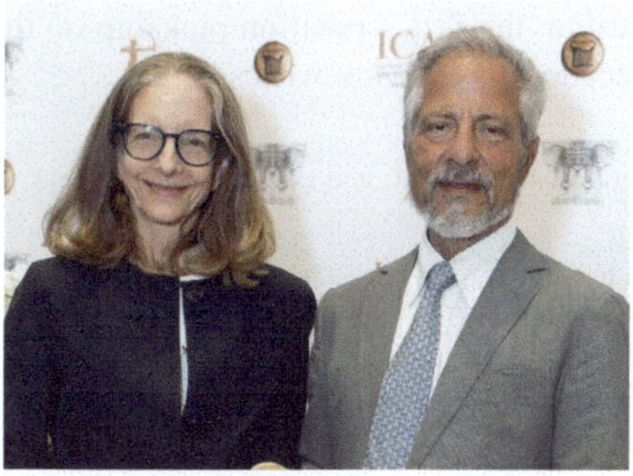

Andrés Duany and Elizabeth Plater-Zyberk in 2013

Photograph by Kristie Blokhim/Shutterstock.com

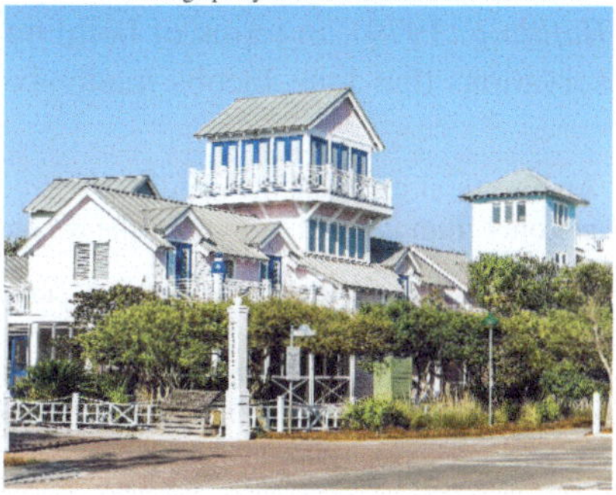

Houses, Seaside, Florida

Photograph by Zstock/Shutterstock.com

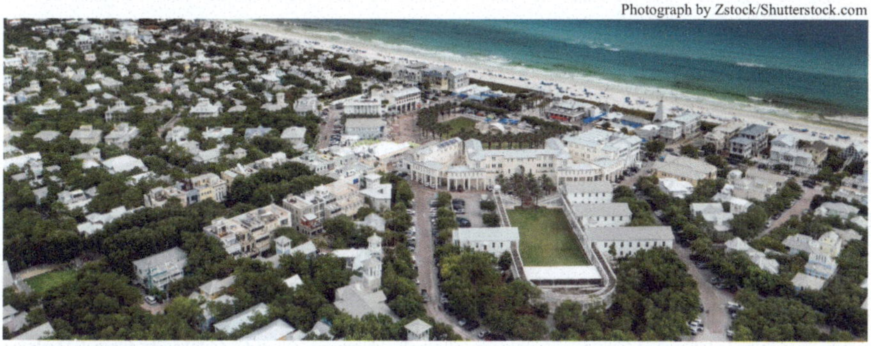

Aerial view, Seaside Florida

Photograph by the author

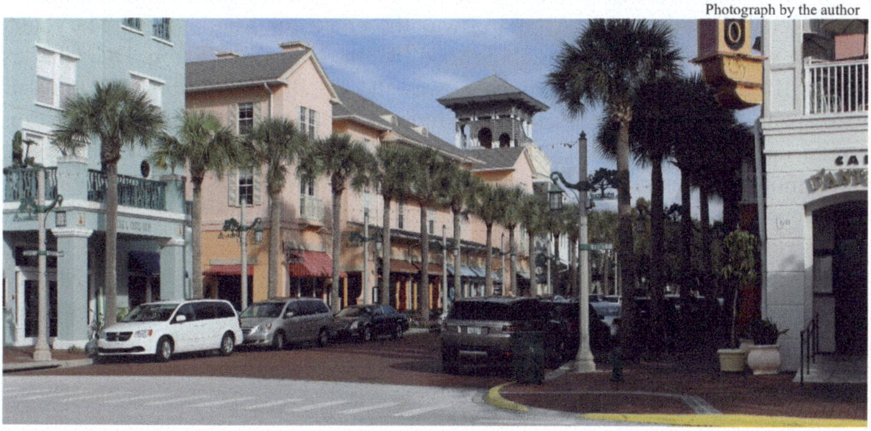

Market Street, Celebration, Florida

Fig. 8.9 Work of DPZ CoDesign

This view echoes that expressed by Christopher Alexander in *The Timeless Way of Building* (1979). In terms of building design, the firm's position picks up on the observations that Jane Jacobs made six decades ago. Duany writes:

> Architectural expression should derive from its urban, cultural, and climatic context, and not only from the will to form of the designer.
>
> That there is a difference between the creativity necessary in the design process and the urge to invent, which may result in transient fame but also in permanent failure. The pursuit of originality condemns cities to incoherence and an architect's work to unwarranted obsolescence. (Duany 2021, originally 1991)

Plater-Zyberk adds that DPZ CoDesign's work:

> … is about dignifying something that was for a long time in American culture. Americans prefer traditional forms for their houses, for their residences … [Architecture] is about the bigger picture rather than just individualistic or branded effort. (cited in Hartman 2022)

DPZ CoDesign's approach to design has earned it many commissions.

8.4.3.2 The Clients

The clients of DPZ CoDesign vary but property developers predominate particularly for urban design work. The Walt Disney Corporation was the client for Celebration in Florida. The Corporation sought to build on its earlier vision for an Experimental Prototype Community for the Future (E.P.C.O.T.). It hired Robert A. M. Stern and Jaquelin T. Robertson (1933–2020) along with DPZ to produce one. A local developer, Joe Alfandre, was the client for Kentlands in Maryland. Alfandre wanted to create a neighbourhood like those in Washington, DC. Charles Osborn III, a former Baptist minister, was the developer of Norton Commons in Prospect, Kentucky. Other clients include civic leaders, institutions, and commercial organizations. For bespoke houses they are individuals. What they have in common is a desire for designs that reflect their locale as well as their own aspirations.

Robert Davis, the property developer of Seaside, the design that gave DPZ a clear identity, is typical. He inherited an 80 acre (32 ha) site on the Gulf of Mexico coast of Florida from his grandfather in 1970. He sought to create a town based on the happy memories of his boyhood vacations and the places that he enjoyed. He noted:

> The idea of Seaside started with the notion of reviving the building traditions of Northwest Florida which had produced wood-framed cottages so well adapted to the climate that they enhanced the sensual pleasure of life by the sea. (Davis 1989, 92)

Davis hired DPZ to do a design for the town and the division of properties into development sites. He then sold the individual plots to purchasers whose architects had to follow the design guidelines established by Duany and Plater-Zyberk.

8.4.3.3 The Work

DPZ CoDesign's projects include new towns, urban districts, and individual buildings. The firm has been much involved in writing the design guidelines for other architects to follow when designing the individual components of a master design plan (Lombard 2005). The objective of the guidelines is to get both a diversity and a sense of unity in a development. At Seaside, Florida, the firm did all three: the urban design, the guidelines, and some individual building design (Katz 1994).

Seaside, a resort community on the Gulf of Mexico remains a good example of DPZ CoDesign's position on urban design and on architecture. During the late twentieth century its conceptual underpinnings formed a new urban design and architectural paradigm, both functionally and procedurally. Today Seaside is a town with a community hall, school, chapel, post office, retail stores, commercial tenants, and a wide variety of housing types. It has a small, but increasing, number of permanent residents. Seaside is important to urban design prescriptive theory because it emphasizes the public over the private realm in its layout.

The master plan was created in 1982. It consists of a hierarchy of street sizes depending on their perceived role in handling traffic flows, particularly pedestrian movements. They are laid out in a concentric network that focuses on the town's centre. Everything that residents want daily is within a five-minute walk of their homes—a hallmark of New Urbanism's prescriptions. At Seaside's heart is a 'semi-octagonal' square with a post office, shops, library, and other communal facilities. This area is linked by a street—the town's main commercial area—to a square. The town hall is located there to the northwest of the site. The public buildings and spaces of Seaside include pavilions at the Gulf of Mexico end of each north–south street (Katz 1994).

While the plan, building codes, and design guidelines for the buildings are clearly the work of Duany and Plater-Zyberk, the ideas were framed during a two-week long charette with designers, local officials, and consultants. Other architects acted as potential property owners and/or designers of individual buildings. The code includes regulations aimed at obtaining a unified overall design while giving architects, many of international repute, the freedom to express their own brands of work. A sense of unity is, nevertheless, clearly achieved because the design guidelines were clearly written, logically presented, and strictly implemented.

The wood-clad vernacular architecture of the region set the precedent for the guidelines. Each residential building is required to have a porch facing the street. It was hoped that residents would sit on them and greet passers-by and so establish acquaintances. Each house must have a roof made either of wood shakes or metal. Residential buildings have white picket fences; each is different from the others on the street. Garages are located at the rear of buildings. Indigenous plants were prescribed for the gardens (Lombard 2005).

The architectural cognoscenti find the town 'too cute' but they are outsiders. Residents find Seaside a good place to live for its sense of community. For firms whose work follows the principles of the New Urbanism Movement, Seaside is a model to imitate. Kentlands, a neighbourhood in Gaithersburg, Maryland (1989+) was designed by DPZ CoDesign soon after their Seaside experience. Celebration in Florida is Seaside writ large (see Ross 1999; Lang 2017, 2021). Each design has been built following local traditions. The design of Norton Commons in Kentucky (2004+) draws on precedents different to those shaping Seaside, Kentlands or Celebration. The competition winning design, by one of the firm's partners, Marina Khoury, for a new urban centre for Dammam, Saudi Arabia (2005) draws on traditional climatically and culturally sensitive Saudi design precedents (Lang 2021).

8.4.3.4 The Functions of Central Attention

The urban design and architecture of the DPZ attempts to be richly functional in the terms described in Chap. 3. Drawing on traditional local building types for precedents enables architects to meet well enough many basic functional needs as defined using Maslow hierarchy. The implicit assumption is that those types that serve as models for DPZ's designs function well on a multiplicity of cultural dimensions. They include safety and survival needs and the opportunity for individual householders to obtain an identity through the places they inhabit. The codes in DPZ's urban designs do allow for each property to have a unique character within an overall established design framework.

The urban design work of DPZ CoDesign has aimed at the creation of territorially based communities. A sense of community—a similarity in values—is established through the unity in building and landscape character. The search, through design, for a genuine sociological local community where people interact with each other in more than a casual manner remains elusive. The communities of the residents of Seaside and Celebration are ones of common interests that extend far and wide. Designing to enable people to sit out on their porches to be comfortable in the summer and engaging passers-by in pleasantries is an intriguing design goal. Times have, however, changed. Air-conditioned houses in Seaside provide greater levels of summer comfort than their antecedents offered. The porches, nevertheless, function as clear thresholds between semi-private and private space and thus add, however marginally, to the psychological comfort of the town's residents.

Denser, walkable precincts function well in providing individuals with activity-rich local environments. Not everybody seeks them. Making walking easy and interesting, nevertheless, 'invites' people to walk and has a positive effect on their health if they do. The cognoscenti see the architectural qualities of DPZ CoDesign's architecture as regressive. It is seen to be aping the past even though its work represents a challenging new urban design paradigm. One of the 'problems' with places such as Seaside and, certainly Celebration, is that their success as attractive alternatives to life in pragmatically laid out suburbs has had the catalytic impact of creating steeply increasing property values. These increases exclude the possibility of the people on modest incomes who work there from living there (Ross 1999). They must commute in from surrounding areas.

8.5 Commentary

Designing a building is a joint venture of a client, often a financial sponsor, and an architect. Their values are shaped by the prevailing political attitudes of the society in which they are working. The brands of work of the nine architects explored here appealed to, or appeals to, some segment of the population that is requesting services. The review shows that the design process involves both programmatic and typological thinking. It also shows that however original design approaches may seem they draw on antecedent ideas.

The difficulty rationalists face is to marry the detailed requirements of a building program and other human aspects of environmental experience with their desire to follow their own intellectual design ideas. The Empiricists' problem is to marry what works well, or has worked well historically, with new ideas that meet the requirements of a changing world. The pragmatists simply plough ahead.

Of the architects whose work has been discussed here, Le Corbusier, Bernard Tschumi, Zaha Hadid and Alejandro Aravena are widely celebrated and discussed in architectural schools. They, and their work, lead the way in socializing students into the contemporary values of the architectural discipline. Of the other architects whose work is addressed in this chapter, Howard Shaw, Richard Seifert, and Duany and Plater-Zyberk had/have major practices. Bubbar, more typically of the profession, had a small, but respected, practice with a clear prescriptive line of thought. It received few accolades. In contrast, Al-Wakil's whose work addressed design concerns within a clear cultural context, has been much awarded in Classical architectural circles. This chapter has shown the diversity of design ideas that exist today, the types of architects who embrace them and the clients, who by their commissions, support them.

AI is beginning to shape architects' thinking, the whole design process, and modes of architectural production. A danger arises in relying on design paradigms that are easy to use as a basis for coding into algorithms. Many lines of architectural thinking then become ignored. AI algorithms, as they are developed, need to recognize the complex relationships among clients, architects, and design attitudes if they are to do more than copy various brands of architectural design. The mainstream of the intellectual elite of the profession admires buildings for what they offer as innovative, artistic expressions. Geometrically innovative building forms attract attention and are straightforward to generate using parametric design procedures. To fully comprehend the full range of architectural possibilities the variety of architectural practices working within specific cultural contexts and philosophical traditions needs to be understood. This chapter has provided an outline of one way to do so.

References

Howard Van Doren Shaw

Cohen SE (2015) Inventing the new American house: Howard Van Doren Shaw, architect. Monacelli Press, New York, NY

Green VA (1999) The architecture of Howard Van Doren Shaw. Chicago Review Press, Chicago, IL

R. Seifert and Partners and Richard Seifert

Bradbury D (2020) Richard Seifert: British brutalist architect. Lund Humphries, London

Harrison EM (2019) Richard Seifert (1910–2001). Archit Rev. architectural-review.com. Accessed 19 Aug 2022

Pawley M (2001) Richard Seifert: workaholic architect whose tall towers changed the skyline of Britain and Europe. The Guardian, 29 Oct 2001. https://www.theguardian.com/news/2001/oct/29/guardianobituaries.arts. Accessed 20 Aug 2022

Elemental and Alejandro Aravena Mori

Hernández D (2012) ELEMENTAL: incremental housing and participatory design manual. Hatje-Cantz, Berlin

Kimmelman M (2016) Alejandro Aravena, the architect rebuilding a country. The New York Times Style Magazine, 23 May 2016. nytimes.com. Accessed 13 Aug 2023

Marinovic GI, Baek J (2016) Lessons of incremental housing: two Chilean case studies: elemental Lo Espejo and Las Higuera. Archit Res 18(4):131–128. Lessons_of_Incremental_Housing_Two_Chilean_Case_St.pdf. Accessed 20 Feb 2024

Le Corbusier and Pierre Jeanneret Partnership and Le Corbusier

Brooks HA (1997) Le Corbusier's formative years: Charles Edouard Jeanneret at La Chaux-de-Fonds. University of Chicago Press, Chicago, IL

Etlin R (1987) A paradoxical avant-garde: Le Corbusier's villas of the 1920s. Archit Rev 181:21–32. architectural-review.com. Accessed 2 June 2024

Le Corbusier (1923) Vers une architecture translated into English by E. Etchells as towards a new architecture. Praeger, New York, NY, 1970

Bernard Tschumi Architects and Bernard Tschumi

Migayrou F (ed) (2014) Bernard Tschumi: concept & notation. Centre Georges-Pompidou, Paris

Tschumi B (1987) Cinégramme Folie: Le Parc de la Villette. Princeton Architectural Press, Princeton, NJ

Tschumi B (1988) Notes towards a theory of architectural disjunction. J Archit Urban 216:11–13

Zaha Hadid Architects and Zaha Hadid

Hadid Z (1982) Randomness vs arbitrariness. AA Files Ann Archit Assoc Sch Archit 2. https://www.scribd.com/document/476575250/Hadid-Randomness-Arbitrariness. Accessed 30 Aug 2022

Hadid Z (1983) The world (89 degrees). Planetary architecture two. Architectural Association, London. https://www.zaha-hadid.com/design/the-world-89-degrees. Accessed 30 Aug 2022

Jodido P (2020) Zaha Hadid: complete works 1979–today. Taschen GmbH, Köln

The Angles Architects and Darshan K. Bubbar

Bubbar DK (2005) The spirit of Indian architecture: Vedantic wisdom of architecture for building harmonious spaces and life. Rupa, New Delhi

Abel-Wahed El-Wakil

al-Asad M (1992) The mosques of Abdel-Wahed El-Wakil. In: Khan H-U (ed) MIMAR 42: architecture in development. Concept Media, London. Academia.edu. Accessed 7 Dec 2023

El Sayed Foundation (2009) Abdel-Wahed El-Wakil or the triumph of the Islamic style. Muslim Heritage. https://muslimheritage.com/abdel-wahed-wakil-architect. Accessed 23 Aug 2023

DPZ CoDesign and Andrés Duany and Elizabeth Plater-Zyberk

Duany A (2021, originally 1991) A general agreement for the practice of contemporary architecture. Common Edge. A General Agreement for the Practice of Contemporary Architecture | ArchDaily. Accessed 21 Nov 2022

Katz P (1994) The new urbanism: toward an architecture of community. McGraw Hill, New York, NY

Lombard J (2005) The architecture of Duany Plater-Zyberk and Company. Rizzoli, New York, NY

Epilogue

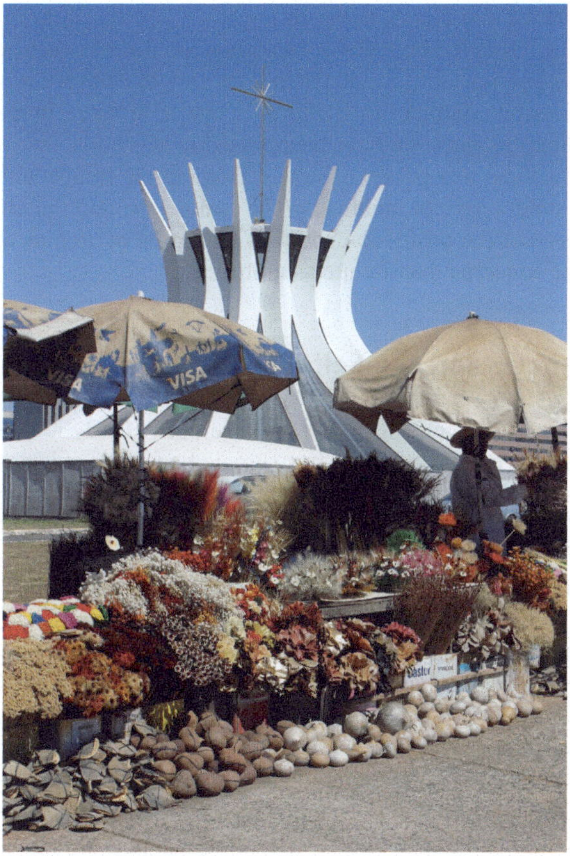

Catedral Metropolitana Nossa Senhora Aparecida de Brasília (1958–1970, renovated 2012)
The State of Brazil was the original client; completed by the Archdiocese of Brasilia

J. Lang, *The Scope of Architectural Theory*, Design Science and Innovation,
https://doi.org/10.1007/978-981-96-3690-7

Oscar Niemeyer, architect; Joaquim Cardoza, structural engineer
Photograph by the author

Some Current Issues

The book brings up but does not resolve current issues that will shape the nature of architectural theory and practice in the future. These issues include the self- and public image of architects, the current advocacies of various pressure groups, the nature of the public interest and how architects deal with it, and the impact of AI on creative thinking. Dealing with these issues will shape the development of both architectural practice and theory in the future.

Over the past century the knowledge base required for architects to practise well has expanded. The responsibilities that architects should embrace is open to debate. Should the profession contract or expand its scope of concern in response? Many issues surrounding this question have been raised in this book. Those addressed in this Epilogue are some that persist. The first is about the image architects have of themselves and others have of architects. The realities behind the image define the domain of concern of the profession and thus the scope of architectural theory. So will the range of concerns that architects embrace.

The Public Image and the Self-image of Architects

Since the mid-nineteenth century architects have been clearly identified as professionals distinct from engineers. They have created an independent profession that controls a place in the marketplace for services (Gutman 1988; Duffy and Hutton 1998, 2017). Registration acts require building designers to demonstrate that they are properly qualified before they can call themselves architects. At the same time many buildings throughout the world are designed well enough by developer/contractors without the help of registered architects. Where does that leave the architect? What is the value added to a building by an architect?

"People purvey and sustain a misleading impression of what it is to be an architect" (Saint 1983, 16). The public sees architects as the designers of buildings and engineers as the people who make them stand up. Architects believe engineers lack a creative spirit despite many widely accepted, aesthetically pleasing, innovative structures having been designed by engineers. Engineers, in turn, regard architects as creative individuals with marginal practical skills (Held 2014). The public image shapes the expectations people have of architects and the self-images of architects shapes their attitudes towards their work. The image architects have of themselves is important. Howard Shaw, Frank Wright, and Ludwig Mies worked under extended professional names; Louis Kahn and Frank Gehry were born with different ones. Alvar Aalto altered his mode of dress after meeting

Frank Lloyd Wright. These changes reflect both the self-images of individuals and contemporary societal attitudes.

The Architect as Artist

The image of an architect to emulate that is offered students in schools of architecture is one of a visual artist who generates unique designs based on inspired, original, creative thought. All buildings and urban designs can, indeed, be viewed as sculptures to be admired (Davies 1994). They are also inhabited places. Architects have difficulty reconciling the creation of works of self-expression with designing environments that afford the multi-functional needs of people's everyday lives. That is the paradox (Tschumi 1998). Similar paradoxes face other professions. Contact sports, for instance, are faced with being displays while reducing the long-term effects of concussions on their players.

Architects model and examine their design proposals in graphic form. This way of looking reinforces the view of buildings as art objects. The same observation is true of the photographs of completed buildings that appear in books, like this one, or in journals. Movement is essential to fully experience the environment. Computer-based graphics that simulate the views of a moving person enable architects and their clients to see buildings from the outside as they move around them and their interiors as they move through them. The sounds, tactile encounters, and odours of a place are yet to form part of this modelling process. Viewers must imagine them.

The image of architects primarily as aestheticians and the sole authors of buildings and public spaces has grown over the past decades since the Modernists' strong views of the social ends and the means for achieving them were questioned. The symbolic function of buildings as artistic expressions is important but sustaining the image of architects as symbol mongers means that they rely on the input of people from many other disciplines to complete their work. The view of architects as artists serves the profession well but is it sufficient given the range of work and practices mentioned in this book?

The Architect as Social Engineer

Many Modernists, implicitly or explicitly, considered themselves to be social engineers. They sought not only to improve the built environments but also to improve people. They were guided by their models of what a good person is and does. The assumption was that the layout of environment and the facilities it contains will lead to well-behaved people (Lipman 1969). Taken to the extreme, police and prisons would not be required (Garnier 1917). This view was taken not only by architects but also by philanthropic industrialists building company towns and by social commentators.

Architects continue to make claims about how their designs will shape people's lives. It helps sell their work. We have recognized the fallacy of many of these claims since the 1960s. The power of designs to change human behaviours is limited. Albert Mayer (1897–1981), an American architect known for co-designing the plan for Chandigarh that Le Corbusier rationalized, noted:

> We all naively thought that if we could eliminate the very bad physical conditions of dwellings and surroundings in slums, the new sanitized conditions would *per se* cure social ails. We know better now. (Mayer 1967, 20)

The nature of the built environment may more powerfully affect the lives of species other than humans. They have less ability to adapt to a changing world. The human social and cultural environments, in contrast are highly deterministic in shaping people's activities and aspirations. Winston Churchill's much quoted, observation: "We shape our buildings and then our buildings shape us" should be taken with caution but certainly not dismissed.

What buildings afford affects the way their users inhabit them as shown in Fred Steele's *Physical Settings and Organizational Development* (1973). The same applies to urban designs. Most patterns of the environment afford a wide variety of activities. As described in Chap. 1, people scan the patterns to see if they can do what they want to do given their competencies. Environments, other than coercive ones, offer many opportunities and will not necessarily be used in the ways architects might expect. Also, if the affordances of the built environment for a particular activity do not exist, that behaviour simply cannot take place.

The Architect as Team Coordinator and Facilitator

Designing buildings is rarely a solitary effort. The collaborative nature of architectural practice has been lost to architectural historians and critics today and, thus, to architectural theory (Coldstream 2003). Developments in structural technology and the demands of society for an evidence-based design and evaluation of buildings and urban designs means that the range of knowledge required to design well is vast and beyond the capacity of any individual to master. If architects are to maintain or expand their services it will have to be in collaboration with others (Lorrimer 2015; Pelsmakers 2018).

Successful collaborative efforts whether between architectural firms or within a firm are led by individuals with an empathy for different attitudes. They are politically adept at getting agreements as compromises are made. They are also willing to make decisions. One role of an architect is that of project coordinator. It is taken on to maintain the integrity of a proposed design. Often, however, the position is usurped by a politically powerful individual among the stakeholders.

The Architect as Entrepreneur

Apart from designing their own houses or those of their parents, architects are seldom property developers. Exceptions exist. Some architects have deep enough pockets to finance their own projects (see Portman and Barnett 1976; Petty 2018). Otherwise, practices require work to flow into them in sufficient amounts and regularity to sustain them. Not only do architects require personal contacts with potential clients but a willingness to show and praise their own work. They display their designs and argue for their high quality in journals and on the internet. Professional photographers take pictures to best show a building's spatial and aesthetic qualities. Philip Johnson's portrayal of architects as "hucksters" is something he regretted but, like other words spoken in jest, it has an element of truth to it.

Practices produce hagiographic books on their own work or solicit others to write them. The books will be on display in architects' offices or be presented as gifts to potential clients. The objective is to show clients the esteem in which the architectural firm and its work is held by others. The books also reinforce the primary image of architects held by the public as fine artists of the built environment.

Conclusion

Architects can play many roles in professional practice. As artists they require colleagues and staff to complete the consequential tasks that bring designs to fruition. Pierre Jeanneret played that role for Le Corbusier and John Rauch for Robert Venturi. Architectural practice has increasingly become collaborative in nature (Seddon 2017; Pelsmakers 2018). Different architects will find their own niches in architectural practice. Some will be leaders and others remain jobbers. The collaborative work that is required to complete and build designs needs to be recognized both in the education of architects and when awards are given for fine work. Doing so will require reimagining the scope of concern of architecture and architects and thus of architectural theory.

The Scope of Concern of the Profession

Professional architectural practice:

> has been progressively desiccated, undermined, and commodified through the instrumental processes of the contemporary development industry and professional practice. (Francis-Jones 2022)

The role of consultants has accompanied—perhaps has led to—the decline in the former all-round role of the architect. The image of the architect as an artist relegates other skills. What tasks should architects be undertaking? Some have to do with substantive issues and others with procedural concerns. Urban designing is an example of the former and designing programs for buildings and running the whole building process are examples of the latter.

Architects and the City

Who takes responsibility for policies and designs that shape the overall built environment that we inhabit? Does the architectural profession and/or individuals within it wish to deal with urban design seriously rather than to simply use it as another opportunity to express their own ideas as city saviours, as expressed by José (Josep) Luis (Lluís) Sert in *Can Our Cities Survive? An ABC of Urban Problems, their Analysis, their Solutions* (Sert and C.I.A.M. 1944)?

The criticism of many urban designs, and thus the declarative theory on which they were based, has come from many sources including people who had supported architects' ideas but then found them lacking (for example, Jane Jacobs when she was an architectural journalist and the architect Peter Blake (1920–2006) in practice (Jacobs 1961; Blake 1977; see also Lawrence 2016). Many architects have been unwilling to extend their knowledge base to include an understanding of how cities and the public spaces within them function. They treat the city as an evolving work of art (Olsen 1986). Other architects faced up to the shortcomings in the profession's thinking about cities. They sought the knowledge base required to create better living environments for humans and, although much less so, for other species.

The consequence has been that urban design has been functioning as a discipline on its own, with its journals and educational programs. Dana Cuff, an American architectural theorist, suggests that instead of narrowing their scope of concern, larger architectural practices should develop their knowledge base and be designing urban infrastructure and design projects (Cuff 2014; Held 2014). If so, the need will be for the scope of declarative architectural theory to be expanded beyond the domain it presently covers to deal with issues of urban morphology.

Programming

The theories of the architectural design process as presented in Chaps. 2 and 7, and the history of the ends and means of buildings as presented in Chaps. 6, 7, and 8, is a history of changing perceptions of the problems that architects address. It is thus one of changing building programs. Programming may be the responsibility of the client but, historically, the architect has been the trusted advisor in establishing what should be done and where the architect has the prerogative to impose his or her own stylistic brand on a design (Van Rensselaer 1893).

Many programs simply specify the spaces required of a building and their sizes. That is all many architects want to receive. Other programs include how buildings should function on multiple dimensions (Cherry 1998; Peña and Parshall 2018).

These dimensions vary from housing activities to the sonic qualities of spaces, to a building's energy performance, to its aesthetic qualities.

Who prepares such programs? Large architectural firms usually have an in-house programming department; others hire consultants. In either case, those architects who are responsible for designing the artefact can avoid getting involved in the politics of creating programs and having to deal with stakeholders' conflicting interests. How aloof should design architects be from the programming process? Do they wait until the program is produced by somebody else before challenging and amending it to suit their own interests? How thorough should the program be anyway? What ends should be left for the architect to decide?

Managing the Building Process

As the timely hiring of contractors and sub-contractors and the procuring of materials on site has become more demanding so the role of project managers has become central to the production of a building. They coordinate activities and see that working drawings are updated as a building's design is changed during implementation. They ensure that safety requirements and building codes are met (Project Management Institute 2013).

Project managers must satisfy the requirements of both clients and architects as circumstances change and unexpected problems and delays occur. They, while managing the day-to-day implementation process, must keep the whole team functioning as a unit and ensure that work proceeds on schedule and within a budget to the extent possible (Horine 2017).

Artificial Intelligence (AI) may have a major role in deciding what the future holds. Its application may well require a change in the self-image of architects and the nature of architectural education. How much narrower does the domain of an architect's control become? Several leading architects worry that the architect's role is becoming much diminished. The alternative is for architects to broaden their skills and their knowledge.

Contentious Projects

Some potential projects are highly contentious. They may be intrusions into existing precincts whose character would be dramatically changed in a manner that locals oppose (Longley 2021). Other proposed buildings may be located on ecologically sensitive sites that should not be developed but are, nevertheless, available for development. They may be projects for authoritarian despots with poor human rights records seeking self-aggrandisement or, more commonly, for a property developer who sees an opportunity for making a profit any cost (Wainwright 2019). Should architects take on such projects?

Much depends on what architects understand to be within their domain of concern and their obligations. One position is that such concerns fall outside the responsibilities of architectural practices. Architects should go ahead if a project is legal. Another is for architects to take on the project in the belief that they have the skills to minimize any harm caused by its development. The goal of the architect, they argue, is simply to do the best job possible. Another response is to not accept the commission. Taking that position has financial consequences for an architectural practice.

The Totality of a Building's Design

The Hufeisensiedlung (1923) in Britz, Berlin was designed by the much-admired architect Bruno Taut and the more politically active Martin Wagner (1885–1957). The units were let furnished down to the art works on the walls (Jager 2007). Taut believed that the renter's tastes in furniture and art would debase the complex's unified design quality.

The Viennese architect Adolf Loos (1870–1933) took strong exception to such views in his criticism of the stance taken by his contemporaries Josef Hoffman (1870–1956) and the Wiener Werkstätte, an association of designers. The Wiener Werkstätte argued that a house should be a total work of art. It was also something sought by advocates of the Gesamkunstwerk movement. In his parable, *The Poor Little Rich Man* (1900), Loos argued that such places were like prisons; the residents had no freedom of action and could not personalize their environments (Heathcote 2021). Les Quartiers Modernes Frugès in Pessac, a contemporary of the Hufeisensiedlung proved easy to adapt. Its first occupants took advantage of that. In doing so they ruined what was also a work of art (Boudon 1972).

The restrictions have long since disappeared at the Hufeisensiedlung. The project is now a UNESCO World Heritage site and is much liked as a place to live. The question remains: How total should the design of a building be? Should it be down to the furnishings? Much depends on the architect-client relationship and the client's values. Many architects still seek to create total works of art. Everything, they believe, should be one composition. Should it? Some believe so; others leave much for the craftspeople and inhabitants to complete.

Questions of Fit

A question asked earlier in the book, and implicit in much of the ensuing discussion in it, remains. How tailor-made should a building and its layout be in meeting the requirements of its first owners and occupiers? How comfortable should the environment be? How challenging should it be physically and mentally?

Some buildings meet the programmatic requirements of their first generation of users well. They continue to serve new users well as circumstances change. The instrumental purposes for which they were built remain unchanged. Theatres have retained their essential form for a long time as the purposes they serve are stable. The demands of the various stakeholders for greater levels of comfort have led to some changes but not to the basic building type. Churches, mosques, and temples have served the same liturgies for centuries and have remained unchanged in general nature. Changes in the liturgy in Catholic churches after the Second Vatican Council's 1964 pronouncements, however, made the traditional basilica type layouts difficult to adapt to fulfil a new priest-congregation relationship (Hackett 2011). Many high-rise commercial buildings are being converted to apartment buildings as market demands change. Does one design for potential changes or does one leave it for future owners to worry about how to make them?

Designing to Enable Easy Future Changes and/or for Easy Demolition

We live in a rapidly changing socio-economic world inhabited by what is said to be societies in which objects are used and discarded as soon as their utility passes. The Metabolists and members of Archigram sought prescriptions for an easy-to-change inhabitable world. Elements could be plugged into an infrastructural framework and easily replaced when no longer needed (Sadler 2005; Arellano 2020; Pernice 2022). They designed hypothetical environments that would be easy to change in a rapidly changing world. The only building built by one of the Metabolists, the Nakagin Capsule Tower in Tokyo designed by Kisho Kurokawa (1934–2007) in 1970 had, however, all its pods and furniture fixed in space. Nothing could be changed without difficulty and expense. The building was demolished in 2022. It became admired for what it was—an aesthetic experiment.

Should architects, as a matter of routine, design with adaptability or the demolition of components of a building or a whole building in mind (Andrade et al. 2019)? Maybe architects should do so. If it does not happen, there is no loss. One wing of the Maison Médicale in Brussels designed by Lucien Kroll (1927–2022), an architect noted for his participatory design work, was designed assuming that successive groups of students would change the layout of their spaces to meet their own preferences. They made few changes; it was much easier to adapt to what was already there (Poletti 2010).

Challenging or Comfortable Environments?

Special questions arise when considering institutionalised populations as noted in Chap. 1. In places such as aged-care centres where residents' physical and/or mental competencies may be in decline, if the environment is made easy for them to negotiate, their competencies will decline even more rapidly than expected to the level demanded by both the social and physical environments (Lawton 1989). The ageing process is one of continual behavioural adaptation to the external environment and to changes in one's competency levels. How easy to negotiate should the social and physical environment be? Similar issues arise when designing for people at other stages in their life cycles.

Children's competencies increase and their lives are full of self-testing activities. How challenging should places such as playgrounds be? Fear of litigation has led to playgrounds in many jurisdictions being designed to be unchallenging on safety grounds. Places such as adventure playgrounds test children's capabilities but are eyesores for many people (Bengtsson 1973). How can one design for safety and adventure in the broader urban environment? Should designers favour one approach or another? Or leave it to public policymakers to decide? What is in the public interest?

Private Concerns and the Public Interest

Public interest concerns in the built environment are represented in master plans, zoning ordinances, and building codes. They establish what can and/or what cannot be built, and the minimum standards that buildings must attain to obtain approval from consent authorities. Such codes date back to well before our common era to, at least, the Code of Hammurabi in 1772 BCE. Rules must be enforced to achieve their intended purpose. Recent disasters around the world have shown that where enforcement is lax the consequences can be dire.

Property developers and their architects want to do things their own way and resent constraints on what they can do (Guy and Henneberry 2002). Few architects oppose codes that are designed to stop buildings being damaged by fire or causing fires to spread or those that ensure that buildings do not collapse because of earthquakes or cyclones. They may believe, correctly or incorrectly, that they can achieve the outcomes sought by codes in other than specified ways. Other purposes that codes serve can be contentious. How far should developers and architects have to go in designing barrier-free environments? Why should they comply with aesthetic requirements when they are after all simply a matter of taste? Why should they have to comply with regulations that specific how buildings should be designed for reducing the impact of global climate change? The requirements may contradict what is desirable for both architect and client in a competitive world.

As noted earlier, some architects oppose all regulations controlling what they can design for their clients (Schumacher 2013; Frearson 2016). They argue that regulations interfere with architects' basic freedoms of expression. Moreover, they restrict their and their clients' wealth creation. Many more architects are opposed to restrictions on what they do when the codes seem to be arbitrary to them and interfere with them producing their brand of work.

Foci of Concern

The design of many schemes is biased by an overriding central idea. It may be to celebrate a structural technique or to be eco-friendly or to fulfil an architect's aesthetic ideal (see Fig. 3.10). The character of the designed product becomes a selling point for a building and an advertisement for its architect. Almost inevitably designing to emphasize some requirements of a design makes the attainment of others more difficult. Property developers, project managers, and architects tend to focus on immediate concerns and on ensuring the maximization of short-term profits. If hesitant, they strive to ensure that no short-term financial loss is incurred. Maintenance and operational costs are not given much attention.

What should the focus of concern of architects be? Should they be worrying about the short-term or long-term life of a project? Should they be designing for a building to have a place in a globalizing financial world or be sensitive to local conditions? Many requirements to make buildings ecologically sound contradict what is currently seen as representative of world-wide aesthetic trends. Today it is often the uniqueness of the geometrical form of a building that is the focus of concern. Such a focus tends to result in much lauded, bold sculptural statements. Multi-functional, well-crafted buildings do not receive the same recognition. They seldom stand out in the crowd. Should the status quo persist or should we change the way we evaluate architectural works?

Buildings have a catalytic effect on their surroundings. Architects generally show little interest in the impact of their designs on their contexts. Buildings throw shadows, shed heat, channel wind flows, and generate traffic. It is beyond the ability of an architect to give equal attention to every concern. What should the priorities be? Whose interests should be of central concern? Who decides?

Clients' Interests

The stakeholders in a project of consequence can be numerous. They may bring diverse attitudes to the discussion of ends and means. The property developer, whether a family for a bungalow or a multi-national corporation for a skyscraper, or a city administration for a public facility that will put it on the world map, are the fundamental stakeholders.

How much attention should architects pay to understanding the needs of non-paying clients such a building's potential users and the neighbours of a building (Ornstein 2022)? Should the natural world always be regarded as a non-paying client? How about other architects and architectural critics? Many critics have been influential in shaping the architectural discourse. Should they be regarded as public-interest clients? Political conservatives say the obligation of the architect to achieve what they think is best is paramount; other views should be disregarded (for a factional example see Rand 1968).

The Geographic, Social, and Visual Context

Some architects consider their works to be independent of their social and physical surroundings. The many eye-catching architectural projects of Zaha Hadid, Daniel Libeskind, and Frank Gehry pay little heed to their surroundings. The designs of well-known architects are not the only ones competing to be 'iconic' landmarks, but their authors' reputations as fine artists give them considerable freedom of action. They argue that their projects serve other intangible cultural needs and the desires of their clients to be part of a prestigious international global community.

The counter argument is that these bold projects could have been eye-catching while fitting in with their built contexts.

Those people who advocate designing with the local context in mind argue that it is important to maintain a sense of a locale because it grounds people in a cultural context (Relph 2008, originally 1976; Norberg-Schulz 1980; Glendinning 2010). Around the world there are many expressions of the loss of a sense of local identity when new property developments are built. Dealing with that loss may be more important at an urban design than at a building scale. If architects design acknowledging the qualities of sunlight and the nature of vegetation in a locale, the scale of the surroundings massing of buildings, the way buildings meet the street, and the behaviour settings that define a culture, they can surely have freedom to do what they want. Is that enough?

Attitudes to Architectural Heritages

The world is full of historically important buildings, urban complexes, and objects such as memorial obelisks and statues. Debates rage over whether a social loss is incurred by memorials and buildings of note being demolished. Differences in opinion exist between cultures and within cultures about what is important to retain and what is not (Weiler and Gutschow 2017). Whose past is being celebrated anyway? What of the past should be preserved, conserved, or restored?

All buildings have a heritage value but those that are regarded as worth keeping are those that represent an era or have witnessed events or housed people of historical significance as perceived by a group of people today (Lynch 1972; Glendinning 2013). Questions arise over the interpretation of the word 'significance'. The works of major architects and/or those that are representative of an architectural era are also part of our heritage.

Architecturally important buildings of a former colonial power may have negative associations for the present population of a country. Fine Japanese era buildings in Korea carry painful memories. In addition, such buildings when much liked may, as in Lutyens' New Delhi, be on sites highly valued in the real estate marketplace. New buildings would generate greater profits and tax revenues than the existing yield. Other questions arise. What can one do with buildings of historic importance whose instrumental functions no longer exist? What are the socially acceptable reuses for existing buildings such as religious edifices? Different architects and different societies take different stances on these issues.

Artificial Intelligence and the Future of Professional Work

The development of Artificial Intelligence—the power of computer-based algorithms to search for information from vast data sets, to learn from experience, and to solve complex problems—promises much. Although initiated in the 1950s, AI only reached the point of challenging human cognitive abilities in the 2010s.

Recent progress has been especially rapid as noted in Chap. 7. Zaha Hadid Architects, under the leadership of Patrik Schumacher has developed parametric algorithms that create designs. The firm uses them for generating optional designs that can form the basis for asking questions about ends and means. Neil Leach, a professor at Florida International University, believes that AI has much more to offer. He notes:

> The most revolutionary change is in … the automation of the entire design package, from developing initial options right through to construction. In terms of strategic thinking and real time analysis, AI is already way beyond what human architects are capable of. This could be the final nail in the coffin of a struggling profession. (Leach 2021 cited in Wainwright 2023)

That day is a still a while in the future. Or is it?

No algorithm is neutral. All are based on the assumptions made by their creators at the time of coding and, in architecture, are founded on a typological approach to design. The values implicit in them are largely opaque to the architects who may want to use them. They can, however, be designed with the values that an architect wants incorporated rather than what is easiest to code. Algorithms can be designed to generate creative expressions by customizing approaches. AI systems can be built with multiple world views in mind. When further developed AI tools will prove to be an invaluable asset for architects (Harrell 2023).

Although AI will revolutionize the nature of architectural practice, many aspects of current operations will persist. Architects will need to have a thorough understanding of what different software packages offer and the values on which they are based. They will need to remain in control of the algorithms not the other way round.

Commentary

This epilogue is neither comprehensive in covering the current architectural issues that are open to debate nor complete in the coverage of those topics that have been identified. What is important is that we understand that many of the positions we take on architectural matters are open to debate. We need to take stands on them based on the evidence available rather than our personal hopes and habits.

Addressing wicked problems well involves specialized knowledge. These areas of knowledge become more numerous as possible ways of building increase and competing stakeholders in a design vie for attention to be paid to their own interest. As a result, the nature and scope of both declarative and prescriptive architectural theories remain open to continuous debate.

Allied to questions of how to deal with problems and issues confronting the profession, is the need to turn architectural education's "obsession with [architects as] form producers, to an understanding of the political and economic forces shaping form" (Cuthbert 2011, xix). This book, by focussing on the scope and nature of architectural theory, is only a partial response to that demand. The way architectural theory has been presented here may be regarded as "pedantic nonsense" by many architects. To them only the prescriptive theories of architectural form, as presented in Chaps. 6 and 8 are of interest.

Students internalize what to do from their exposure to whatever examples they are shown in lectures or from reading magazines online. They learn from their instructors' responses to the work they produce in studios. They will also know what leading architects—Jane Jacobs' "saints"—advocate as good design. With that as a base and from their own professional experiences, architects develop a set of beliefs about how the built environment functions and how to go about designing it. Other knowledge is regarded as extraneous and gets in the way of creative thinking. The users and observers of buildings and places are adaptable and can work out how to accommodate themselves to almost any environment. They will be happy enough especially if the financial and emotional cost to them is low. Such a position is short-sighted if the profession wants to do more than provide the world with satisficing, geometrically innovative, building forms.

In 1943 Kurt Lewin (1890–1947), a psychologist, noted that "nothing is so practical as a good theory" (Lewin 1951). The goal here has been to outline the scope of architectural theory. Whether it has any utility remains to be seen. A broad understanding of both declarative and prescriptive theories will, nevertheless, enhance architects' understanding of what they do and could do. Students will have a better understanding of what they are learning and the nature of design than current graduates possess.

References

Held J (2014) Deskilling and reskilling. The Business of Architecture. ACA—Association of Consulting Architects, Australia. Accessed 16 Feb 2022

Leach N (2021) Architecture in the age of artificial intelligence. Bloomsbury, London

Lorrimer L (2015) 10 challenges facing architecture and ways to turn them into opportunities. Architect Des, 25 Feb 2015. [ACA series]. architectureanddesign.com.au. Accessed 16 Feb 2023

Pelsmakers S (2018) Changing practice: new roles. RIBA J, 15 Feb 2018. Accessed 22 June 2023

Saint A (1983) The image of the architect. Yale University Press, New Haven

Credits

Except for the following items, all the illustrations, tables and diagrams are the work of the author or he is their copyright holder. Other than those so identified all the Wikimedia Commons images are in the public domain as are the U.S. Government publications. Credits have been labelled 'Collection of the author' in cases where it has been impossible to identify or trace the copyright holder. The sub-figure number are from left to right and the top to bottom of each page.

Prologue Photograph by Carol Highsmith/Wikimedia Commons

Chapter 1

Fig. 1.1b Photograph by Gren/Wikimedia Commons
Fig. 1.2b Drawing by Walter Moleski. Source: US Department of the Army (1976, 3–9): US Government publication: In the public domain
Fig. 1.3a Adapted from Lang and Moleski (2010, 42): Courtesy of Taylor and Francis
Fig. 1.3b Adapted from Lang and Moleski (2010, 58): Courtesy of Taylor and Francis

Chapter 2

Table 2.1 Adapted from Fischer (2004) by the author: Courtesy of Gerhard Fischer
Fig. 2.1 Adapted from various sources by the author
Fig. 2.2 Adapted from Lang (1987) by the author
Fig. 2.3 Adapted from Lang (2017) by Molly O'Neill Robertson: Courtesy of Molly O'Neill Robertson

Chapter 3

Fig. 3.1 Adapted from Maslow (1987), Huitt (2004) and Lang and Moleski (2010, 56) by Alix Verge: Courtesy of Alix Verge and Taylor and Francis

J. Lang, *The Scope of Architectural Theory*, Design Science and Innovation,
https://doi.org/10.1007/978-981-96-3690-7

Fig. 3.2	Adapted from Lewis (1977) and Lang and Moleski (2010) by the author: Courtesy of Nigel Lewis and Taylor and Francis
Fig. 3.3	Adapted from Izumi (1968) by the author: Courtesy of Taylor and Francis
Fig. 3.4d	Source: Newman (1975). US Government publication: In the public domain
Fig. 3.5a	Photograph by Nancy Marshall: Courtesy of Nancy Marshall
Fig. 3.6	Adapted from Altman (1975) and Lang and Moleski (2010, 146) by the author: Courtesy of Taylor and Francis
Fig. 3.7a	Photograph by Kirill Neiezmaklov/Shutterstock.com
Fig. 3.7b	Photograph by High Contrast/Wikimedia Commons
Fig. 3.7c	Photograph by Bill Price III/Wikimedia Commons
Fig. 3.7d	Photograph by Trungydang/Wikimedia Commons
Fig. 3.7e	Photograph by John O'Neill/Wikimedia Commons
Fig. 3.8	Source: Lang and Moleski (2010, 42): Courtesy of Taylor and Francis
Table 3.1	Source: Lang and Moleski (2010, 216). Adapted from Mann (1979) by the author: Courtesy of Taylor and Francis
Fig. 3.9e	Photograph by Alix Verge: Courtesy of Alix Verge
Fig. 3.9f	Photograph by the author: Courtesy of Taylor and Francis
Fig. 3.10a	Adapted from Palladio (1570) by the author: In the public domain
Fig. 3.10b	Photograph by Marcok/Wikimedia Commons
Fig. 3.10c	Adapted from various sources by the author
Fig. 3.10d	Photograph by Carol M. Highsmith/Wikimedia Commons
Fig. 3.10e	Photograph by Marsupium/Wikimedia Commons
Fig. 3.10f	Photograph by Mondeo79/Wikimedia Commons

Chapter 4

Fig. 4.1a	Collection of the author. Various sources: In the public domain
Fig. 4.1b	Collection of the author. Various sources: In the public domain
Fig. 4.2	Source: Patti (1782, Plate 3): In the public domain
Fig. 4.3	Source: Durand (1809, Plate 2): In the public domain
Fig. 4.4b	Photograph by M. Kebra/Shutterstock
Fig. 4.4c	Photograph by Hansueli Krapf/Wikimedia Commons
Fig. 4.4d	Photograph by iofoto/Shutterstock.com
Fig. 4.4e	Photograph by Ahmed Hussein/Wikimedia Commons
Fig. 4.5	Source: Bohl and Plater-Zyberk (2008): Courtesy of DPZ CoDesign
Fig. 4.6a	Source: Krier (2005, 9): Collection of the author
Fig. 4.6b	Source: Krier (2005, 33): Collection of the author
Fig. 4.7e	Photograph by Mn000/Wikimedia Commons
Fig. 4.7f	Photograph by Brit in Seoul/Wikimedia Commons
Fig. 4.8a	Photograph by Wdzinc/Wikimedia Commons
Fig. 4.8b	Photograph by Michael Reed/Wikimedia Commons
Fig. 4.8c	Photograph by Alexander Cuthbert: Courtesy of Alexander Cuthbert
Fig. 4.8e	Photograph by Sardaka/Wikimedia Commons

Fig. 4.8f	Photograph by Romany V. Sansky/Wikimedia Commons
Fig. 4.8g	Photograph by No Swan so Fine/Wikimedia Commons
Fig. 4.8h	Photographer unknown/Wikimedia Commons
Fig. 4.9	© Estate of Bernd and Hilla Becher, represented by Max Becher. Courtesy of the Paula Cooper Gallery, New York
Fig. 4.10b	Photograph by Ian Watt/Wikimedia Commons
Fig. 4.10c	Photograph by Rovanienen taidemuseo/Wikimedia Commons
Fig. 4.10d	Photograph by Cobrecht/Wikimedia Commons
Fig. 4.10e	Photograph © by Iwan Baan: Courtesy of the photographer
Fig. 4.10f	Courtesy of Bernard Tschumi Architects
Fig. 4.11	Source: Schielke (2023): Courtesy of Siobhan Rockcastle and Marilyne Anderson
Fig. 4.12a	© Le Corbusier/Perriand/Jeanneret-Oeuvres ADAGP/Copyright Agency, 2024
Fig. 4.12b	© Le Corbusier/Perriand/Jeanneret-Oeuvres ADAGP/Copyright Agency, 2024
Fig. 4.12c	Source: Sert and C.I.A.M (1944): Courtesy of the Frances Loeb Library, GSD, Harvard University
Fig. 4.13a	Source: Lang (2021, 216). Originally Perry (1929): In the public domain
Fig. 4.13b	Adapted from Lang (2021, 216) by the author: Courtesy of Taylor and Francis
Fig. 4.14a	Adapted from Richards (1962) and Lang (2021, 178) by the author
Fig. 4.14c	Source: A. Savin/Wikiphotospace: In the public domain
Fig. 4.14d	Adapted from Hilberseimer (1927) by the author
Fig. 4.14e	Unknown photographer/Shutterstock.com
Fig. 4.14g	Photograph by Erhard Pfeiffer: Courtesy of Erhard Pfeiffer
Fig. 4.14h	Photograph by Abdullah/Wikimedia Commons

Chapter 5

Fig. 5.1a	Photograph by Anaïs Napoleon/Wikimedia Commons
Fig. 5.1b	Adapted from US Library of Congress/Wikimedia Commons
Fig. 5.1c	Shawn Beckwith Collection: Collection of the author
Fig. 5.1d	Photograph by William Dunlap/Edith Farnsworth House Historical Site
Fig. 5.1e	Photograph by Michiel Hendryckx/Wikimedia Commons
Fig. 5.1f	Photograph by Kathy Hutchins/Shutterstock.com
Fig. 5.1g	Photograph by President.az/Wikimedia Commons
Fig. 5.2b	Photograph by Teemu08/Wikimedia Commons
Fig. 5.3c	Watercolour by J. McNeven/Wikimedia Commons
Fig. 5.3d	Source: Lang and Moleski (2010, 126): Courtesy of Taylor and Francis
Fig. 5.3f	Photograph by Alix Verge: Courtesy of Alix Verge

Chapter 6

Fig. 6.2a	Source: B. Jankuloski/Wikimedia Commons
Fig. 6.2b	Attributed to various artists/Wikimedia Commons
Fig. 6.2c	Adapted from an 1846 painting by Leo von Klenze/Wikimedia Commons
Fig. 6.2d	Adapted from an 1882 etching by I. Buhlmann/Wikimedia Commons
Fig. 6.3b	Photograph by Iman Fakhri/Wikimedia Commons
Fig. 6.3c	Photograph by Nino Barbieri/Wikimedia Commons
Fig. 6.3f	Photograph by Ninarus/Wikimedia Commons
Fig. 6.4a	Photograph by Commonists/Wikimedia Commons
Fig. 6.4b	Illustration by Straffarello Gustavo/Wikimedia Commons
Fig. 6.4d	Etching by Guiseppe Vasi/Wikimedia Commons
Fig. 6.4e	Collection of the Urbino National Gallery/Wikimedia Commons
Fig. 6.5b	Photograph by Karel/Wikimedia Commons
Fig. 6.5c	Photograph by Dan/Wikimedia Commons
Fig. 6.5d	Photograph by Kawneer/Wikimedia Commons
Fig. 6.5e	Mr and Mrs Charles Wrightsman Gift, MMA/Wikimedia Commons
Fig. 6.6b	Photograph by Moneycantbuy/Wikimedia Commons
Fig. 6.6c	Photograph by Marc Ryckaert/Wikimedia Commons
Fig. 6.6d	Photograph by Joe Ravi/Wikimedia Commons
Fig. 6.6f	Photograph by Benh Lieu Song/Wikimedia Commons
Fig. 6.7a	Photography by Reading Tom/Wikimedia Commons
Fig. 6.7b	Photographer unknown/Wikimedia Commons
Fig. 6.7c	Photograph by Stephen Richards/Wikimedia Commons
Fig. 6.7d	Photograph by Baycrest/Wikimedia Commons
Fig. 6.7e	Source: Geological Survey, US Department of the Interior: In the public domain
Fig. 6.8a	Source: Sullivan (1891). In the public domain
Fig. 6.8b	Source: Mujica (1929, Plate CXXIV): In the public domain
Fig. 6.8c	Source: Ferris (1929, 63): In the public domain
Fig. 6.8d	Courtesy of the Chicago Architectural Center
Fig. 6.8e	Photograph by Limno/Wikimedia Commons
Fig. 6.8h	Photograph by Chris Simpson/Wikimedia Commons
Fig. 6.9a	Source: Wagner (1912): In the Public Domain
Fig. 6.9b	Source: Garnier (1917)/Wikimedia Commons
Fig. 6.9c	© Le Corbusier/Perriand/Jeanneret-Oeuvres ADAGP/Copyright Agency, 2024
Fig. 6.9d	Photograph by Carsten Jansen/Wikimedia Commons
Fig. 6.9e	Photograph by Svetlov Artem/Wikimedia Commons
Fig. 6.9g	Photograph by Wladyslaw/Wikimedia Commons
Fig. 6.11a	Source: Scott Leonard for the Kellogg Dolittle Estate: Courtesy of kelloggdolittlehouse.com
Fig. 6.11c	Photograph by Colros/Wikimedia Commons
Fig. 6.11d	Photograph by Gary Knight/Wikimedia Commons
Fig. 6.11e	Photograph by Fredericus Magus/Wikimedia Commons
Fig. 6.12a	Photograph by D. Conventini/Wikimedia Commons

Fig. 6.12b	Photograph by J. Schnake/Wikimedia Commons
Fig. 6.12c	Photograph by Alex Sergeev/Wikimedia Commons
Fig. 6.12d	Photograph by John Pavelka/Wikimedia Commons
Fig. 6.12f	Photograph by Elsabornfree/Wikimedia Commons

Chapter 7

Fig. 7.1	Source: Alexander et al. (1968, 54): Courtesy of the Center for Environmental Structure, University of California at Berkeley
Fig. 7.2a	Courtesy of Zaha Hadid Architects
Fig. 7.2b	Courtesy of Zaha Hadid Architects
Part III	Photograph by Pivari.com/Wikimedia Commons

Chapter 8

Fig. 8.1a	Various sources. In the public domain
Fig. 8.1b	Source: Henry Fuerman, American Country Houses (1904)/ Wikimedia Commons
Fig. 8.1c	Photograph by Andrew Jameson/Wikimedia Commons
Fig. 8.1d	Photograph by Teemu08/Wikimedia Commons
Fig. 8.1e	Adapted from various source by the author
Fig. 8.1f	Photograph by Richard Drew/Wikimedia Commons
Fig. 8.2a	© National Portrait Gallery London: Courtesy of the Nationals Portrait Gallery
Fig. 8.2b	Photograph by Philafrenzy/Wikimedia Commons
Fig. 8.2d	Photograph by Colin/Wikimedia Commons
Fig. 8.3a	Photograph by Centro de Politicas UC/Wikimedia Commons
Fig. 8.3b	Photograph by Gabriela Alcanta/Shutterstock.com
Fig. 8.3c	Photograph by V. Morande/Wikimedia Commons
Fig. 8.3d	Photograph adapted from various sources by the author
Fig. 8.3e	Photograph adapted from various sources by the author
Fig. 8.3f	Photograph adapted from various sources by the author
Fig. 8.3g	Photograph adapted from various sources by the author
Fig. 8.4a	Photograph by Jeet Mehrotra/Wikimedia Commons
Fig. 8.4c	Photograph by Shakespeare/Wikimedia Commons
Fig. 8.4d	Photograph by Dijana Alic: Courtesy of the photographer
Fig. 8.4e	Source: Mumford (1954). Collection of the author
Fig. 8.4f	Adapted by the author from a drawing in the British Library/ Wikimedia Commons
Fig. 8.5a	Photograph by Columbia GAAPP/Wikimedia Commons
Fig. 8.5b	Courtesy of Bernard Tschumi Architects
Fig. 8.5c	Drawing by Yin Yin: Courtesy of Yin Yin
Fig. 8.5d	Courtesy of Bernard Tschumi Architects
Fig. 8.6a	Photograph by Dimitri Ternovoy/Wikimedia Commons
Fig. 8.6b	Photograph by Yale/Wikimedia Commons

Fig. 8.6c	Photograph by Bob Boom/Wikimedia Commons
Fig. 8.6d	Photograph by Brenac/Wikimedia Commons
Fig. 8.6e	Photograph by Khalilov/Wikimedia Commons
Fig. 8.6f	Image by Flying Architecture: Courtesy of Zaha Hadid Architects
Fig. 8.7a	Courtesy of Chandrika Bubbar
Fig. 8.7b	Source: Lang et al. (2022, 30): Courtesy of CEPT University Press
Fig. 8.7c	Source: Bubbar (2005, 169): Courtesy Chandrika Bubbar
Fig. 8.7d	Source: Bubbar (2005, 118): Courtesy Chandrika Bubbar
Fig. 8.7e	Source: Bubbar (2005, 180): Courtesy Chandrika Bubbar
Fig. 8.8a	Photographer unknown: Collection of the author
Fig. 8.8b	Photograph by Najah Alsalami/Wikimedia Commons
Fig. 8.8c	Photograph by Leonardo Martin/Shutterstock.com
Fig. 8.8d	Photographer unknown: Collection of the author
Fig. 8.8f	Photograph by HSTASYF/Alamy
Fig. 8.8h	Photograph by Motacilla/Wikimedia Commons
Fig. 8.9a	Photographer unknown. Collection of the author
Fig. 8.9b	Photograph by Kristie Blokhim/Shutterstock.com
Fig. 8.9c	Photograph by Zstock/Shutterstock.com
Bibliography	Photograph by Alix Verge: Courtesy of Alix Verge

References and Bibliography

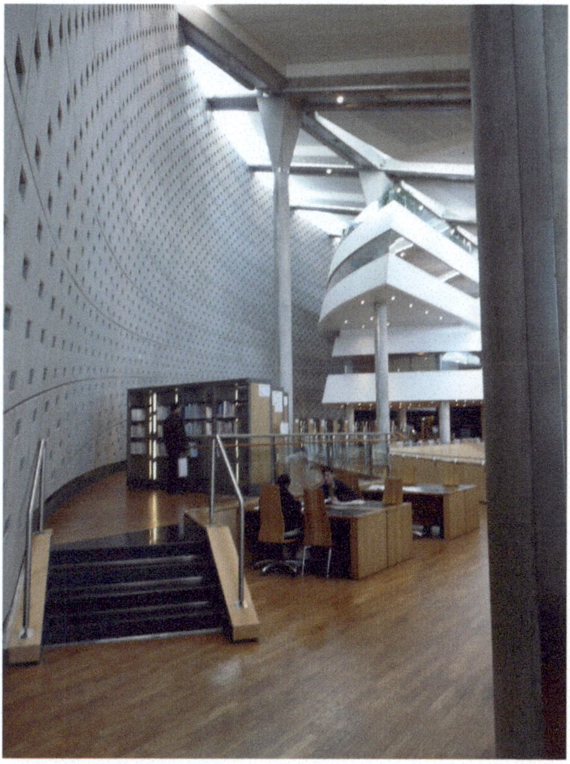

Alexandrina Library, Alexandria, Egypt
Government of Egypt and other international, organizations, clients
Snøhetta Hamza Consortium, architects, Hamza Associates engineers
Photograph by Alix Verge; Courtesy of Alix Verge

J. Lang, *The Scope of Architectural Theory*, Design Science and Innovation,
https://doi.org/10.1007/978-981-96-3690-7

Bibliography

Aburawa A (2012) Hassan Fathy's "New Gourna" model village lies shambles. Cities, 6 Aug 2012. Green Prophet. Accessed 6 Aug 2023

Ackoff R (1978) The art of problem solving. Wiley, New York, NY

Adajania N (2018) The politics of acknowledgement: exploding the male architectural canon. Tekton 5(10):74–81. Academia.edu. Accessed 19 Feb 2024

Aksamija A (2016) Integrating innovation in architecture, design, methods and technology for progressive practice and research. Wiley, Hoboken, NJ

al-Asad M (1992) The mosques of Abdel-Wahed El-Wakil. In: Uddin H (ed) Mimar 42: architecture in development. Concept Media, London. Academia.edu. Accessed 7 Dec 2023

Alberti LB (1452) De re aedificatoria: Libri decem. In: Rykwert J, Tavernor R, Leach N (trans) On the art of building in ten books. Translated from the Latin. MIT Press, Cambridge, MA, 1988

Alexander C (1964) Notes on the synthesis of form. Harvard University Press, Cambridge, MA

Alexander C (1969) Major changes in environmental form required by social and psychological demands. Ekistics 78–85

Alexander C (1979) The timeless way of building. Oxford University Press, New York, NY

Alexander C (2000–2005) The nature of order, 4 vols. Center for Environmental Structure, University of California, Berkeley, Berkeley, CA

Alexander C, Chermayeff S (1964) Community and privacy: toward a new architecture of humanism. Doubleday Anchor, New York, NY

Alexander C, Ishikawa S, Silverstein M (1968) A pattern language which generates multi-service centers. Center for Environmental Structure, Berkeley, CA

Alexander C, Ishikawa S, Silverstein M (1977) A pattern language: towns, buildings, construction. Oxford University Press, New York, NY

Allport G (1954) The nature of prejudice. Addison Wesley, Cambridge, MA

Alter L (2022) Arabia's plans for a 105-mile mirrored city. Treehugger, 29 July 2022. https://www.treehugger.com/the-line-saudi-arabia-mirror-city-6279123. Accessed 19 July 2023

Altman I (1975) Environment and social behavior: privacy, personal space, territory, crowding. Brooks/Cole, Monterey, CA

Anderson C (2014) Feng Shui: organizing the home and office: Feng Shui rules explained. Speedy Publishing, Newark, DE

Andrade J, Bragança L (2019) Assessing buildings' adaptability at the early design stages. IOP Conf Ser Earth Environ Sci. https://www.researchgate.net/publication/331313898_Assessing_buildings'_adaptability_at_early_design_stages. Accessed 10 June 2023

Anthony KH (2012) Design juries on trial: the renaissance of the design studio. Amazon Digital Services. Originally published in 1991 by Van Nostrand Reinhold, New York

Appleyard D (1982) Toward an urban design manifesto. Institute for Urban and Environmental Development, Berkeley, CA

Appleyard D, Lynch K, Myer JR (1965) The view from the road. MIT Press, Cambridge, MA

Archer LB (1970) An overview of the structure of the design process. In: Moore GT (ed) Emerging methods in environmental design and planning. MIT Press, Cambridge, MA, pp 285–307

J. Lang, *The Scope of Architectural Theory*, Design Science and Innovation, https://doi.org/10.1007/978-981-96-3690-7

Ardalan N, Candilis G, Doshi BV, Safdie M, Sert JL (1976) Habitat bill of rights. Hamdami Foundation, Tehran

Ardizzola P (2017) Architectural practice and theory: the case of Bruno Taut's house in Berlin-Dahlewitz. VITRUVIO 2(1):45. researchgate.net. Accessed 28 July 2023

Arellano M (2020) Archigram and the dystopia of small-scale living spaces. ArchDaily. Translated from the Italian by Johnson M. https://www.archdaily.com/948954/archigram-and-the-dystopia-of-small-scale-living-spaces. Accessed 10 Oct 2020

Argan GC (1963) On the typology of architecture. Translated from the German by Rykwert J. Archit Des 33:564–561. scribd.com. Accessed 2 Nov 2022

Arndt I, Tautz J (2014) Animal architecture. Harry N. Abrams, New York, NY

Arnheim R (1977) The dynamics of architectural form. University of California Press, Berkeley and Los Angeles

Asch RG, Birke AM (1991) Princes, patronage, and the nobility: the court at the beginning of the modern age c. 1450–1650. The German Historical Institute and Oxford University Press, New York, London

Asher CB (1992) Architecture of Mughal India, part 1, vol 4. Cambridge University Press, Cambridge

Asimov M (1962) Introduction to design. Prentice Hall, Englewood Cliffs, NJ

Astbury J (2022a) Bernard Tschumi is the deconstructivist architect with big ideas. Dezeen, 11 May 2022. dezeen.com. Accessed 16 Aug 2023

Astbury J (2022b). Zaha Hadid's Vitra Fire Station is "ready to explode into action at any moment". Dezeen, 24 May 2022. https://www.dezeen.com/2022/05/24/vitra-fire-station-zaha-hadid-deconstructivism/. Accessed 6 Dec 2023

Attoe W, Logan D (1989) American urban architecture: catalysts in the design of cities. University of California Press, Berkeley and Los Angeles

Audi R (2015) The Cambridge dictionary of philosophy, 3rd edn. Cambridge University Press, Cambridge

Bahrami A, Dagli CH (1993) Models of design processes. In: Parsaei HR, Sullivan WG (eds) Concurrent engineering: contemporary issues and modern design tools. Chapman and Hall, New York, NY, pp 113–126

Baldwin J (1997) Buckyworks: Buckminster Fuller's ideas for today. Wiley, New York, NY

Banham R (1967) Theory and design in the first machine age. Praeger, New York, NY

Banham R (1969) Architecture of the well-tempered environment. University of Chicago Press, Chicago, IL

Banham R (1976) Megastructures: urban futures of the recent past. Thames and Hudson, London. Republished by Monacelli Press, New York in 2020

Barber DA (2020) Modern architecture and climate design before air conditioning. Princeton University Press, Princeton and Oxford

Barker RG (1968) Ecological psychology: concepts and methods for studying human behavior. Stanford University Press, Stanford, CA

Barnes C (2011) Understanding disability and the importance of design for all. J Accessibil des All 1(1):55–80

Barnett J (1974) Urban design as public policy: practical methods for improving cities. Architectural Record Books, New York, NY

Barnett K (2023) Frank Lloyd Wright colors. The Decorologist. Accessed 1 Aug 2023

Barron FX (1969) Creative person, creative process. Holt, Reinhart & Winston, New York, NY

Batchelor P, Lewis D (eds) (1984) Urban design in action. School of Architecture, North Carolina State University, Charlotte, NC

Bay JH, Ong BL (eds) (2006) Tropical sustainable architecture: social and environmental dimensions. Elsevier, London

Bayer P (1999) Art deco architecture: design, decoration and detail from the twenties and thirties. Thames and Hudson, London. Originally published in 1992 by H. N. Abrams, New York

Bazjanac V (1974) Architectural design theory: models of the design process. In: Spillers WR (ed) Basic questions of design theory. North Holland, Amsterdam, pp 3–20

Beatley T, James CL, Rainey R (eds) (2018) Healthy environments, healing spaces: practices and directions in health, planning and design. University of Virginia Press, Charlottesville, VA
Beaton R (2021) The Greeks: a global history. Faber and Faber, London
Becher B, Becher H (1991) Pennsylvanische Kleinbergwerke. Schimmer-Mosel, München
Bednar J (ed) (1977) Barrier-free environments. Dowden, Hutchinson & Ross, Stroudsburg, PA
Behne A (1926) The modern functional building. Translated from the German by Robinson M. Republished, The Getty Research Institute for History of Arts and Humanities, Santa Monica, CA, 1996
Beltramini G (2006) Carlo Scarpa: architecture and design. In: Zannier I (ed) Rizzoli, New York, NY
Bengtsson A (1973) Adventure playgrounds. Crosby Lockwood Staples, London
Benjamin A (1988) Deconstruction and art/the art of deconstruction. In: Norris C, Benjamin A (eds) What is deconstruction? St Martin's Press, New York, NY, pp 13–56
Ben-Joseph E (2005) The code of the city: standards and the hidden language of placemaking. MIT Press, Cambridge, MA
Bergman D (2012) Sustainable design: a critical guide. Princeton Architectural Press, New York, NY
Berlyne DE (1970) Novelty, complexity, and hedonic value. Percept Psychophys 8:279–286
Betjeman J (1974) A pictorial history of English architecture. Penguin, Harmondsworth
Bhatia G (1994) Punjabi baroque and other memories of architecture. Penguin, New Delhi
Bhatia G (2003) Laurie Baker, life, work, writings. Penguin, New Delhi
Birren F (1965) Color psychology and color therapy: a factual study of the influence of color on human life. University Books, New Hyde Park, NY. Republished by Pickle Partners in 2016
Blake P (1977) Form follows fiasco. Little, Brown, Boston, MA
Blakenbehler B (2015) How Greek temples correct visual distortion. Correcting optical illusions in Greek temples. Search (bing.com). Accessed 28 May 2023
Blanciak F (2008) SITELESS: 1001 building forms. MIT Press, Cambridge, MA
Blau E, Troy NJ (eds) (1997) Architecture and cubism. MIT Press and Canadian Centre for Architecture, Cambridge MA
Boguslaw R (1965) The new utopians: a study of system design and social change. Prentice Hall, Englewood Cliffs, NJ
Bohl CC, Plater-Zyberk E (2006) Building community across the rural-to-urban transect [the transect]. Places 18(1). https://placesjournal.org/assets/legacy/pdfs/building-community-across-the-rural-to-urban-transect.pdf. Accessed 5 May 2019
Booker L (1966) Teaching engineering design. Institute of Engineering Designers, London
Borngässer B, Bednorz A (2008) History of architecture from classical to contemporary. In: Toman R (ed) Translated from the German by Basedow M, Dobrian M, Marianiello L, Schreiber S. Parragon, Bath
Borràs È (2024) 'We don't need air con': how Burkina Faso builds schools that stay cool in 40°C heat. The Guardian, 29 Feb 2024. Architecture. Accessed 2 Mar 2024
Boudon P (1972) Lived in architecture: Le Corbusier's Pessac. Revisited translated from the French by Onn G. MIT Press, Cambridge, MA
Bourdieu P (1984) Distinction: a social critique of the judgement of taste translated from the French la distinction: critique sociale du jugement by Nice R. Harvard University Press, Cambridge, MA
Bradbury D (2020) Richard Seifert: British brutalist architect. Lund Humphries, London
Breeze C (2003) American art deco: modernist architecture and regionalism. W. W. Norton, New York, NY
Bressani M (2014) Architecture and the historical imagination: Eugène Viollet-le-Duc, 1814–1879. Routledge, London
Brewer DD (2020) Essentials of scientific research: a practical guide. Evidence Guides, Seattle, WA

Britton J (1807) The architectural antiquities of Great Britain. Longman, Hurst, Rees, and Orme, London. https://archive.org/details/BrittonJTheArchitecturalAntiquitiesOfGreatBritainVol11807/page/n15/mode/2up. Accessed 19 Feb 2024

Broadbent G, Bunt R, Jencks C (1980) Signs, symbols, and architecture. Wiley, New York, NY

Brolin B (1976) The failure of modern architecture. Van Nostrand Reinhold, New York, NY

Bromwich J (2006, originally 1996) Roman remains of Southern France: a guidebook. Routledge, London and New York

Brooks HA (1997) Le Corbusier's formative years: Charles-Édouard Jeanneret at La Chaux-de-Fonds. University of Chicago Press, Chicago, IL

Brown J (1996) Lutyens and the Edwardians: an English architect and his clients. Viking, New York, NY

Bubbar DK (2005) The spirit of Indian architecture: vedantic wisdom of architecture for building harmonious spaces and life. Rupa & Co., New Delhi

Buggy G (2012/13) Pierre Jeanneret under the solar clock: the role of Pierre Jeanneret in the design and construction of Chandigarh, 1951–65. Academia.edu. Accessed 23 Aug 2022

Bunszel A (2024) How architect add AI to their digital toolbox. ArchDaily, 29 May 2024. Accessed 29 May 2024

Burnette CH (2009) Intentionality and design. In: Durling D, Shackleton J (eds) Originally in common ground: design research society international conference 2002. Academia.edu. Accessed 12 July 2024

Burton E, Mitchell L (2006) Inclusive urban design: streets for life. Routledge, London

Butina-Watson G, Bentley I (2007) Identity by design. Routledge, London

Çalişkan O (2012) Design thinking in urbanism: learning from the designers. Urban des Int 17(4):272–296

Çalişkan O (2017) Parametric design in urbanism: a critical reflection. Plan Pract Res 32(4):417–443

Campbell C (1715) Vitruvius Britannicus. The Author, London. Partially available at https://archive.org/details/gri_33125008447589/page/n33/mode/2up. Accessed 2 Mar 2024

Canter D (1970) The need for a theory of function in building. Archit J 151(5):299–302. Republished in Readings in the psychology of space: selected works of David Canter published by Routledge in 2023

Carson R (1962) Silent spring. Houghton Mifflin, Boston. http://library.uniteddiversity.coop/More_Books_and_Reports/Silent_Spring-Rachel_Carson-1962.pdf. Accessed 3 June 2019

Carthian JP (1979) The Ecole des Beaux arts: modes and manners. J Art Educ 33(2):7–17

Chakrabarti V (1998) Indian architectural theory: contemporary uses of Vastu Vidya. Curzon Press, Richmond. Vibhuti Sachdev. Academia.edu. Accessed 4 Feb 2023

Chandavarkar P (2022) Latitude as identity. In: Musings on architecture and urbanism. wordpress.com. Accessed 28 Jan 2023

Charitonidou M (2020) Simultaneously space and event: Bernard Tschumi's conception of architecture. ARENA J Archit Res 5(1):5. arena-architecture.eu. Accessed 24 Nov 2022

Charles V, Carl KH (2010). Rococo. Parkstone International, New York, NY

Charter of Machu Pichu (1979) J Archit Res 7(2):5–9

Chatel M (2019) Spotlight: Alejandro Aravena. Archdaily, 22 June 2019. Accessed 24 July 2022

Chelkoff G (2003) For an ecological approach to architecture: perception and design. Academia.edu. Accessed 19 Feb 2024

Cheney E (1998) The evolution of the grand tour: Anglo-Italian cultural relations since the renaissance. Cass, London

Cherry E (1998) Programming for design: from theory to practice. Wiley, New York, NY

Cherry K (2022) Maslow's hierarchy of needs. verywellmind.com. Accessed 8 Sept 2022

Cherry K (2023a) What are Jungian archetypes? Verywellmind (11 March). 12 Archetypes: definition, theory, and types. verywellmind.com. Accessed 18 July 2023

Cherry K (2023b) How habituation in psychology works and affects relationships. Verywellmind, 23 Dec 2023. https://www.bing.com/search?q=Cherry+of+habituation+levels&FORM=AWRE. Accessed 17 July 2024

Cherry E, Petronis J (2016) Architectural programming. WBDG—whole building design guide. Accessed 19 May 2022
Chisholm H (1911) Apollodorus. In: Encyclopedia Britannica, 2, 11th edn. Cambridge University Press, Cambridge
Churchman CW (1967) Guest editorial: wicked problems. Manage Sci 14(4):141–142. wordpress.com. Accessed 28 Sept 2023
Clarke PW (2004) The economic currency of architectural aesthetics. In: Cuthbert A (ed) Designing cities: critical readings in urban design. Blackwell, Oxford, pp 28–45
Cohen SE (2015) Inventing the new American house: Howard Van Doren Shaw, architect. Monacelli Press, New York, NY
Coldstream N (2003) The architect, history and architectural history. Trans R Hist Soc 13:219–226
Coles RT (2016) Architecture + advocacy. Compiled and edited by Siener WH, Coles S. Buffalo Arts Publishing, Buffalo, NY
Colquhoun A (1967) Typology and design method. ARENA J Archit Assoc 83(913):11–14
Colquhoun I (2003) Design out crime: creating safe and sustainable communities. Routledge, London and New York
Congress for the New Urbanism (2013) In: Talen E (ed) The charter of the new urbanism. McGraw Hill, New York. Available at https://www.cnu.org/who-we-are/charter-new-urbanism. Accessed 2 Mar 2024
Conn S (2014) Americans against the city: anti-urbanism in the twentieth century. Oxford University Press, Oxford
Conrads U (ed) (1970) Programs and manifestoes on twentieth century architecture. Translated from the German by Bullock M. MIT Press, Cambridge, MA. Conrads_Ulrich_ed_Programs_and_Manifestoes_on_20th-Century_Architecture_1970.pdf. monoskop.org. Accessed 9 June 2023
Conway H, Roenisch R (2005) Understanding architecture: an introduction to architecture and architectural history. Psychology Press, London. Aseel Rababah. Academia.edu. Accessed 24 Jan 2023
Cooper C (1974) House as image of self. In: Lang J, Burnette C, Moleski W, Vachon D (eds) Designing for human behavior: architecture and the behavioral sciences. Dowden, Hutchinson & Ross, Stroudsburg, PA, pp 130–146
Cooper TE (1996) Mecenatismo or clientelismo? The character of renaissance patronage. In: Wilkins DG, Wilkins RL (eds) The search for a patron in the middle ages and the renaissance. Edwin Mellen, Lewiston, NY, pp 19–32
Coyne R (2005) Wicked problems revisited. Des Stud 26(1):5–17. ScienceDirect. Accessed 17 May 2023
Cozens P, Love T (2015) A review and current status of crime prevention through urban design (CPTED). J Plan Lit 30(4):393–412. https://doi.org/10.1177/0885412215595440. Accessed 25 Feb 2023
Crandall CS, Silver PJ, N'Gbala AN, Tsang J-A, Dason K (2007) Balance theory. Unit relations, and attributions: the underlying integrity of Heiderian theory. Rev Gen Psychol 11(1):12–30. researchgate.net. Accessed 15 Feb 2019
Craven J (2019) Mies van der Rohe gets sued: the battle with Farnsworth. In: Mies van der Rohe and the battle with Farnsworth. ThoughtCo, 4 Nov 2019. thoughtco.com. Accessed 18 May 2023
Crawford M (1995) Building the workingman's paradise: the design of American company towns. Verso, London and New York
Cremer-Petersen CL, Ahmed-Kristensen S (2016) Argumentation and reasoning in design: an empirical analysis of the effects of verbal reasoning on idea value in group idea generation. In: International design conference—design 2016. Search (bing.com). Accessed 23 July 2023
Critchlow K (1976) Islamic patterns: an analytical and cosmological approach. Schocken, New York, NY. Republished in 1999 by Inner Traditions
Croft W (2002) Typology and universals. Cambridge University Press, Cambridge

Crosbie MJ, Watson D (eds) (2005) Time-saver standards for architectural design: technical data for professional practice, 8th edn. McGraw-Hill, New York, NY

Cross N (1984) Developments in design methodology. Wiley, New York, NY

Cuff D (1992) Architecture: the story of practice. MIT Press, Cambridge, MA

Cuff D (2014) Architecture's undisciplined urban desire. Archit Theory Rev 19(1):92–97. https://www.academia.edu/106215798/Architectures_Undisciplined_Urban_Desire. Accessed 20 Feb 2024

Cullen G (1961) Townscape. Architectural Press, London

Cuthbert AR (2011) Understanding cities: method in urban design. Routledge, London and New York

Cuthbert AR (2012) Sustainable theory: sustainable form—the future of urban design. In: Haas T (ed) Sustainable urbanism and beyond: rethinking cities for the future. Rizzoli, New York, NY, pp 180–183

Dagit CE Jr. (2017) Louis I. Kahn—architect: remembering the man and those who surrounded him. Routledge, New York, NY

Dahinden J (1972) Urban structures for the future. Translated from Stadtstrukturen für Morgen by Onn G. Praeger, New York, NY

Davey P (1980) Arts and crafts architecture: the search for earthly paradise. Architectural Press, London

Davidoff P (1965) Advocacy and pluralism in planning. J Am Inst Plann 31(4):331–338

Davies S (1994) Is architecture art? In: Mitrovic B (ed) Philosophy and architecture. Rodopi, Amsterdam and Atlanta, pp 31–47. https://www.academia.edu/15110875/Is_Architecture_Art. Accessed 20 Feb 2024

Davis R (1989) Seaside, Florida. Gahomes 27:96–133

de Bono E (1992) Serious creativity: using the power of lateral thinking to create new ideas. HarperBusiness, New York, NY

de Bono E (2015) Serious creativity: how to be creative under pressure and turn ideas into action, Reprint. Random House, New York, NY

De Chiara J, Callendar JH (2020) Time saver standards for building types, 2nd edn. McGraw Hill, New York, NY

De Grandis L (1986) Theory and use of colour. Translated from the Italian Teoria e uso del Colore by Gilbert J. Blandford Press, Poole, Dorset

de Sola Morales I (1989) Weak architecture. Ottagono 92(Summer):87–129. Architecture theory since 1968. cornell.edu. Accessed 20 Feb 2024

Department of the Army (1976) Design guidelines: recreation centers. The Authors, Washington, DC

Desai M (2017) Women architects and modernism in India: narratives and contemporary practices. Routledge, London and New York

Desai M, Desai M, Lang J (2012) The bungalow in twentieth century India: the cultural expression of changing ways of life and aspirations in the domestic architecture of colonial and post-colonial society. Ashgate, Farnham

Descartes R (1637) Discourse on the method of rightly conducting one's reason and of seeking truth in the sciences. Translated from the French by Woolaston A. Penguin, Harmondsworth, 1960. https://ia801906.us.archive.org/17/items/rmcg001/Descarte-Discourse1.pdf. Accessed 20 Feb 2023

Diaconu M (2006) Reflections on an aesthetics of touch, smell and taste. Contemp Aesth 4. http://hdl.handle.net/2027/spo.7523862.0004.008. Accessed 9 June 2023

Dickinson D (2023) Architecture always reflects the values of its current culture. ArchDaily, 18 Aug 2023. Accessed 19 Aug 2023

Domański B, Gwosdz K (2010) Multiplier effects in local and regional development. Quæst Geogr 29(2):27–37. Multiplier_Effects_in_Local_and_Regional_Developme.pdf. Accessed 18 July 2023

Dowling EM (2004) New classicism: the rebirth of traditional architecture. Rizzoli, New York, NY

Drexler A (ed) (1977) The architecture of the École des Beaux arts. Museum of Modern Art, New York, NY

Dreyfuss H (2003, originally 1955) Designing for people. Allworth Press, New York, NY

Duany A (2021) A general agreement for the practice of contemporary architecture. Common Edge. ArchDaily. Accessed 21 Nov 2022. Originally published in Common Edge in 1991

Duany A, Talen E (2002) Transect planning. J Am Plann Assoc 68(3):245–266. https://urbandesign.uchicago.edu/wp-content/uploads/2019/01/TransectPlanning-min-min.pdf. Accessed 13 Feb 2024

Duany A, Plater-Zyberk E, Speck J (2000) Suburban nation: the rise of sprawl and the decline of the American dream. North Point Press, Berkeley, CA

Duffy F, Hutton L (1998) Architectural knowledge: the idea of a profession. Spon, London and New York. Republished by CRC Press in 2017

Duran R (2017) Crystal city: Frank Lloyd Wright's long-lost D.C. masterpiece. Curbed, 5 Oct. https://dc.curbed.com/2017/10/5/16416038/crystal-city-frank-lloyd-wright-roy-thurman. Accessed 10 Jan 2019

Durand J-N-L (1809) Précis des Leçons de Architecture données à l'École Royale Poltechnique. Chez l'Auteur et Barnard, Paris. https://archive.org/details/prcisdesleon02dura/page/n7/mode/2up. Accessed 3 Mar 2024

Eastman C (1970) On the analysis of the intuitive design process. In: Moore GT (ed) Emerging methods of environmental design and planning. MIT Press, Cambridge, MA, pp 21–37

Eaton L (1969) Two Chicago architects and their clients: Frank Lloyd Wright and Howard van Doren Shaw. MIT Press, Cambridge, MA

El Sayed Foundation (2009) Abdel-Wahed El-Wakil or the triumph of the Islamic style. Muslim Heritage. https://muslimheritage.com/abdel-wahed-wakil-architect. Accessed 23 Aug 2022

Elemental (2008) Quinta/ELEMENTAL. ArchDaily, 31 Dec 2008. Quinta Monroy/ELEMENTAL | ArchDaily. Accessed 12 Aug 2023

Elemental (2013) Villa Verde housing. ArchDaily, 13 Nov 2013. https://www.archdaily.com/447381/villa-verde-housing-elemental. Accessed 12 Aug 2023

Elliot AJ, Maier MA (2013) Color psychology: effects of perceiving color on psychological functioning in humans. Annu Rev Psychol 65:95–100. https://doi.org/10.1146/annurev-psych-010213-115035. Accessed 1 July 2023

Ellis R, Cuff D (eds) (1989) Architect's people. Oxford University Press, New York, NY

Ellis RL, Heath DD, Spalding J (eds) (1996) The collected works of Francis Bacon. Routledge, London and New York

El-Wakil A-W (2015). The benefits of traditional construction in contemporary architecture. A talk at the University of Miami. http://archidatum.com/articles/abdel-wahed-el-wakil-the-benefits-of-traditional-construction-in-contemporary-architecture/. Accessed 23 Aug 2022

Emerson M (2018) Greek sanctuaries & temple architecture: an introduction, 2nd edn. Bloomsbury, London

Emmitt S (2018) Barry's introduction to construction of buildings, 4th edn. Wiley-Blackwell, Hoboken, NJ

Etlin R (1987) A paradoxical avant-garde: Le Corbusier's villas of the 1920s. Archit Rev 181:21–32. architectural-review.com. Accessed 2 June 2024

Evans F (2023) How the Colosseum was built—and why it was an architectural marvel. History. Accessed 11 Aug 2023

Fairs M (2015) Zaha Hadid is "resentful and wronged" says Stephen Bailey. Dezeen, 5 Oct 2015. dezeen.com. Accessed 16 Aug 2023

Fathy H (1969) Architecture for the poor: an experiment in rural Egypt. University of Chicago Press, Chicago, IL. Republished in 2010

Faucheu J, Weiland B, Juganaru-Mathieu M, Witt A, Cornuault P-H (2019) Tactile aesthetics: textures that we like or hate to touch. Acta Psychol 201. PubMed (nih.gov). Accessed 12 Sept 2023

Ferris H (1929) The metropolis of tomorrow. Ives Washburn, New York, NY. https://archive.org/details/mettomo00ferr. Accessed 28 Apr 2017

Festinger L (1962) A theory of cognitive dissonance. Stanford University Press, Stanford, CA

Filippi C, Hooper H (2019) The architect/the advocate: the role of advocacy and influence in architecture and design. Architecture Now. Accessed 10 Sept 2022

Finkel K (2020) The rise and fall of Southwark. The PhillyHistory Blog. Accessed 21 July 2023

Fischer G (2004) Reflective practitioners and unselfconscious cultures of design. In: CHI'2004 workshop. Fischer.pdf (uci.edu). Accessed 19 June 2023

Fisher T (2008) Public-interest architecture: a needed and inevitable change. In: Wakeford K, Bell B (eds) Expanding architecture: design as activism. Metropolis Books, New York. Berkeley prize essay competition. Accessed 22 June 2023

Fletcher SB (1954) A history of architecture on the comparative method, Sixteenth. Batsford, London

Flinchum RA, Meyer RO (2022) Henry Dreyfuss: designing for people. State University of New York Press, Albany, NY

Florian M-C (2023) The curb cut effect: how accessible architecture is benefiting everybody. ArchDaily, 17 Aug 2023. Accessed 18 Aug 2023

Frai M (2022) The design first approach: building products with people at the center. Greylock. https://greylock.com/archive/the-design-first-approach/. Accessed 18 May 2023

Fraker H (2007) Where is the urban design discourse? Places 19(34):61. https://www.academia.edu/80623485/Where_is_the_Urban_Design_Discourse. Accessed 10 July 2010

Frampton K (2020) Modern architecture: a critical history, 5th edn. Thames and Hudson, London

Francescato G (1994) Types and the possibility of an architectural scholarship. In: Franck KA, Schneekloth LH (eds) Ordering space: types in architecture and design. Van Nostrand Reinhold, New York, NY, pp 253–270

Francis-Jones R (2022) Truth and lies in architecture. ORO Editions, Novato, CA

Franck KA, Schneekloth LH (eds) (1994) Ordering space: types in architecture and design. Van Nostrand Reinhold, New York, NY

Frankl P (2000) Gothic architecture. Revised by Crossley P. Yale University Press, New Haven, CT

Frearson A (2016) Patrik Schumacher calls for social housing and public space to be scrapped. Dezeen, 18 Nov 2016. https://www.dezeen.com/2016/11/18/patrik-schumacher-social-housing-public-space-scrapped-london-world-architecture-festival-2016/. Accessed 25 Feb 2019

Gallace A, Spence C (2011) Tactile aesthetics: towards a definition of its characteristics and neural correlates. Soc Semiot 21(4):569–589. Academia.edu. Accessed 26 Jan 2024

Gallo T, Fidino M (2018) Biodiversity: making wildlife welcome in urban areas. Insight | eLife. elifesciences.org. Accessed 19 Nov 2023

Gans H (1962) The urban villagers: group and class in the life of Italian-Americans. The Free Press, New York, NY

Gans H (1999, originally 1975) Popular culture and high culture: an analysis and evaluation of taste. Basic Books, New York, NY

Garnier T (1917) Une Cité Industrielle: The Cité Industrielle. Massin, Paris, 1930. Translated from the French by Ellis A. Reprint edited by Marianni R. Rizzoli, New York, NY, 1990

Gattupalli A (2023) How culture shapes kitchens: exploring influences and design differences between Japan and the United States. ArchDaily, 24 May 2023. Accessed 25 May 2023

Geddes P (1904) City development: a study of parks, gardens, and culture-institutes; a report to the Carnegie Dunfermline Trust. Geddes & Co., Edinburgh

Geddes P (1915) Cities in evolution: an introduction to the town planning movement and to the study of civics. Williams, London

Ghirardo DY (1996) Architecture after modernism. Thames and Hudson, New York, NY

Ghisleni C (2023) Types of urban blocks: different ways of occupying the city. Translated by Duduch T. ArchDaily, 29 July 2023. Accessed 30 July 2023

Gibson JJ (1979) The ecological approach to visual perception. Houghton Mifflin, Boston, MA. Republished by Psychology Press in 2014

Gibson D (2009) The wayfinding handbook: information design for public places. Princeton Architectural Press, New York, NY

Giedion S (1963, originally 1941) Space, time and architecture: the growth of a new tradition, 5th edn. Harvard University Press, Cambridge, MA
Givoni B (1998) Climate considerations in building and urban design. Wiley, New York, NY
Glassey H (1968) Patterns in the material culture of the eastern United States. University of Pennsylvania Press, Philadelphia, PA
Glendinning M (2010) Architecture's evil empire? The triumph and tragedy of global modernism. Reaktion Books, London
Glendinning M (2013) The conservation movement: a history of architectural preservation: antiquity to modernity. Routledge, London and New York
Goffman E (1961) Asylums: essays on the social situations of mental patients and other inmates. Anchor, Garden City, NY. Published online by Cambridge University Press in 2018
Golany GS (1983) Design for arid regions. Van Nostrand Reinhold, New York, NY
Gold JR (2019) The Athens charter (C.I.A.M.) 1933. In: Orum AM (ed) The Wiley Blackwell encyclopedia of urban and region studies. Wiley-Blackwell, Chichester. https://www.academia.edu/34353829/The_Athens_Charter_CIAM_1933. Accessed 15 Jan 2024
Goldsmith S (1976) Designing for the disabled. Royal Institute of British Architects, London
González de Canales F, Ray N (2015) Rafael Moneo: building teaching writing. Yale University Press, New Haven, CT
Gordon DLA (1997) Politics and planning on the New York waterfront. Routledge and Gordon & Breach, London
Grant J (2006) Planning the good community: new urbanism in theory and practice. Routledge, London
Graves CP (2020) The use of typologies in designing architecture and urban design. Looking@Cities, 23 Oct 2020. lookingatcities.info. Accessed 8 Nov 2022
Gray J (2022) A powerful brand: Marion Mahony's original form of graphic representation. The Whirling Arrow, 14 Feb 2022. Frank Lloyd Wright Foundation. Accessed 29 May 2023
Green VA (1999) The architecture of Howard Van Doren Shaw. Chicago Review Press, Chicago, IL
Green J (2016) Which way to a better future for Baltimore? Huffpost, 6 Apr 2016. Accessed 21 Feb 2017
Greenberg M (1995) Poetics of cities: designing neighborhoods that work. Ohio State University Press, Columbus, OH
Greenbie B (1976) Designing for diversity. Elsevier, Amsterdam
Groat L, Canter D (1979) A study of meaning: does post-modern architecture communicate? Prog Arch 60(12):84–87
Gropius W (1955, originally 1953) Scope of total architecture. Harper and Row, New York, NY and Evanston, IL. monoskop.org. Accessed 12 Feb 2023
Güney Y (2007) Type and typology in architectural discourse. Mimarlık kuramlarında tip ve tipoloji kavramları. DocShare.tips. Accessed 2 Nov 2022
Gupta R, Mehrotra R, Shetty P (2011) Architecture and contemporary Indian identity (75) 1. Aca demia.edu. Accessed 12 Sept 2023
Gutman R (1988) Architectural practice: a critical review. Princeton University Press, Princeton, NJ
Guy S, Henneberry J (eds) (2002) Development and developers: perspectives on property. Wiley-Blackwell, Hoboken, NJ
Habrakan NJ (2021, originally 1962) Supports: an alternative to mass housing. Routledge, New York and London
Hackett SP (2011) The architecture of liturgy: liturgical ordering in church design: the Australian experience in perspective. Unpublished doctoral dissertation, University of New South Wales, Sydney
Hadid Z (1982) Randomness vs arbitrariness. AA Files Ann Archit Assoc Sch Archit 2. https://www.scribd.com/document/476575250/Hadid-Randomness-Arbitrariness. Accessed 30 Aug 2022

Hadid Z (1983) The world (89 degrees). Planetary architecture two. Architectural Association, London. https://digitalcommons.risd.edu/specialcollections_books_architecture/2/. Accessed 30 Aug 2022

Haider H (2011) Calligraphic architecture: stroke to form, space and surface. Fields J Huddersfield Stud Res 7(1). fieldsjournal.org.uk. Accessed 12 Feb 2024

Hakim BS (1986) Arabic-Islamic cities: building and planning principles. Kegan Paul International, London

Hall ET (1969) The hidden dimension. Anchor, New York, NY. https://www.academia.edu/43785083/The_Hidden_Dimension_Edward_Hall. Accessed 26 Feb 2023

Halpin DW, Senior BA, Lucko G (2017) Construction management, 5th edn. Wiley, Hoboken, NJ

Hamm C (ed) (2015) Evolution of lightweight structures: analyses and technical applications. Springer, Dordrecht

Haneman JT (1923) A manual of architectural composition. Architectural Book Publishing, New York. Republished by Dover in 2011

Hansell MH (2004) Animal architecture. Oxford University Press, Oxford

Hansen CT, Dorst K, Andreasen MM (2009) Problem formulation as a discursive design activity. In: International conference on engineering design. ICED'09. Stanford University. https://opus.lib.uts.edu.au/bitstream/10453/11338/1/2009003490.pdf. Accessed 3 Nov 2022

Hardwick CS, Cook J (eds) (1977) Semiotics and significs: the correspondence between Charles S. Peirce and Victoria Lady Welby. Indiana University Press, Bloomington and London

Harrell DF (2023) AI can help shape society for the better—but humans and machines must work together. The Guardian, 18 Aug 2023. Accessed 19 Aug 2023

Harrison EM (2019) Richard Seifert (1910–2001). Archit Rev, 17 Sept 2019. architectural-review.com. Accessed 19 Aug 2022

Hartley LP (1953) The go-between. Hamish Hamilton, London

Hartman JC (ed) (2022) The women who changed architecture. Princeton Architectural Press, New York, NY

Harvey JH (1971) The master builders: architecture in the middle ages. McGraw Hill, New York, NY

Hascher R, Arnold T, Jeska S, Klauck B (eds) (2002) Office buildings: a design manual. Princeton Architectural Press, New York, NY

Hatch CR (ed) (1984) The scope of social architecture. Van Nostrand Reinhold, New York, NY

Hayden D (1980) What would a non-sexist city look like? Speculations on housing, urban design, and human work. Signs J Women Cult Soc 5(3):S170–S187

Hayden D (1981) The grand domestic revolution: a history of feminist design for American homes, neighborhoods, and cities. MIT Press, Cambridge, MA. https://monoskop.org/images/a/a7/Hayden_Dolores_The_Grand_Domestic_Revolution_A_History_of_Feminist_Designs_for_American_Homes_Neighborhoods_and_Cities_1981.pdf. Accessed 5 June 2019

Hays KM (ed) (1998) Architecture/theory/since 1968. MIT Press, Cambridge, MA

Heathcote E (2021) Adolf Loos's spaces of intimacy, darkness and mystery. Financial Times, 8 Jan 2021. ft.com. Accessed 12 Feb 2023

Hebron M (2018) Patronage in the renaissance. In: Sgarbi M (ed) Encyclopedia of renaissance philosophy. Springer, Cham. https://doi.org/10.1007/978-3-319-02848-4_1150-1. Accessed 2 July 2023

Heft H (1996) The ecological approach to navigation. A Gibsonian approach. In: Portugal J (ed) The construction of cognitive maps. Kluwer Academic, Dordrecht, pp 105–132

Hegemann W, Peets E (1922) The American Vitruvius: an architects' handbook of civic art. The Architectural Book Publishing Co., New York, NY. Republished in an amended version by Dover in 2010

Heider F (1958) The psychology of interpersonal relationships. Wiley, New York, NY

Held J (2014) Deskilling and reskilling. The business of architecture. ACA—Association of Consulting Architects Australia. Accessed 16 Feb 2022

Hensley J, Kitch B (2023) What is co-design: a primer on participatory design. Mural, 1 Feb 2023. https://www.mural.co/blog/co-design-method. Accessed 16 June 2023

Hernández A (1969) J.N.L. Durand's architectural theory: a study in the history of rational building design. Perspecta 12:153–160
Hernández D (2012) ELEMENTAL: incremental housing and participatory design manual. Hatje Cantz, Berlin
Herring SR, Jones BR, Bailey B (2009) Idea generating techniques among creative professionals. In: Proceedings of the 42nd Hawaii international conference on system sciences. Idea generation techniques among creative professionals. illinois.edu. Accessed 24 July 2023
Heschong L (1979) Thermal delight in architecture. MIT Press, Cambridge, MA
Hewitt MA (2023) Away from old architecture: what Le Corbusier really meant. ArchDaily, 5 May 2023. https://www.archdaily.com/1000499/away-from-old-architecture-what-le-corbusier-really-meant. Accessed 5 Dec 2023
Heyda P (2019) What about typology: an update for late capitalism. J Archit Educ. https://www.academia.edu/41561143/What_About_Typology_An_Update_for_Late_Capitalism. Accessed 22 Feb 2022
Hilberseimer L (1927) Großstadt Arkitektur—metropolis architecture. Julius Hoffman, Stuttgart
Hill M, Werczberger E (1978) Goal programming and the goals-achievement matrix. Int Reg Sci Rev 3(2):2023. https://doi.org/10.1177/016001767800300205
Hillier B (1968) Art deco of the 20s and 30s. Schocken, New York, NY
Horine GN (2017) Project management: absolute beginner's guide, 4th edn. QUE, Indianapolis, IN
Horowitz H (1967) The program's the thing. Am Inst Archit J 47:94–100
Howard E (1902) Garden cities of tomorrow. Swan Sonnenschein, London
Hughes R (1980) The shock of the new: art and the century of change. British Broadcasting Corporation, London
Huitt WG (2004) Maslow's hierarchy of human needs. Educational psychology interactive. Valdosta State University, Valdosta, GA. http://chiron.valdosta.edu/whuitt/col/regsys'maslow.html. Accessed 10 Apr 2006
Huppatz DJ (2015) Revisiting Herbert Simon's "science of design". Des Issues 31(2):29–40. Design Methods | Design. scribd.com. Accessed 18 May 2023
Hutchings M, Jarvis P (2012) The relationship between practice, theory and research. In: Higgs J (ed) Practice-based education: perspectives and strategies. Sense Publishers, Rotterdam, pp 175–186. https://doi.org/10.1007/978-94-6209-128-3_13. Accessed 3 Sept 2022
Huxtable AL (2004) Frank Lloyd Wright. Lipper/Viking, New York, NY
Imani N, Zafarmandi S (2017) Origins of taste in architecture. Sci J NAZAR 14(53):33–42. www.bagh-sj.com/jufile?ar_sfile+1025272. Accessed 3 Mar 2023
Imrie R, Hall P (2001) Inclusive design: designing and developing accessible environments. Spon, London
Irving RG (1981) Indian summer: Lutyens, Baker and Imperial Delhi. Yale University Press, New Haven, CT
Itten J (1964) Design and form: the basic design course at the Bauhaus. Translated from Mein Vorkurs am Bauhaus. Gestaltungs—und Formenlehre by Maass J. Reinhold, New York, NY
Izumi K (1968) Some psycho-social considerations of environmental design. mimeographed
Jacobs J (1961) The death and life of great American cities. Random House, New York, NY
Jacobs A, Appleyard D (1987) Toward an urban design manifesto. J Am Inst Plann 53(1):112–120
Jager M (2007) Housing estates in the Berlin modern style. In: Haspel J, Jaeggi A (eds). Deutscher Kunstverlag, Berlin
Jagtap S (2019) Design and poverty: a review of contexts, roles of poor people, and methods. Res Eng Des 30(5):41–62. Accessed 31 July 2023
James J (1982) The master masons of Chartres. West Grinsted Publications, London. Reprinted by Boydell and Brewer in 1990
Jencks C (1977) The language of post-modern architecture. Academic Editions, London
Jencks C, Kropf K (eds) (2006) Theories and manifestoes of contemporary architecture, 2nd edn. Wiley, Hoboken, NJ. Originally published in 1997 by Wiley Academic
Jenkins F (1961) Architect and patron: a survey of professional relations and practice in England from the sixteenth century to the present day. Oxford University Press, London and New York

Jenkins S (2024) A short history of British architecture: from Stonehenge to the Shard. Penguin, Harmondsworth

Jenkyns R (2004) Westminster Abbey: a thousand years of national pageantry. Profile Books, London

Jodidio P (2018) Green architecture. Taschen, Köln

Jodidio P (2020) Zaha Hadid: complete works 1979–today. Taschen, Köln

Johnson P-A (1994) The theory of architecture: concepts, themes & practices. Van Nostrand Reinhold, New York, NY

Jones JC (1992, originally 1972) Design methods: the seeds of human futures. Wiley, New York, NY

Jung CG (1968, originally 1961) Man and his symbols. Dell, New York, NY

Kahneman D (2011) Thinking fast and slow. Farrar, Straus and Giroux, New York

Katz P (1994) The new urbanism: toward an architecture of community. McGraw Hill, New York, NY

Kepes G (1944) Language of vision. Paul Theobald, Chicago, IL

Kilroy-Ewbank L, Graham H (2023) Why commission artwork during the renaissance? Khan Academy. Accessed 7 Aug 2023

Kimmelman M (2016) Alejandro Aravena, the architect rebuilding a country. The New York Times Style Magazine, 23 May 2016. nytimes.com. Accessed 13 Aug 2023

Kirsch K (1990) The Weissenhofsiedlung: experimental housing built for the Deutscher Werkbund, Stuttgart, 1927. Rizzoli, New York, NY

Kitchin R, Freundschuh S (eds) (2018) Cognitive mapping: past, present and future. Routledge, London and New York

Klaus SL (2004) A modern Arcadia: Frederick Law Olmsted Jr. and the plan for forest hill gardens. University of Massachusetts Press, Amherst and Boston, MA

Klawitter J, Klesen F, Wolff A (2021) Algorithms for floor planning with proximity requirements. In: Gerber D, Pantazis E, Bogosian B, Nahmad A, Miltiadis C (eds) Computer-aided architectural design. Design imperatives: the future is now. CAAD futures 2021. Springer, Singapore, pp 151–171. https://doi.org/10.1007/978-981-19-1280-1_10. Accessed 24 Jan 2023

Knight C (2001) Frank Lloyd Wright. Parragon, Bath

Koberg D, Bagnall J (1981) The all new universal traveller: a soft system guide to creativity, problem-solving, and the process of reaching goals. W. Kaufmann, Los Altos, CA

Koenigsberger OH (1994) Personal interview. London

Korporaal G (2015) Making magic: the Marion Mahony Griffin story. BookPOD, Canberra, ACT

Krahn T (2019) Essential rammed earth construction: the complete step-by-step guide. New Society Publishers, Gabriola, BC

Krier R (1988) Architectural composition. Rizzoli, New York, NY

Krier R (2005, originally 1979) Urban space. Rizzoli, New York, NY. Originally published as Stadtraum. Translated from the German by Czechowski C, Black G. Academy Editions, London, 1976

Kristo S, Dhiamandi J (2016) Urban catalyst as the tool for public space transformation. In: Proceedings of the first international scientific conference on professional sciences. researchgate.net. Accessed 3 Mar 2023

Kroemer KHE, Grandjean E (1997) Fitting the task to the human: a textbook of occupational ergonomics. CRC Press, Boca Raton, FL

Krueger R, Casey MA (2014) Focus groups: a practical guide for applied research, 5th edn. Sage, Newbury Park, CA

Kruger C, Cross N (2006) Solution driven versus problem driven design: strategies and outcomes. Des Stud 27(5):527–548. SolutionVProblemStrats.pdf.open.ac.uk. Accessed 3 Nov 2022

Kuhn TS (1962) The structure of scientific revolutions. University of Chicago Press, Chicago, IL. https://archive.org/details/ThomasS.KuhnTheStructureOfScientificRevolutions/page/n3/mode/2up. Accessed 21 Feb 2024

Lai RT (1988) Law in urban design and planning: the invisible web. Van Nostrand Reinhold, New York, NY

Lam WMC (1986) Sunlighting as formgiver for architecture. Van Nostrand Reinhold, New York, NY

Lambert J, Pearson J (1974) Adventure playgrounds. Jonathan Cape, London

Lang J (1987) Creating architectural theory: the role of the behavioral sciences in environmental design. Van Nostrand Reinhold, New York, NY

Lang J (2017) Urban design: a typology of procedures and products, 2nd edn. Routledge, London and New York

Lang J (2021) The Routledge companion to twentieth and early twenty-first century urban design: a history of shifting manifestoes, paradigms, generic solutions and specific designs. Routledge, London and New York

Lang J (2024) Shifting urban design paradigms in the development of two New York projects: Battery Park City and Hudson Yards. Ekistics New Habit 84(1). Available at: https://ekisticsjournal.org/index.php/journal/issue/view/348. Accessed 20 Dec 2023

Lang J, Marshall N (2016) Urban squares as places, links and displays. Routledge, London and New York

Lang J, Moleski W (2010) Functionalism revisited: architectural theory and practice and the behavioral sciences. Ashgate, Farnham. Republished by Routledge in 2016

Lang J, Desai M, Desai M (2022) Architecture and independence: the search for identity—India 1880 to 1980, 2nd edn. CEPT University Press, Ahmedabad. Originally 1997. Oxford University Press, New Delhi

Latour A (ed) (1991) Louis I. Kahn: writings, lectures, interviews. Rizzoli, New York, NY

Laugier M-A (1755) An essay on architecture. Translated from Essai sur l'Architecture by Wolfgang and Hermann A. Hennessey & Ingalls, Los Angeles, CA, 1977

Lawrence PL (2016) Becoming Jane Jacobs. University of Pennsylvania Press, Philadelphia, PA

Lawson B (2005) How designers think: the design process demystified, 4th edn. Architectural Press, Oxford. https://www.academia.edu/29106712/Bryan_lawson_how_designers_think. Accessed 21 Feb 2024

Lawton MP (1989) Behavior-relevant ecological factors. In: Warner Schaie K, Schooler C (eds) Social structure and aging: psychological processes. Lawrence Erlbaum, Hillsdale, NJ, pp 57–78

Le Corbusier (1923) Vers une architecture. Translated from the French by Etchells F as Towards a new architecture. Praeger, New York, NY, 1970

Le Corbusier (1925a) Urbanisme. Translated from the French by Etchells F as The city of tomorrow and its planning. Architectural Press, London, 1927

Le Corbusier (1925b) L'arte décoratif d'aujourd'hui. Translated from the French by Dunnett JI as The decorative art of today. Architectural Press, London, 1987

Le Corbusier (1931) Polychromie Architecurale. Editions de l'Architecture d'Aujourdhui, Paris. Edited and republished by Birkhäuser in 2016

Le Corbusier (1933) Charte d'Athènes. Translated from the French by Eardley A as The Athens charter. Grossman, New York, 1973. https://portal.uur.cz/pdf/charter-of-athens-1933.pdf. Accessed 21 Feb 2024

Le Corbusier (1951) Le Modulor. Translated from the French by de Francia P, Bostock A as The modulor. MIT Press, Cambridge, MA, 1968

Le Corbusier (1960) My work with an introduction by Maurice Jardot. Translated from the French by Palmer J. Architectural Press, London

Le Corbusier (1967) The radiant city: elements of a doctrine of urbanism to be used as the basis of our machine age civilization. Translated from the French, La Ville Radieuse (1933) by Knight P, Levieux E, Coltman D. Orion, New York, NY

Le Corbusier's five points of architecture. Undated. http://www.arch2o.com/why-was-villa-savoye-hated-by-the-savoyes/. Accessed 6 Sept 2022

Leach N (2021) Architecture in the age of artificial intelligence: an introduction to AI for architects. Bloomsbury Publishing, London

Lee RM (2019) Unobtrusive methods. In: Liamputtong P (ed) Handbook of research methods in health social sciences. Springer, Singapore, pp 491–507. https://doi.org/10.1007/978-981-10-5251-4_85. Accessed 19 May 2022

Lefaivre L, Tzonis A (2003) Critical regionalism: architecture and identity in a globalized world. Prestel, Munich

Leighton AH (1959) My name is legion: foundations for a theory of man in relation to culture. Basic Books, New York, NY

Lennertz B, Lutzenhiser A (2014) The charrette handbook, 2nd edn. Routledge, London and New York

Levi D (1974) The Gestalt psychology of expression in architecture. In: Lang J, Burnette C, Moleski W, Vachon D (eds) Designing for human behavior: architecture and the behavioral sciences. Dowden, Hutchinson & Ross, Stroudsburg, PA, pp 111–119

Lewin K (1951) Field theory in social science: selected theoretical papers. In: Cartwright D (ed). Harper and Row, New York, NY

Lewis NC (1977) A procedural framework attempting to explore the relationship of human factors to the physical design process. Unpublished student paper, University of Pennsylvania

Lewis NC (2020) Design and order: perceptual experience of built form: principles in the planning and making of place. Wiley-Blackwell, Oxford

Lidwell W, Holden K, Butler J (2007) Universal principles of design. Essential tools for architects, artists, developers, engineers, inventors and makers. Rockport, Gloucester, MA

Lipman A (1969) The architectural belief system. Br J Sociol 20(2):190–204. Accessed 28 June 2023

Lippmann W (1955) Essays in public philosophy. Little Brown, Boston, MA

Livingstone K, Parry L (eds) (2005) International arts and crafts. Victoria & Albert Publishing, London

Lobos D, Donath D (2021) O problema da distribuição espacial na arquitetura: Pesquisa e reflexões. Arquitetura Rev 6(2):136–161. https://doi.org/10.4013/arq.2010.62.05. Accessed 9 June 2023

Lockard MS (2017) The nature of design: principles processes and the purview of architects. ORO Editions and Gordon de Vries Studio, Granville, PQ

Lombard J (2005) The architecture of Duany Plater-Zyberk and Company. Rizzoli, New York, NY

Longley R (2021) Gentrification: why is it a problem? ThoughtCo, 23 Apr 2021. thoughtco.com. Accessed 25 June 2023

Lorrimer L (2015) 10 challenges facing architecture and ways to turn them into opportunities. Archit Des, 25 Feb 2015. ACA series. architectureanddesign.com.au. Accessed 16 Feb 2023

Loth C (2012) The Tempietto, grandfather of domes: classical comments. ICCA, 24 Sept 2012. Institute of Classical Architecture & Art. classicist.org. Accessed 11 Aug 2023

Low SM (2003) Behind the gates, life, security and happiness in fortress America. Routledge, London and New York

Lubell S (2004) Libeskind's World Trade Center guidelines raise doubts. Archit Rec 192(6):47

Lutz A (2016) American malls are dying faster than you think—and it's about to get even worse. Business Insider Australia, 1 Sept 2016. https://finance.yahoo.com/news/american-malls-dying-faster-think-153026830.html. Accessed 18 Nov 2017

Lynch K (1960) The image of the city. MIT Press, Cambridge, MA

Lynch K (1972) What time is this place? MIT Press, Cambridge, MA

Lynes R (1983, originally 1949) The tastemakers: the shaping of American popular taste. Dover, New York, NY

MacKinnon DW (1962) The nature and nurture of creative talent. Am Psychol 17(7):484–495

Madanipour A (2003) Public and private spaces of the city. Routledge, London and New York

Makstutis G (2018) Design process in architecture: from concept to completion. Laurence King, London

Mann DA (1979) Architecture, aesthetics, and pluralism: theories of taste as a determinant of architectural standards. Stud Art Educ 20(3):15–29

Margolis L, Chaouni A (eds) (2015) Out of water: design solutions for arid regions. Birkhäuser, Basel. https://www.academia.edu/13282196/Out_of_Water_Design_Solutions_for_Arid_Regions. Accessed 21 Feb 2024
Marinovic GI, Baek J (2016) Lessons of incremental housing: two Chilean case studies: elemental Lo Espejo and Las Higuera. Archit Res 18(4):131–128. Lessons_of_Incremental_Housing_Two_Chilean_Case_St.pdf. Accessed 20 Feb 2024
Marmot AF (1981) The legacy of Le Corbusier and high-rise housing. Built Environ 7(2):82–95
Marshall S (ed) (2011) Urban coding and planning. Routledge, London and New York
Martienssen RD (1956) The idea of space in Greek architecture with special reference to the Doric temple and its setting. University of Witwatersrand Press, Johannesburg. Published online by Cambridge University Press in 2013
Martin JR (1977) Baroque. Harper and Row, New York, NY
Maslow A (1943) A theory of human motivation. Psychol Rev 50(4):370–396
Maslow AH (1987) Motivation and personality, 3rd edn. Revised by Frager R, Fadiman J, McReynolds C, Cox R. Harper and Row, New York, NY. https://www.academia.edu/1613921/Motivation_and_personality. Accessed 21 Feb 2024
Mayer A (1967) The urgent future; people, housing, city, region. McGraw Hill, New York, NY
McClure WR, Bartuska TJ (eds) (2007) The built environment: a collaborative inquiry into design and planning, 2nd edn. Wiley, Hoboken, NJ
McHarg IL (1969) Design with nature. Natural History Press, New York, NY
McManus J (2018) Harvard researchers detail the 9 factors that make a healthy building. ArchDaily, 28 June 2018. Accessed 11 Sept 2023
McNeill D (2008) The global architect. Firms, fame and urban form. Routledge, New York and London
Meagher BR, Marsh KL (2017) Seeking the safety of sociofugal space: environmental design preferences following social ostracism. J Exp Soc Psychol 68:192–199
Mele C (2000) Selling of the lower east side: culture, real estate, and resistance in New York City. University of Minnesota Press, Minneapolis, MN
Michell G (1977) The Hindu temple: an introduction to its meaning and forms. Paul Elek, London
Michelson W (1968) Most people don't want what architects want. Transactions 5(8):37–43
Middleton R (ed) (1982) The Beaux-arts and nineteenth-century French architecture. Thames and Hudson, London
Middleton R, Watkin D (1980) Neoclassical and 19th century architecture. Harry N. Abrams, New York, NY
Migayrou F (ed) (2014) Bernard Tschumi: architecture: concept & notation. Centre Pompidou, Paris
Mikellides B (1980) Architecture for people: explorations in a new humane environment. Holt, Rinehart and Winston, New York, NY
Mirza M (2016) The Egyptian architect who has designed Saudi Arabia's most famous mosques. Spirituality. https://destinationksa.com/the-egyptian-architect-who-has-designed-saudi-arabias-most-famous-mosques/. Accessed 14 Aug 2023
Mitchell JR, Stebenne DL (2007) New city upon a hill. A history of Columbia, Maryland. Arcadia, New York, NY
Mitchell L, Burton E, Raman S (2004) Dementia-friendly cities: designing intelligible neighbourhoods for life. J Urban des 9(1):89–101
Mitchell V, Ross T, Sims R, May A, Parker CJ (2015) Empirical investigations of the impact of co-design methods when generating travel solutions. CoDesign 12(4):205–222. https://doi.org/10.1080/15710882.2015.1091894. Accessed 16 Nov 2022
Mitzkus S (2022) Idea generation techniques: how to inspire new ideas that drive innovation? Digital Leadership. http://gdyuelve.com/index-74.html. Accessed 19 May 2023
Moneo R (1978) On typology. Oppositions 18:23–45. moneo_on-typology_oppositions.pdf. wordpress.com. Accessed 1 July 2023
Mönninger M (2010) Coop Himmelb(l)au: complete works: 1968–2010. In: Gossel P (ed). Taschen America, Los Angeles, CA

Moore CH (1905) Character of renaissance architecture. MacMillan, London and New York. Republished in 2019 by Wentworth Press, New York

Morris T (2019) A sublime example of organic architecture in Joshua Tree. Archit Digest, Sept 2019. Accessed 30 Aug 2023

Moser W (2015) How the conservatively radical architect Howard Van Doren Shaw changed the American house. Chicago Magazine, 24 Apr 2015. Accessed 17 Jan 2024

Moss-Eccardt J (1973) Ebenezer Howard. Bloomsbury, London

Mujica F (1929) History of the skyscraper. Archaeological and Architectural Press, Paris and New York. Reprinted by DaCapo Press in 1977

Mumford L (1954) Nowicki's work in India. Archit Rec 116(3):115–119

Nadal M, Vartanian O (eds) (2019) The Oxford handbook of empirical aesthetics. Oxford University Press, Oxford

Nasar JL (1988) Architectural symbolism: a study of house-style meanings. In: Lawrence D, Wasserman B (eds) Paths to co-existence. Proceedings of EDRA 19. EDRA, Riverside, CA, pp 163–172

Nesbitt K (ed) (1996) Theorizing a new agenda for architecture: an anthology of architectural theory 1965–1995. Princeton University Press, Princeton

Neuman R (2013) Baroque and Rococo art and architecture. Pearson Education, London. Partially available at (71). Academia.edu. Accessed 2 Mar 2024

Newman O (1972) Defensible space: crime prevention through urban design. MacMillan, New York, NY

Newman O (1975) Creating defensible space. US Department of Housing and Urban Development, Washington, DC

Nicolau H, Guerreiro T, Jorge JAP (2009) Design guidelines for blind people. https://www.researchgate.net/publications/235005410_Designing_Guides_for_Blind_People. Accessed 7 Oct 2022

Nicolin P (2020) Architettura in Quarantena. Skira, Milan

Niland J (2023) David Adjaye previews his Asaase III earthwork installation for COUNTERPUBLIC in St. Louis. Archinect, 2 June 2023. Accessed 28 July 2023

Norberg-Schulz C (1965) Intentions in architecture. MIT Press, Cambridge, MA

Norberg-Schulz C (1980) Genius loci: towards a phenomenology of architecture. Rizzoli, New York, NY

Olds K (2001) Globalization and urban change: capital, culture, and the Pacific Rim mega-projects. Oxford University Press, Oxford

Olgyay V (2015) Design with climate: bioclimatic approach to architectural regionalism. New and expanded version. Princeton University Press, Princeton, NJ

Oliver P (ed) (1997) Encyclopedia of vernacular architecture of the world. Cambridge University Press, Cambridge

Oliver P (2003) Dwellings: the vernacular house world wide. Phaidon, London

Olsen DJ (1986) The city as a work of art: London, Paris, Vienna. Yale University Press, New Haven and London

Ornstein SW (2022) Towards a pro-user architecture and city planning: potential chronologies. Acad Lett. Article 43635. https://www.academia.edu/67134660/TOWARDS_A_PRO_USER_ARCHITECTURE_AND_URBAN_PLANNING_Potential_Chronologies. Accessed 15 Feb 2024

Orwell G (1954) Politics and the English language. On shooting an elephant. Harcourt and Brace, New York, pp 77–92

Osmond H (1966) Some psychiatric aspects of design. In: Holland LB (ed) Who designs America? Doubleday, New York, NY, 281–318

Palladio A (1581) I quattro libri dell'architettura. Translated and edited by Hart V, Hicks P as Palladio's Rome, a translation of Palladio's two guidebooks to Rome. Yale University Press, New Haven, 2006

Pallasmaa J (2012, originally 1994) The eyes of the skin: architecture and the senses. Wiley, Chichester

Panero J, Zelnick M (1979) Human dimensions & interior space: a source book of design reference standards. Watson Guptill, New York, NY

Passini R, Rainville C, Marchand N, Joanette Y (1998) Wayfinding and dementia; some research findings, and a new look at design. J Archit Plan Res 15(2):133–151

Patti P (1782) Essai sur l'Architecture Théatrale. Chez Moutard, Paris. Republished by Kessinger in 2009

Pätynen J, Tervo S, Robinson PW, Lokki T (2014) Concert halls with strong lateral reflections enhance musical dynamics. Proc Natl Acad Sci PNAS. Accessed 6 Dec 2022

Pawley M (2001) Richard Seifert: workaholic architect whose tall towers changed the skyline of Britain and Europe. The Guardian, 29 Oct 2001. https://www.theguardian.com/news/2001/oct/29/guardianobituaries.arts. Accessed 20 Aug 2022

Peckham A, Schmiedeknecht T (eds) (2013) The rationalist reader: architecture and rationalism in Western Europe 1920–1940/1960–1990. Routledge, London and New York

Pelletier J (2019) Do architects advocate for the client or for the architect? Board and Vellum. https://www.boardandvellum.com/blog/do-architects-advocate-for-the-client-or-the-architect/. Accessed 22 Feb 2024

Pelsmakers S (2018) Changing practices: new roles. RIBA J, 15 Feb 2018. Accessed 22 June 2023

Peña WM, Parshall SA (2012) Problem seeking: an architectural programming primer, 5th edn. Wiley, Hoboken, NJ

Penner R, Adams L, Robson S (2013) Hotel design, planning and development, 2nd edn. Routledge, London and New York

Pérez Oyarzún F, Mori AA, Chala JQ (1999) Los Hechos de Arquitectura. Ediciones ARQ, Santiago

Pernice K (2013) Charettes (design sketching): ½ inspiration, ½ buy in. Design charrettes & team sketching: ½ inspiration, ½ buy-in (nngroup.com). Accessed 5 Feb 2023

Pernice R (ed) (2022) The urbanism of metabolism: visions, scenarios and models for the mutant city of tomorrow. Routledge, London and New York

Perry CA (1929) The neighborhood unit: a scheme of arrangement for the family-life community. In: Regional survey of New York and environs, vol VII, neighborhood and community planning. Reprinted. Routledge Thoemmes Press, New York, NY, 1998

Petty J (2018) Architect & developer: a guide to self-initiating projects. CreateSpace Independent Publishing Platform, New York, NY

Pocock D, Hudson R (1978) Images of the urban environment. Columbia University Press, New York, NY

Poletti R (2010) Lucien Kroll: utopia interrupted. Domus, 30 June 2010. domusweb.it. Accessed 22 June 2023

Polshek JS (2014) Build, memory. Monacelli, New York, NY

Portman J, Barnett J (1976) The architect as developer. McGraw, New York, NY

Powers A (ed) (2011) Robin Hood gardens. Twentieth Century Society, London. Poplar: 'an exemplar—a demonstration of a more enjoyable way of living'. Municipal Dreams. wordpress.com. Accessed 25 Mar 2024

Prak NL (1984) Architects: the noted and the ignored. Wiley, New York, NY

Preiser WFE (ed) (1977) Facilities programming: methods and applications. McGraw Hill, New York, NY

Preiser WFE, Rabinowitz HZ, White ET (2015) Post-occupancy evaluation. Routledge, London and New York. Originally published by Van Nostrand Reinhold in 1988

Pressman N (ed) (1995) Northern cityscapes: linking design to climate. Winter Cities Association, Yellowknife, NWT

Prina F (2011) The story of gothic architecture. Prestel, Munich

Prina F (2012) The story of Romanesque architecture. Prestel, Munich

Prisco J (2022) Everglades jetport: the 'world's greatest airport' that never was. CNN. Accessed 15 May 2023

Project Management Institute (2013) A guide to project management body of knowledge (PMBOK ® guide), 5th edn. The Authors, Newtown, PA

Provensal H (1904) L'art de demain: Vers l'Harmonie Intégrale. Perrin, Paris. Republished by Hachette Livre in 2017. Free download, borrow, and streaming: internet archive. Accessed 22 Feb 2024

Pugin AWN (1841) The true principles of pointed or Christian architecture. J. Weale, London

Ramani N (2007) Bombay art deco architecture: a visual journey, 1930–1953. Roli, New Delhi

Rand A (1968) The fountainhead. Bobbs-Merrill, Indianapolis, IN

Rapoport A (1969) House form and culture. Prentice Hall, Englewood Cliffs, NJ

Rapoport A (1990) History and precedent in environmental design. Plenum Press, New York and London

Rapoport A, Watson N (1972) Cultural variability in physical standards. In: Gutman R (ed) People and buildings. Basic Books, New York, NY, pp 33–53. Reprinted by Routledge in 2009

Rasmussen SE (1959) Experiencing architecture. MIT Press, Cambridge, MA

Ray N (2023) Thinking through twentieth century architecture. Routledge, London

Reid HH, Farber JC (1980) Palladio's architecture and its influence. Dover, Mineola, NY

Reisch KH (2007) The lost vanguard: Soviet architecture. 1922–32. Monticelli Press, New York, NY

Relph EC (1976) Place and placelessness. Pion, London. Republished in 2008 by Sage

Requena-Ruiz I (2014) Experiments on thermal comfort and modern architecture: the contribution of André Missenard and Le Corbusier. European Architectural History Network, Turin. https://www.academia.edu/7446374/Experiments_on_Thermal_Comfort_and_Modern_Architecture_The_Contributions_of_Andr%C3%A9_Missenard_and_Le_Corbusier?email_work_card=thumbnail. Accessed 10 Sept 2022

Richards JM (1962) An introduction to modern architecture. Penguin, Harmondsworth

Ricœur P (1977) The rule of metaphor: the creation of meaning in languages. University of Toronto Press, Toronto. Republished by Psychological Press in 2003

Ridley J (2003) Edwin Lutyens: his life, his wife, his work. Pimlico, London

Risselada M, van den Heuvel D (eds) (2005) Team 10: 1953–81: in search of a utopia of the present. NAi Uitgevers, Rotterdam

Rittel H (1971) Some principles for the design of an educational system for design. J Archit Educ 25(1–2):16–26

Rodgers N (2013) The architecture of ancient Rome: an illustrated guide to the glorious classical heritage of the Roman Empire. Lorenz, Dayton, OH

Rodley L (1994) Byzantine art and architecture: an introduction. Cambridge University Press, Cambridge

Roggema R (ed) (2013) The design charrette: ways to envision sustainable futures. Springer, Dordrecht

Rogoff B, Morelli GA, Chavajay P (2010) Children's integration in communities and segregation from people of differing ages. Perspect Psychol Sci 5(4):431–440

Roozenburg NFM, Dorst K (1998) Describing design as a reflective practice: observations on Schön's theory of practice. In: Frankenberger E, Badke-Schaub P, Birkhofer H (eds) Designers: the key to successful product development. Springer, London, pp 29–41. https://doi.org/10.1007/978-1-4471-1268-6_3. Accessed 14 Apr 2024

Rosenheim J (2022) Bernd & Hilla Becher. Metropolitan Museum of Art, New York, NY

Ross A (1999) The celebration chronicles: life, liberty and the pursuit of property values in Disney's new town. Ballantine Books, New York, NY

Rossi A (1982) The architecture of the city. Revised American edition. Translated from L'architettura della cìtta by Ghirardo D, Ockman J. MIT Press, Cambridge, MA

Rothschild J (ed) (1999) Design and feminism: re-visioning spaces, places and everyday things. Rutgers University Press, New Brunswick, NJ and London

Rowe C (1976) The mathematics of the ideal villa and other essays. MIT Press, Cambridge, MA. https://monoskop.org/images/9/96/Rowe_Colin_1947_1976_The_Mathematics_of_the_Ideal_Villa.pdf. Accessed 15 Jan 2023

Rowe C (1983) Program vs paradigm. Cornell J Archit 2:8–12. https://issuu.com/cornellaap/docs/cja002-opt. Accessed 25 Jan 2023

Rowe PG (1987) Design thinking. MIT Press, Cambridge, MA
Rowe C, Kotter F (1979) Collage city. MIT Press, Cambridge, MA. scribd.com. Accessed 31 Dec 2023
Rudofsky B (1964) Architecture without architects. A short introduction to non-pedigreed architecture. University of New Mexico Press, Albuquerque, NM
Rudofsky B (1969) Streets for people; a primer for Americans. Doubleday, Garden City, NY
Ruprecht LA (2021) Quatremère de Quincy's moral consideration on the place and purpose of works of art. Lexington, Lanham, MD
Ruskin J (1849) The seven lamps of architecture. Smith, Elder & Co., London. The project Gutenberg eBook. Accessed 25 Jan 2024
Ruskin J (1853) The stones of Venice. Volume II: the sea stories. Smith, Elder & Co., London. The project Gutenberg eBook. Accessed 25 Jan 2024
Sabastian SK, Shankar R, Al Qeisi S (2018) Design approach of Zaha Hadid form vocabularies and design techniques. JETIR 5(6). https://www.academia.edu/68080832/Design_Approach_of_Zaha_Hadid_Form_Vocabularies_and_Design_Techniques. Accessed 16 June 2023
Sachdev V (2011) Paradigms for design: the Vastu Vidya codes of India. In: Marshall S (ed) Urban coding and planning. Routledge, London and New York. https://www.academia.edu/8087629/Paradigms_for_Design_The_Vastu_Vidya_Codes_of_India. Accessed 4 Mar 2023
Sadler S (2005) Archigram: architecture without architecture. MIT Press, Cambridge, MA
Saint A (1983) The image of the architect. Yale University Press, New Haven, CT
Sakr YM (1987) The mosque between modernity and tradition: a study of recent designs of mosque architecture in the Muslim world. Unpublished master's thesis, Massachusetts Institute of Technology. The_mosque_between_modernity_and_traditi.pdf. Accessed 7 Jan 2024
Salama A (1995) New trends in architectural education: designing the design studio. Tailored Text and Potential Publishing, Raleigh, NC
Salingros N (2007) Architectural cannibalism in Athens. Orthodoxy Today, 20 Nov 2007. Architecture's New Scientific Foundations. architexturez.net. Accessed 16 Aug 2023
Samuel F (1999) A Profane annunciation: the representation of sexuality at Ronchamp. J Archit Educ 53(2):74–90
Santayana G (1896) The sense of beauty: being the outline of aesthetic theory. Reprinted. Dover, New York, NY, 1955. https://archive.org/details/sensebeautybein00santgoog/page/n12/mode/2up. Accessed 23 Feb 2024
Santayana G (1935) Why I am not a Marxists. Mod Mon 9:77–79
Savić S, Savićic G (eds) (2013) Unpleasant design: designing out unwanted behaviour. G.L.O.R.I.A., Belgrade. researchgate.net. Accessed 12 Sept 2023
Schaathun HG (2022) Where Schön and Simon agree: the rationality of design. Des Stud 79. ScienceDirect. Accessed 17 May 2023
Schielke T (2023) 10 typologies of daylighting: from expressive dynamic patterns to diffuse light. ArchDaily, 27 Mar 2023. Accessed 27 Mar 2023
Schiler M, Valmont E (undated) Microclimatic impact: glare around the Walt Disney concert hall. Academia.edu. Accessed 17 July 2023
Schön DA (1983) The reflective practitioner: how professionals think in action. Basic Books, New York, NY. Achmad Aldo. Academia.edu. Accessed 23 Feb 2024
School of Architecture, University of Notre Dame (2023) Abel-Wahed El-Wakil 2009 laureate. https://driehausprize.nd.edu/laureates/abdel-wahed-el-wakil/. Accessed 7 Dec 2023
Schumacher P (2008) Parametricism as style—parametricist manifesto. Presented and discussed at the dark side club, 11th architecture Biennale, Venice. http://www.patrikschumacher.com/Texts/Parametricism%20as%20Style.htm. Accessed 24 Feb 2017
Schumacher P (2013) Free-market urbanism—urbanism beyond planning. In: Verebes T (ed) Masterplanning the adaptive city—computational urbanism in the twenty-first century. Routledge, London and New York. https://www.patrikschumacher.com/Texts/Free%20Market%20Urbanism%20-%20Urbanism%20beyond%20Planning.html. Accessed 2 June 2019
Schumacher P (ed) (2016) Parametricism 2.0: rethinking architecture's agenda for the 21st century. Wiley, Hoboken, NJ

Schuyt K (2019) Scientific integrity: the rules of academic research. Translated from the Dutch by Gherman K. Leiden University Press, Leiden. https://scholarlypublications.universiteitleiden.nl/access/item%3A2732754/view. Accessed 14 Apr 2024

Scott GG (1857) Remarks on secular & domestic architecture, present and future. John Murray, London. https://books.google.com.au/books?id=N4IEAAAAYAAJ&redir_esc=y. Accessed 13 Dec 2022

Scott Brown D (2004) The redefinition of functionalism. In: Venturi R, Scott Brown D (eds) Architecture as signs and systems for a mannerist time. Belknap Press, Cambridge, MA, pp 142–174

Scruton R (2013) The aesthetics of architecture, 2nd edn. Princeton University Press, Princeton, NJ

Scully VJ Jr (1962) Louis I Khan. George Braziller, New York, NY

Seddon RC (2017) The changing role of the architect in the building process. Open Dissertation Press, Maidenhead

Seguin P (2014) Le Corbusier & Pierre Jeanneret: Chandigarh, India. Editions Galerie Seguin, Paris

Sen A (2000) East and west: the reach of reason. The New York Times Review of Books, 20 July 2000, pp 33–38

Serraino P (2016) The creative architect: inside the great midcentury personality study. Monacelli Press, New York, NY

Sert JL, C.I.A.M. (1944) Can our cities survive? An ABC of urban problems, their analysis, their solutions. Harvard University Press, Cambridge, MA

Shakya H (2023) Exploring different types of urban blocks with examples. Architectures Styles, 23 Jan 2023. https://architecturesstyle.com/types-of-urban-blocks/. Accessed 2 Jan 2024

Shandrow KL (2015) How the color of your office impacts productivity. Entrepreneur. Infographic. Accessed 13 Sep 2023

Shane G (1976) Theory vs practice. Archit Des. Recombinant urbanism. cargo.site. Accessed 24 Nov 2023

Sharp D (1978) The rationalists: theory and design in the modern movement. Architectural Press, London

Shaw FW (1926) Concerning Howard Shaw in his home. https://marktown.org/pdf/concerning_howard_van_doren_shaw.pdf. Accessed 15 Aug 2023

Sherwood R (1978) Modern housing prototypes. Harvard University Press, Cambridge, MA. https://archive.org/details/ModernHousingPrototypes. Accessed 23 Mar 2017

Short TL (2009) Peirce's theory of signs. Cambridge University Press, Cambridge

Sikora S (2018) Willey house stories part 5—the best of clients. The Whirling Arrow. Frank Lloyd Wright Foundation. Accessed 28 Aug 2023

Simon HA (1957) Models of man. Wiley, New York, NY

Simon HA (1969) The sciences of the artificial. MIT Press, Cambridge, MA

Singh M (2023) US buildings kill up to a billion birds a year: these architects want to save them. The Guardian, 28 Dec 2023. Accessed 29 Dec 2023

Sitte C (1889) Der Städtebau nach seinen Künstlerischen Grundsätzen. Translated from the German by Collins GR, Collins CC as City planning according to its artistic principles. Phaidon Press, London, 1965. https://archive.org/details/derstadtebaunach00sitt/page/62/mode/2up. Accessed 23 Feb 2024

Skaburskis A (2008) The origin of wicked problems. Plan Theory Pract 9(2):277–280

Skinner J (2008) Public spaces, universal spaces: taking accessibility to the next level. Planning 74(7):10–13

Sklar L (2000) The nature of scientific theory. Routledge, London and New York

Skogan WC (1990) Disorder and decline: crime and the spiral of decay in American neighborhoods. Free Press, New York

Slunecko O (2018) What are served and servant spaces? Constructing Architect. https://www.constructingarchitect.com/what-are-served-and-servant-spaces/. Accessed 14 June 2023

Smith K, Guitart M (2012) Introducing architectural theory: debating a discipline. Routledge, London and New York

Smith S, Mark S (2011) Marktown: Clayton Mark's planned worker community in northwest Indiana. South Shore J 43. archive.org. Accessed 18 Sept 2022

Smith VM, Stewart BL (2006) Feng Shui: a practical guide for architects and designers: corporate, retail, residential, hospitality, health care. Kaplan, Chicago, IL

Smithson A (1968) Team 10 primer. MIT Press, Cambridge, MA

Soth L (1983) Le Corbusier's clients and their Parisian houses of the 1920s. Art Hist 6(2):188–198

Southworth M (1969) The sonic environment of cities. Environ Behav 1(1):49–70

Souza E (2023) Le Corbusier's color theory: embracing polychromy in architecture. ArchDaily, 18 July 2023. https://www.archdaily.com/1003880/le-corbusiers-color-theory-embracing-polychromy-in-architecture. Accessed 19 July 2023

Spirn AW (1984) The granite garden: urban nature and human design. Basic Books, New York, NY

Staff Writer (2013) Chinese heritage group boycotts RIBA award for Hadid's Galaxy SOHO. Middle East Architect. Accessed 1 June 2023

Steele FI (1973) Physical settings and organization development. Addison-Wesley, Reading, MA

Steil L (1988) On imitation. Archit Des 58(9/10):8–9. https://www.academia.edu/7322485/On_Imitation. Accessed 23 Feb 2024

Stern RAM (1975) George Howe: toward a modern American architecture. Yale University Press, New Haven, CT

Stern RAM (1981) The Anglo-American suburb. Architectural Design, London

Sternberg E (2002) What makes buildings catalytic? How cultural facilities can be designed to spur surrounding development. J Archit Plan Res 19(1):30–43

Sternberg ME (2009) Healing spaces: the science of place and wellbeing. Harvard University Press, Cambridge, MA

Sternberg M (2012) Monastic paradigms in modernism: Le Thoronet and the romantic legacy. J Archit 17(6):925–949. https://www.academia.edu/2949368/Monastic_Paradigms_in_Modernism_Le_Thoronet_and_the_Romantic_Legacy. Accessed 25 Jan 2024

Steuteville R (2022) The new urbanist developer king. Public Square, 12 Sept 2022. https://www.cnu.org/publicsquare/2022/09/12/new-urbanist-developer-king. Accessed 19 Nov 2023

Steuteville R (2024) A pattern language applied to a suburban strip. Public Square, 4 Apr 2024. CNU. Accessed 6 Apr 2024

Steyn G (2012) Le Corbusier's town planning ideas and the ideas of history. S Afr J Art Hist 27(1):83–106. https://www.academia.edu/71590528/Le_Corbusiers_town_planning_ideas_and_the_ideas_of_history?email_work_card=thumbnail. Accessed 15 Sept 2022

Steyn G (2014) West African influence in various projects by Le Corbusier. S Afr J Art Hist 29(2):120–136. https://www.scribd.com/document/294005272/Gerald-Steyn-West-African-Influence-on-Various-Projects-by-Le-Corbusier. Accessed 23 Feb 2024

Stouhi D (2024) 5 art movements that influenced architecture. ArchDaily, 17 Feb 2024. https://www.archdaily.com/972816/5-art-movements-that-influenced-architecture. Accessed 18 Feb 2024

Strauss AL (1961) Images of the American city. The Free Press, New York, NY. Republished by Routledge in 2017

Studer R (1969) The dynamics of behaviour-contingent physical systems. In: Ward A, Broadbent G (eds) Methods in architecture. Lund Humphries, London, pp 53–70

Sullivan LH (1891) The advantage of regulation on the height of commercial buildings. The Graphic, Dec 1891. See Tafuri (1973, 47)

Sussman A, Chen K (2017) The mental disorders that gave us modern architecture. Common Edge. Accessed 20 Aug 2022

Suswanto A (2000) A model of living cultures. In: Soegijapranata Catholic university workshop on urban design and conservation, 26 Nov 2000

Swain J, French S, Barnes C, Thomas C (eds) (2014) Disabling barriers—enabling environments, 3rd edn. Sage, Newbury Park, CA

Swinburne HH (1967) Change is the challenge. AIA J 83–90

Tafuri M (1968) Teorie e Storia dell'architettura. Translated by Verrecchia G as Theories and histories of architecture. Grenada, London, 1980

Tafuri M (1973) Progetto e Utopia. Translated by Taylor R, Pisani D, Ortiz M as Architecture and utopia: design and capitalist development. MIT Press, Cambridge, MA

Talen E (2000) The charter of the new urbanism. Bull Sci Technol Soc. Academia.edu. Accessed 31 May 2024

Talen E (2012) City rules: how regulations affect urban form. Island Press, Washington, DC

Taschen (ed) (2022) Architectural theory: pioneering texts on architecture from the renaissance to today. Taschen, Cologne

Taylor J, Lang J (2016) The great houses of Calcutta: antecedents, precedents and portents. Niyogi, Delhi

Thakar A (2021) How Louis Khan's Institute influenced a generation of architects. ArchDaily, 19 Oct 2021. https://www.archdaily.com/970564/how-louis-kahns-salk-institute-influenced-a-generation-of-architects. Accessed 29 July 2023

Tjallingii SP (1996) Ecological conditions: strategies and structures in environmental planning. Wageningen, DLO Institute for Forestry and Nature Research

Todd JA, Lindsay G (2016) Planning and conducting integrated design (ID). Whole Building Design Guide. https://wbdg.org/resources/planning-and-conducting-integrated-design-id-charrettes. Accessed 21 June 2023

Toker U (2022) Parametric design for beginners: architecture applications. Routledge, London and New York

Toman R (2002) Romanesque architecture, sculpture, painting. Translated from the German by Hulse F, Macmillan I. Feierabend Verlag, Berlin

Tovar E (2023a) Generative space design: exploring 8 transformative tools in architecture. ArchDaily, 29 June 2023. https://www.archdaily.com/1001757/generative-space-design-exploring-8-transformative-tools-in-architecture. Accessed 30 June 2023

Tovar E (2023b) How to guide people in architectural spaces with tactile paving surfaces? ArchDaily, 11 July 2023. Accessed 12 July 2023

Troy J (2021) Designing with people in mind: an approach to creating contemporary communities that put people first. Form + Place, 27 Oct 2021. https://www.formandplace.com/perspectives/designing-with-people-in-mind-an-approach-to-creating-contemporary-communities-that-put-people-first. Accessed 23 Feb 2023

Trudeau D (2013) A typology of new urbanism neighbourhoods. J Urban Int Res Plan Urban Sustain 6(2):113–188. https://www.academia.edu/27943392/A_typology_of_New_Urbanism_neighborhoods. Accessed 28 May 2023

Tschumi B (1977) The pleasure of architecture. Archit des 47(3):214

Tschumi B (1981) The Manhattan transcripts: theoretical projects. Academy Editions, London

Tschumi B (1987) Cinégram Folie: le Parc de la Villette. Princeton Architectural Press, Princeton, NJ

Tschumi B (1988) Notes towards a theory of architectural disjunction. J Archit Urban 216:11–15

Tschumi B (1996) Architecture and disjunction. MIT Press, Cambridge, MA

Tschumi B (1998) The architectural paradox. In: Hays KM (ed) Oppositions reader: selected essays, 1973–1984. Princeton Architectural Press, New York, NY, pp 324–327

Turner JFC (1976) Housing by people: towards autonomy in building environments. Marion Boyers, London. https://www.academia.edu/40573748/Turner_John_F_C_Housing_by_people. Accessed 3 Dec 2023

Turner PV (1977) The education of Le Corbusier. Garland, London and New York

Tyng AG (1975) Simultaneous randomness and order: the Fibonacci-divine proportion as a universal forming principle. Unpublished doctoral dissertation, University of Pennsylvania, Philadelphia, PA

Tzonis A (2014) Creativity real and imagined in architectural education. Front Archit Res 3:331–333. Academia.edu. Accessed 28 Sept 2023

Ulrich RS (1991) Effects of interior design on wellness: theory and recent scientific research. J Health Care Interior Des 3:97–109. researchgate.net. Accessed 13 Sept 2023
Universitat Bielefeld (2015) Bielefeld University: public university in Germany. www.uni-bielefeld.de. Accessed 23 Mar 2015
US Dept of Housing and Urban Development (2000) Principles for inner city neighborhood design. The Author, Washington, DC
Van den Berg A (2005) Health impacts of healing environments: a review of the evidence for the benefits of nature, daylight, fresh air and quiet on health care. University Hospital Groningen, Groningen. P391 Boekje Agnes Definitief.qxp. agnesvandenberg.nl. Accessed 11 Sept 2023
Van der Linden V, Dong H, Heylighen A (2018) Architects' attitudes towards users: a spectrum of advocating and envisioning future use(rs) in design. Open Ed J 2:196–216. openedition.org. Accessed 1 June 2023
Van Geert E, Wagemans J (2020) Order, complexity, and aesthetic appreciation. Psychol Aesth Creat Arts 14(2):135–154
Van Rensselaer S (Mariana Griswold Van Rensselaer) (1893) Client and architect. N Am Rev 151. Reprinted in Places, Feb 2013. https://placesjournal.org/article/client-and-architect/. Accessed 14 Aug 2023
van Schaik L (1985) Rationalism and contextualism. Architecture SA 52–56
Van Zanten D (1978) Félix Duban and the buildings of the École des Beaux arts, 1832–1840. J Soc Archit Hist 37(3):161–174
Van Zanten D (2011) Marion Mahoney reconsidered. University of Chicago Press, Chicago, IL
Venturi R (1966) Complexity and contradiction in architecture. Museum of Modern Art, New York, NY
Venturi R, Brown DS, Izenour S (1977) Learning from Las Vegas; the forgotten symbolism of architectural form, Revised. MIT Press, Cambridge, MA
Verbeek P-P (2008) Morality and design: design ethics and the morality of technological artefacts. In: Vermaas PE, Kroes PA, Light A, Moore SA (eds) Philosophy and design: from engineering to architecture. Springer, Dordrecht, pp 91–103
Verbruggen S (2017) Changes trumps tradition. The Atlantis project of Léon Krier, 1986–1992. J Archit 22(7):1230–1256
Vilchis Esquivel LC (2014) Design methodology. Theoretical fundamentals. Palibrio, Bloomington, IN
Villari S (1990) J.N.L. Durand (1760–1834) art and science of architecture. Rizzoli, New York, NY. Originally published as J.-N.-L. Durand (1760–1834) Arte e Scienza dell'architettura in 1987
Vio E (2000) The Basilica of St Marks in Venice. Translated from the Italian by Evans H. Riverside Press, Boston, MA
Vitruvius Pollo M (ca 10 BCE) De Architectura, Libri Decem. Translated from the Latin by Morgan MH as Vitruvius: the ten books on architecture. Harvard University Press, Cambridge, MA, 1914. Free download, borrow, and streaming: internet archive. Accessed 16 July 2023
von Frisch K (1974) Animal architecture. Harcourt Brace Jovanovich, New York, NY
von Goethe JW (1982, originally 1816) Italian journey 1786–1788. Translated from the German by Auden WH. Penguin, Harmondsworth
Waddell G (2017) Greek and Roman architecture. Createspace, Scotts Valley, CA
Wainwright O (2019). How money shapes architecture. ArchitectureAU, 21 Mar 2019. Accessed 24 Feb 2023
Wainwright O (2020) The despot dilemma. Should architects work for repressive regimes? The Guardian, 27 Jan 2020. Architecture. Accessed 22 July 2023
Wainwright O (2021) I have picked people up off the street: the secret life of Alvar Aalto. The Guardian, 22 Mar 2021. https://www.theguardian.com/artanddesign/2021/mar/21/secret-life-architect-alvar-aalto. Accessed 5 Jan 2023
Wainwright O (2023) 'It's already way beyond what humans can do': will AI wipe out architects. The Guardian, 9 Aug 2023. Architecture. Accessed 18 Aug 2023

Wainwright O (2024) Like a sylvan spa! Indies Zurich's staggering revolutionary new hospital for kids. The Guardian, 9 Oct 2024. https://www.theguardian.com/artanddesign/2024/oct/09/zurich-hospital-kinderspital-herzog-de-meuron?utm_term=670662358ba4ff50a147b967237 95e62&utm_campaign=GuardianTodayUS&utm_source=esp&utm_medium=Email&CMP=GTUS_email. Accessed 10 Oct 2024

Waite R (2011) Falling to pieces: Hadid's Guangzhou opera house. Archit J, 11 July 2011. architectsjournal.co.uk. Accessed 12 June 2023

Walker E, Tschumi B (2006) Tschumi on architecture: conversations with Enrique Walker. Monacelli Press, New York, NY

Walters D, Brown LL (2004) Design first: design-based planning for communities. Architectural Press, Oxford. https://thearchiblog.files.wordpress.com/2011/02/architecture-ebook-design-first-design-based-planning-for-communities.pdf. Accessed 10 June 2023

Ware WR (1904) The American Vignola: arches and vaults, roofs and domes, doors and steps and staircases. International Textbook Co., Scranton, PA

Watkin D (1977) Morality and architecture: the development of a theme in architectural history and theory from the gothic revival to the modern movement. Clarendon, Oxford

Watkin D (1996) A new order for office buildings. City J (Spring). https://www.city-journal.org/html/new-order-office-buildings-11733.html. Accessed 11 Apr 2017

Watson H (2005) Hotel revolution—21st century hotel design. Wiley, Hoboken, NJ

Watson D, Plattus AJ, Shibley RG (2003) Time saver standards for urban design. McGraw Hill, New York

Weaver M (2006) Iconic buildings often hated by users. Guardian Weekly, 20–26 Oct 2006, p 9

Weaver M (2023) Wembley introduces dementia-friendly measures as part of stadium scheme. The Guardian, 12 June 2023. https://www.theguardian.com/society/2023/jun/12/wembley-dementia-friendly-measures-stadium-scheme. Accessed 14 June 2023

Weber R (1995) On the aesthetics of architecture: a psychological approach to the structure and order of perceived architectural space. Brookfield, Philadelphia, PA

Weber W, Yannas S (eds) (2014) Lessons from vernacular architecture. Routledge, London and New York

Webster H (2007) The analytics of power: re-presenting the design jury. J Archit Educ 60(3):21–27. https://www.academia.edu/2465939/The_Analytics_of_Power_Re_presenting_the_Design_Jury. Accessed 19 May 2023

Weiler K, Gutschow N (eds) (2017) Authenticity in architectural heritage conservation: discourses, opinions, experiences in Europe, South and East Asia. Springer, Cham

Westerman L (2019) Surrealist furniture: Salvador Dali's visionary design. Financial Times, 4 Oct 2019. ft.com. Accessed 21 Dec 2023

Wheeler C (2006) The public interest. We know it's important but do we know what it means? AIAL Forum No. 48. https://www5.austlii.edu.au/au/journals/AIAdminLawF/2006/2.pdf. Accessed 13 May 2023

White MG, White L (1962) The intellectual versus the city from Thomas Jefferson to Frank Lloyd Wright. New American Library, New York, NY

Whyte WH (1980) The social life of small urban spaces. The Conservation Foundation, Washington, DC

Wigley M (1993) The architecture of deconstruction: Derrida's Haunt. MIT Press, Cambridge, MA. monoskop.org. Accessed 14 Feb 2024

Willis C (1995) Form follows finance: skyscrapers and skylines in New York and Chicago. Princeton Architectural Press, New York, NY

Wolf N (2013) Art deco. Prestel, Munich

Wolfe T (1981) From Bauhaus to our house. Farrar Strauss Giroux, New York, NY

Wotton SH (1624) The elements of architecture. John Bill, London. https://archive.org/details/architectureelem00wott/page/10/mode/2up. Accessed 24 Feb 1624. Also reproduced in Springfield, MA by Bassette FA, ca 1897

Wozniak M (2016) How to improve architectural education: learning (and unlearning) from the Beaux arts method. ArchDaily. Accessed 18 May 2023

Wright FL (1945) When democracy builds. University of Chicago Press, Chicago

Wright FL (1969) Skyscraper regulation, 1926. The Frank Lloyd Wright Foundation, Scottsdale, AZ

Wu Y, Lu J, Dijk E, Li H, Schnall S (2018) The colour red is implicitly associated with social status in the United Kingdom and China. Front Psychol 9:1902–1908. https://europepmc.org/article/MED/30344502#free-full-text. Accessed 24 Feb 2024

Yaneva A (2009) Made by the office of metropolitan architecture: an ethnography of design. NAi010 Publishers, Rotterdam

Yarrow T (2018) How conservation matters: ethnographic explorations of historic building renovation. J Mater Cult 24(1). sagepub.com. Accessed 30 June 2023

Yorke T (2017) Gothic revival architecture. Bloomsbury, London

Younés S, Westfall CW (2022) Architectural type and character: a practical guide to the history of architecture. Routledge, London and New York

Zeisel J (2006) Inquiry by design. Tools for environment-behavior research. W. W. Norton, New York, NY. Originally published by in Montrey, CA by Brooks/Cole in 1981

Zevi B (1950) Towards an organic architecture. Translated from Verso un'architettura organica (1945) by Gendel M. Faber and Faber, London

Zimring C, Bosch S (2008) Building the evidence for evidence-based design. Editors' introduction. Environ Behav 40(1):147–150

Index

A
Aalto, Aino, 189
Aalto, Alvar, 33, 143, 157, 189, 242
Aalto, Elissa, 189
Aberdeen, Scotland
  Music Hall, 76
Adam, Robert, 136, 193
Adjaye, David, 95
Advocacy Design, 175, 176
Agamy, Egypt
  Halawa House, 230
Agra
  Akbar's tomb, 53
  Taj Mahal, 49
Alberti, Leon Battista, 63, 66, 91, 101, 129, 131, 132, 135
Alexander, Christopher, 14, 25, 33, 34, 44, 46, 146, 172–174, 204, 234
Alexandria, Egypt
  Bibliotheca Alexandrina, 54
Alfeld an der Leine
  Fagus Shoe-Last Factory, 153, 155
American Vitruvius, An Architect's Handbook of Civic Art, The, 63
Andre, Louis-Jules, 139
Antwerp, Belgium
  Havenhuis, 217, 218
Aravena Mori, Alejandro, 96, 141, 199
Archer, L. Bruce, 13, 178
Archigram, 211, 215, 249
Architecture
  animal, 110
  Empiricist, 97–99, 101–103, 140, 146, 152
  Pragmatic, 98, 110, 113, 143, 144
  Rationalist, 102, 153
  unselfconscious and self-conscious, 14
Architecture of the City, The, 68
Architecture without Architects, 116
Architettura in quarantena, 96
Argan, Giulio, 29, 30, 65, 72, 73
Aristotle, 47, 99, 101
Ashgabat, Turkmenistan, 164
Ashley, Fredrick M., 148
Asimov, Morris, 16, 17
Athens, Greece
  Acropolis, 120, 122, 215
Austin, John C., 148, 201, 204
Austin, Texas
  St Edwards University, 204

B
Bailey, Stephen, 221
Baker, Laurie, 22, 58, 116
Baku, Azerbaijan
  Heydar Aliyev Centre, 94
Balance Theory, 49
Barcelona, Spain
  Block, 68
  Cerda's plan, 200
  CIAM proposal, 102
  Sagrada Familia, 92
Bardi, Lina Bo, 131, 151
Bauhaus, 30, 33, 46, 81, 94, 96, 143, 146, 150, 151, 154, 157, 205
Bechers, Bernd and Hilla, 76
Beijing, China
  Galaxy SOHO, 221
Bélidore, Bernard Forest de, 63
Berlin, Germany
  Berliner Philharmonie, 76
  Hufeisensiedlung, 248
  Jewish Museum, 163
  Unité d'habitation, 210
Bernini, Gian Lorenzo, 131
Bethune, Louise Blanchard, 137

J. Lang, *The Scope of Architectural Theory*, Design Science and Innovation,
https://doi.org/10.1007/978-981-96-3690-7

Birmingham, England
  Alpha Tower, 198
Blake, Peter, 96, 140, 154, 246
Blatteau, John, 163, 228
Bofill, Ricardo, 158, 162
Borromini, Francesco, 131
Bourdieu, Pierre, 50
Brasília
  National Art Gallery, 8
  Pilot Plan, 157, 210
Breuer, Marcel, 151, 199
Brunelleschi, Filippo, 129, 131, 132
Brussels, Belgium
  Hôtel de Ville, 126
  Maison Médicale, 249
Bubbar, Darshan K. and The Angels, Architects, 43, 190, 209, 222–224, 226, 227, 237
Buenos Aires, Argentina
  Edificio Kavanagh, 94
Buffalo, New York
  Hotel Lafayette, 137
Burgee, John, 160
Burnham, Daniel, 137

C
Calatrava, Santiago, 58, 113
Çalişkan, Olgu, 13, 169, 181
Campbell, Colen, 135
Candilis, George, 102, 211
Can our Cities Survive? An ABC of Urban Problems, their Analysis, their Solutions, 246
Caracas, Venezuela
  23 de enero housing, 157
Carson, Rachel, 105, 213
Cathedrals, 56, 126–128, 133, 164, 198, 228
Celebration, Florida, 165, 234–236
Cerdà i Sunyer, Ildefons, 69
Chandigarh, India
  High Court, 109
  Le Corbusier's plan, 30, 209
  Mayer-Nowicki plan, 210
  Panjab University Campus, 206
Charte d'Athènes, 206
Charter of Machu Pichu, The, 102, 152
Chicago, Illinois
  Breasted House, 193
  Lakeside Press Building, 193
  Mentor Building, 193, 194
Churchman, C. West, 15, 16
C.I.A.M., 81, 246
Cincinnati, Ohio
  Rosenthal Museum, 220
  Wexner Center, 54
Cité Industrielle, 150
Clay, Grady, 28
Clients
  Empiricist, 100
  Pragmatic, 103, 140, 163, 201
  Rationalist, 100, 152
Code Civil des Français, 101
Columbia, Maryland, 63, 142, 148, 212
Columbus, Indiana
  Fire Station No. 4, 161
Complexity and Contradiction in Modern Architecture, 91, 103, 158, 189
Constitución, Chile
  Villa Verde, 201, 203
Cook, Peter, 215
Cooper, Alex, 18, 45, 92, 132
Coop-Himmelb(l)au, 160, 162
Costa, Lúcio, 157, 210
Creating Architectural Theory, xv, xviii, xxix, xxxvii
Cret, Paul Phillippe, 146
Cuff, Dana, 13, 157, 246

D
Dali, Salvador, 9
Davidoff, Paul, 175
Da Vinci, Leonardo, 99, 129
De architectura; Libre decem, 91, 120
Death and Life of Great American Cities, The, 99
Defensible Space, 42, 99, 204
De re aedificatoria: Libre decem, 91, 101
Derrida, Jacques, 95, 158, 161, 163, 171, 212, 216
Descartes, René, 13, 14, 101, 174
Designing for People, 10, 20, 177, 249
Design methodology, 13, 177
Design Methods: The Seeds of Human Futures, 178
Design with Nature, 105
De Silva, Minette, 151
De Sola Morales, Ignasi, 102
Doha
  Skyline, 103
Donatello, 131
Doshi, B. V., 96, 102, 152
DPZCoDesign, 100, 190, 222, 232, 234–236
Dresden, Germany
  UFA-Kristallpalast, 162
  Zwinger Palace, 133
Dreyfus, Henry, 37, 177

Duany, Andrés, 66, 91, 222, 230, 232, 234, 235, 237. *See also PDZ CoDesign*
Dubai, UAE, 164
Dublin, Ireland
  Leinster House, 138
Durand, J-N-L, 30, 62

E
East S Louis, Missouri
  Pruitt Igoe, 143, 144, 157
Eckstut, Stanton, 18
Ėcole des Beaux Arts, 30, 63, 135–137, 191, 195
Eisenman, Peter, 81
Ellis, Sir Clough William, 157
El-Wakil, Abdel-Wahed, 190, 222, 227
Essai sur l'Architecture, 62
Essay Concerning Human Knowledge, An, 99

F
Farrell, Tony, 160
Fathy, Hassan, 228
Ferriss, Hugh, 145
Fischer, Gerhard, 14
Fisker, Kay, 157
Florence, Italy
  Palazzo Ruchellai, 132
  Santa Maria Novella, 131, 132
Foster, Norman, 88
Fournier, Colin, 95, 215
Fraker, Harrison, 72, 97
Franzen, Ulrich, 58
Fuller, Buckminster, 20, 113, 215
Functionalism Revisited, xxix, xxxvii

G
Gabriel, Ange-Jacques, 136
Gans, Herbert, 44, 50
Garden Cities of Tomorrow, 100
Garnier, Charles, 136
Garnier, Tony, 150, 153, 205
Gaudi, i Cornet, Antoni, 92
Geddes, Patrick, 148, 150
Gehry, Frank, 160, 163, 203, 218, 242, 251
Genius of Architecture, or, The Analogy of that Art with our Sensations, The, 132
Gibson, James J., 7
Giedion, Sigfried, 7, 151
Gigi, Anton, 133
Ginsburg, Moisei, 151
Goethe. Wolfgang von, 133
Goldberger, Paul, 28
Graves, Michael, 81, 160, 162
Greenberg, Allan, 44, 163, 228
Griffin, Walter Burley, 188
Groningen, The Netherlands
  Glass Video Gallery, 161, 213, 216
Gropius, Walter, 33, 91, 150, 151, 153, 154, 175, 195
Guangzhou, China
  Opera House, 136, 154, 213, 221
Guimard, Hector, 146

H
Habitat Bill of Rights, 102, 152
Habrakan, John, 176, 203
Hadid, Zaha, 76, 94, 95, 110, 160, 161, 163, 170, 171, 188, 190, 201, 205, 211, 215, 217–221, 237, 253
Hall, Peter, 154
Haussmann, Baron Georges-Eugène, 138
Hayden, Dolores, 107
Heider, Fritz, 49
Henderson, Tony, 188
Hilberseimer, Ludwig, 150
Hoffman, Josef, 248
Hood, Raymond, 146
Housing for People: Towards Autonomy in Building Environments, 203
Howard, Ebenezer, 100, 140, 145–147
Hunt, Richard Morris, 137
Huxtable, Ada Louise, 28

I
Innsbruck, Austria
  Helbing Haus, 133
Intentions in Architecture, 33
Introduction to Design, 16
I quattro libri dell'architettura, 131, 135
Iquique, Chile
  Quinta Monroy, 203
Istanbul, Türkiye
  Hagia Sophia, 124, 126, 127
Itten, Johannes, 46
Izenour, Steven, 56, 103
Izumi, Kiyo, 35

J
Jacobs, Jane, 99, 179, 234, 246, 254
Jacobson, Arne, 151
Jeanneret-Gris, Pierre, 152, 188, 205, 207, 209, 245

Jeddah, Saudi Arabia
  Al-Khobar Mosque, 229
Jencks, Charles, 97
Jenny, William Le Baron, 191
Johannesburg, South Africa
  Kerk Street Mosque, 231
  Soweto housing, 144
Johnson, Philip, 24, 160, 188, 245
Jones, J. Christopher, 178
Joshua Tree, California
  Doolittle House, 161
Jung, Carl, 47

K
Kahn, Louis I., 29, 58, 85, 146, 170, 179, 242
Kentlands, Maryland, 234
Kolkata, India
  Great Houses, 129
  Neo-classical architecture, 72, 108, 138
Krier, Leon, 94, 100, 160, 162
Krier, Rob, 30, 69, 161
Kroll, Lucien, 249
Kropf, Karl, 97
Kurokawa, Kisho, 249

L
Ladovsky, Nikolai, 151
Lake Forest, Illinois
  Ragdale, 193
Lamb, William, 142
Langhans, Carl Gottfried, 136
Laugier, Abbé-Antoine, 62
La Ville Contemporaine, 150
Leach, Neil, 253
Learning from Las Vegas, 57
Le Corbusier, 29, 30, 81, 82, 91, 92, 95, 101, 102, 109, 128, 140, 143, 150–154, 157, 188, 190, 205, 206, 208–211, 213, 215, 222, 224, 237, 244, 245
Ledoux, Claude-Nicholas, 96
Leighton, Alexander, 34
L'Enfant, Major Charles, 140
Leptis Magna, Libya, 121, 122
L'Esprit Noveau, 102, 152, 208
Levitt, William and the Levittowns, 142, 143
Lewin, Kurt, 254
Liebeskind, Daniel, 211
Lippmann, Walter, 19
Lissitzky, Lazar Markovich, 152
Livres d'Ornemens Inventés & Dessinés, 132
Locke, John, 99
London, England
  1 Poultry Street, 159
  236 Gray's Inn Road, 197
  Centre Point, 196
  Crystal Palace, 112
  Hoover House, 149
  Paternoster Square, 8
  Richmond Waterfront, 53
  Robin Hood Gardens, 154
  Secret Intelligence Services Building, 162
  Space House, 198
  St Paul's Cathedral, 133
  Strawberry Hill House, 137
  Tower 42 (NatWest Tower), 197
  Westminster Abbey, 126, 127
  Westminster Cathedral, 128
  Wilton Music Hall, 76
Loos, Adolf, 248
Los Angeles, California
  Griffith Observatory, 147, 148
  Our Lady of the Angels, 164
  Walt Disney Concert Hall, 28, 55, 163
Los Hechos de la Arquitectura, 200
Lutyens, Edwin, 92, 146–148, 191
Lynch, Kevin, 43, 45, 68, 99, 252

M
Mackintosh, Charles Rennie, 153, 191
Madrid, Spain
  El Museo d'Arte, 112
  Prado, 164
Mahony, Marion Lucy, 188, 189
Malevich, Kazimir, 95, 171, 215, 217
Manchester, England
  Gateway House, 143
  Midland Bank Building, 148, 149
Marktown, Indiana, 193, 194
Marne la Vallée, France
  Architectural School, 52, 100, 136, 172, 181, 191, 200, 213, 216, 237
Marseilles
  Unité d'habitation, 30, 82, 143, 152, 154
Marsh, George, 42, 188, 198
Maslow, Abraham, 34
Mayne, Thomas, 110
McHarg, Ian, 105
McKim, Meade and White, 96, 136
Medina, Saudi Arabia
  Masjid al-Qiblatayn, 229
Meissonier, Juste-Aurèle, 132, 133
Melbourne, Australia
  Federation Square, 112
Mendelsohn, Erich, 150
Metabolists, 20, 113, 152, 249

Metropolis of Tomorrow, The, 145
Mézières, Nicolas Le Camus, 132
Michelangelo, 131–133
Mies van der Rohe, Ludwig, 109, 151, 154
Mill Run, Pennsylvania
  Falling Water, 147, 148
Modern Functional Building, The, 91
Moleski, Walter, 22, 43, 44, 46, 68
Moneo, Rafael, 29, 65, 164
Montpellier, France
  Quartier Antigone, 162
Moore, Charles, 131, 158, 161
Morris, William, 145, 161
Moscow, Russia
  Narkomfi, 155
Mujica, Francisco, 145
Mumbai, India
  Art Deco, 73, 100, 101, 109, 140, 142, 143, 145–148, 150
  Chhatrapati Shivaji Terminus, 139
Munich, Germany
  Amalienborg Pavilion, 133
  Glyptothek, 48
Murcia, Spain
  City Hall, 164

N

Napolean III, 137, 138
Nashville, Tennessee
  Schermerhorn Center, 76, 163
Nervi, Pier Luigi, 113
New Canaan, Connecticut
  Glass House, 160
New Delhi
  Lutyens' Delhi, 252
  Rashtrapati Bhavan, 148
New Gourna, Egypt, 106
Newman, Oscar, 42, 55, 99, 204
New Orleans, Louisiana
  Piazza d'Italia, 161
New York
  550 Madison Avenue, 160
  Battery Park City, 18, 27, 165
  Blue Condominium, 213, 216
  Chrysler Building, 143, 146–148
  Empire State Building, 142
  Forest Hill Gardens, 82, 145
  4th Avenue, 8
  High Line Park, 70
  Hudson Yards, 27
  Seagram's Building, 188
  Theater District, 72
Nicolin, Pierluigi, 96, 163
Niemeyer, Oscar, 151, 157, 199
Nimes, France
  Pont du Gard, 122
Noisy-le-Grand, France
  Les Espaces d'Abraxsus, 162
Nolan, John, 147
Norberg-Schulz, Christian, 252
Notes on the Synthesis of Form, 14
Nouvel, Jean, 76

O

Oak Park, Illinois
  Arthur Heurtley House, 149
  Malcolm Willey House, 92, 147
  Nathan Moore House, 54
Office for Metropolitan Architecture (OMA), 14, 77
Oliver, Paul, 116
Olmsted, Frederick Law, Jr, 145
Orwell, George, 189
Osmond, Humphrey, 42
Ove Arup Associates, 154
Oxford, England
  Centre for Islamic Studies, 230, 231

P

Paimio, Finland
  Sanotorium, 189
Palladio, Andrea, 58, 63, 131
Parametricism as Style – Parametricist Manifesto, 179
Paris, France
  Castel Beranger, 146
  Grande Arche, 213
  Haussmann's, 162, 163, 165
  Hotel de Soubise, 132, 133
  Musée d'Orsay, 213
  Muséum national d'histoire naturalle, 139
  Notre-Dame, 128
  Opera House, 154, 213
  Parc de la Villette, 95, 163, 213, 215, 216
  Plan Voisin, 153
  Pompidou Centre, 159
  Promenade Plantée, 70
  Sante-Chapelle, 127
  St Denis church, 125
Patte, Pierre, 62, 64
Pei, I. M., 85
Peirce, Charles Saunders, 49, 97
Perriand, Charlotte, 188
Perry, Clarence, 82, 145
Pessac, France

Le Quartier Moderne, 208
Pevsner, Nikolaus, xiv
Philadelphia, Pennsylvania
First Bank, 138
housing projects, 30, 85, 143, 154, 210
PSFS Building, 153
Vana Venturi House, 160, 162
Physical Settings and Organizational Development, 244
Piano, Renzo, 162
Plano, Illinois
Farnsworth House, 58
Plato, 101
Poissey, France
Villa Savoye, 58, 92, 94, 208–210
Politics and the English Language, xviii
Polshek, James Stuart, 95, 96, 200
Polychromie Architecturale, 109
Ponti, Gio, 199
Pöppelmann, Matthäus Daniel, 133
Portland, Oregon
Municipal Services Building, 162
Provensal, Henry, 91
Pugin, Augustus, 136
Pune, India
Shashital House, 226

Q
Quatremère de Quincy, Antoine-Chrysostome, 63

R
Raphael, 131
Reich, Lilly, 154
Rittel, Horst, 15, 170
Rock, David, 9, 152, 163
Rogers, Richard, 92, 162
Rome
Colosseum, 121, 122
Forum, 120
insulae, 122
Palazzo Nuevo, 53
Pantheon, 105, 120–122, 133, 231
Piazza de Campidoglio, 133
Piazza del Populo, 140
San Carlo alle Quattro Fontane, 131
St Peter's Basilica, 134
Tempietto, 133, 139
Trevi Fountain, 87
Ronchamp
Notre Dame du Haut, 154
Rossi, Aldo, 68
Rotterdam, The Netherlands
De Hofbogen, 70
Routledge Companion to Twenty and Early Twenty-first Century Urban Design, xxix
Rowe, Colin, 13, 28–30, 131, 151, 158, 169, 171, 209
Rublyovo-Arkhangelskoye proposal, 220
Rudofsky, Bernard, 82, 116
Rudolph, Paul, 81
Ruskin, John, 136, 227

S
Saarinen, Eero, 152, 154
Saarinen, Eliel, 146
Safdie, Moshe, 102
Sangallo, Antonio, 131
Santayana, George, 52
Santiago, Chile
Las Torres Siamesas, 202
U.C. Innovations Centre, 201
Scarpa, Carlo, 164
Scharoun, Hans, 76
Schinkel, Karl Friedrich, 136
Schumacher, Patrik, 110, 178, 179, 181, 188, 220, 250, 253
Schwarz, David M., 76, 163
Sciences of the Artificial, 13
Scope of Total Architecture, 91, 150
Scott Brown, Denise, 96, 103, 154, 158, 160, 189
Scott, Sir George Gilbert, 22, 33, 103, 115, 116
Seaside, Florida, 100, 103, 158, 232, 234–236
Seiffert, Richard and Partners, 141
Seoul, Korea
Seoullo, 70
Sert, José Luis, 81, 102, 210, 246
Seven Lamps of Architecture, The, 136
Shastras, 96, 209, 222–224, 227
Shaw, George Bernard, 12
Shaw, Howard Van Doren, 143, 190, 191, 220
Shustar, Iran, 68
Silsbee, Joseph Lyman, 146, 188
Simon, Herbert, 13, 16, 22
Simpson, John, 163, 228
Singapore
One North proposal, 218, 220
Sitte, Camillo, 69, 145
Sixtus V, Pope, 140
Smithsons, Alison and Peter, 83, 198
Soane, John, 136
Space, Time and Architecture, 7

Spirit of Indian Architecture, The, 222, 223, 227
Spirn, Anne Whiston, 105
Steele, Fred, 33
Steil, Lucien, 30
Stephens, Frederick William, 139
Sterling, James, 160
Stern, Robert A. M., 222, 234
Stonorov, Oscar, 85
St Petersburg, Russia, 137, 138
Strauss, Anslem, 66
Stuttgart, Germany
  Weissenhofsiedlung, 152
Style and Epoch, 151
Suburban Nation: The Rise and Sprawl and the Decline of the American Dream, 232
Sullivan, Louis, 146, 150
Supports: Alternative to Mass Housing, 176, 203
Swinburne, Herbert, 16
Sydney
  Opera House, 154
  Seidler House, 154

T
Tafuri, Manfredo, 99, 115
Taliesin, Wisconsin
  Unity Chapel, 188
Tallmadge, Thomas, 191
Taut, Bruno, 109, 151, 248
Team 10, 101, 102, 152
Team 10 Primer, 152
Tel Aviv, Israel, 148, 150
Terry, Quinlan, 160, 162
Thoronet, Var, France
  L'abbeye, 205
Timeless Way of Building, The, 146, 174, 234
Tokyo, Japan
  Nakagin Capsule Tower, 249
  National Theatre, 213
Towards an Organic Architecture, 102, 158
Tschumi, Bernard, 29, 77, 95, 114, 160, 161, 163, 171, 190, 205, 211–213, 215–217, 223, 237, 243
Turner. John F. C., 176
Tyng, Anne, 47, 85, 108, 115, 124, 164, 167, 189

U
Universal Traveler, The, 177
Urban Villagers, The, 44
Utzon, Jørn, 152, 154

V
Valadier, Giuseppe, 140
Valencia, Spain
  Museu de les Ciencies, 112
Vallabha Vidya Nagar
  Berger Paints, 224, 226
Van Alen, William, 143, 146, 147
Van Rensselaer, Mariana, 92, 246
Venice, Italy
  Basilica San Marco, 125
  Doge's Palace, 126, 136
Venturi, Robert, Venturi and Rauch, 56, 91, 103, 158, 160–162, 189, 245
Versailles, France
  Petit Trianon, 136, 137
Vers Une Architecture, 91, 150, 151, 153, 206
Vicenza, Italy
  Villa Capra, 58, 133
Vienna, Austria, 145, 218
Vignola, Giacomo Barozzi da, 63
Villanova, Pennsylvania
  Camp-Woods, 193
Ville Contemporaine, 150, 210
Ville Radieuse, 150, 157, 210
Viollet le-Duc, Eugène-Emmanuel, 135
Vision for Britain: A personal View of Architecture, A, 230
Vitruvius Pollio, Marcus, xxiii

W
Walpole, Horace, 137
Washington, DC
  Capitol, 137, 139, 210
  Crystal Heights proposal, 150
  Lincoln Memorial, 139
  Plan, 140
  White House, 109, 138
Weil-am-Rhein, Germany, 220, 221
  Arts Workshop, 137
  Vitra Fire Station, 58, 94, 218, 220, 221
When Democracy Builds, 145
White, William H., 25, 96, 108, 109, 147
Wotton, Sir Henry, xxiii
Wren, Sir Christopher, 133, 195, 196
Wright, Frank Lloyd
  Oak Park houses, 92, 148, 193
  Skyscraper Regulations, 145

Y
Yamamoto, Riken, 96
Ypres, Belgium
  Cloth Hall, 128

Z

Zevi, Bruno, 102, 158, 161, 162
ZHA Architects, *see* Hadid, Zaha

The manufacturer's authorised representative in the EU is Springer Nature Customer Service Centre GmbH, Europaplatz 3, 69115 Heidelberg, Germany. If you have any concerns regarding our products, please contact ProductSafety@springernature.com

Printed and bound by CPI Group (UK) Ltd, Croydon, CR0 4YY
01/10/2026
02225850-0014